AF785470

ALLE ZEIT WACH 1842

Ultraprecision in Manufacturing Engineering

Proceedings of the International Congress for Ultraprecision Technology May 1988, Aachen, FRG

Edited by
Manfred Weck and Robert Hartel
Forschungsgemeinschaft Ultrapräzisionstechnik e.V.

With 216 Figures

Springer-Verlag Berlin Heidelberg NewYork
London Paris Tokyo

Prof. Dr. Manfred Weck
Chairman of the board

Dipl.-Ing. Robert Hartel
Secretary

Forschungsgemeinschaft Ultrapräzisionstechnik e.V.
Steinbachstraße 17
D-5100 Aachen

ISBN-13:978-3-642-83475-2 e-ISBN-13:978-3-642-83473-8
DOI: 10.1007/978-3-642-83473-8

This work is subject to copyright. All rights are reserved, whether the whole or part of the material is concerned, specifically the rights of translation, reprinting, re-use of illustrations, recitation, broadcasting, reproduction on microfilms or in other ways, and storage in data banks. Duplication of this publication or parts thereof is only permitted under the provisions of the German Copyright Law of September 9, 1965, in its version of June 24, 1985, and a copyright fee must always be paid. Violations fall under the prosecution act of the German Copyright Law.

© Springer-Verlag, Berlin, Heidelberg 1988
Softcover reprint of the hardcover 1st edition 1988

The use of registered names, trademarks, etc. in this publication does not imply, even in the absence of a specific statement, that such names are exempt from the relevant protective laws and regulations and therefore free for general use.

2161/3020 543210

Foreword

The International Conference "Ultraprecision in Manufacturing Engineering" in Aachen, FRG, is the first meeting in Western Germany to present the state-of-the-art in this area of engineering. Many advanced technology products depend entirely on one or more components being manufactured to tolerances or dimensions in the micro- or even nanotechnology range. Examples and summaries of the mechanism of high precision material removal processes, ultra precision machine systems with their advanced servo motion control and the dimensional measurement of finished components are topics of papers presented on this conference. The progress in research work of ultra precision engineering should help the machine tool industry in the demand to meet higher workpiece accuracy.

The Forschungsgemeinschaft Ultrapräzisionstechnik e.V. is indebted to thank the program committee for collecting these advanced papers from authors of many different countries. The authors presenting their papers are welcomed in Aachen. In addition we thank our colleagues and collaborators for organizing the conference and take care of printing this book.

Aachen, May 1988

Manfred Weck
Robert Hartel

Table of Contents

SESSION I - "MANUFACTURING TECHNOLOGY"
Chair: W.König, FRG

W.J. WILLS-MOREN:
Experiences in the Precision Machining of Brittle Materials 3

T.G. BIFANO, T.A. DOW and R.O. SCATTERGOOD:
Ductile-Regime Grinding of Brittle Materials 22

M. MIYASHITA:
Review on Ultraprecision Grinding Technology 41

W. KÖNIG and M. POPP:
Precision Machining of Advanced Ceramics - A Challenge in Production Technology 58

K.H. FIEDLER:
Precision Grinding of Brittle Materials 72

G. SPUR, D. SIMPFENDÖRFER and I. SABOTKA:
Form Accuracy and Surface Integrity in Plane Lapping 78

T.C.G. LO-A-FOE, J.H. DAUTZENBERG and A.C.H. VAN DER WOLF:
Cutting Forces and Their Influences upon the Surface Integrity in Single-Point Diamond Turning 110

N. IKAWA, S. SHIMADA and G. OHMORI:
Recent Trends in Diamond Tool Technology 126

W.J. BEEKMANS:
Prediction of Tolerances in Precision Cutting for Mass Production 143

SESSION II - "MACHINE DEVELOPMENT AND IMPROVEMENT"
Chair: E. Heynacher, FRG

M. WECK, R. HARTEL and T. BISPINK:
Limits of Workpieces Accuracy Caused by the Geometrical and Dynamical Behaviour of Ultraprecision Diamond Turning Machines 153

M. TSUTSUMI, K. UNNO, M. YOSHINO and F. YAMAUCHI:
New Material Application to Ultraprecision Lathe 171

H. HASHIMOTO:
Machining of Aspherical Surfaces 184

P.J. FALTER and T.A. DOW:
A Diamond Turning Apparatus for Fabrication on Non-Rotationally Symmetric Surfaces........................187

D.H. YOUDEN:
The Design and Construction of a Sub-Microinch Resolution Lathe..202

K. BECKER and K. BECKSTETTE:
M400 and P400 - a Pair of Machines for Computer-Controlled Fine Correction of Optical Surfaces........................212

P.H. KNOL, D. SZEPESI and M. HUIJBERS:
New Design Features of a Fully Hydrostatic Ultra Precision CNC-Turning Machine, Type Super-Mikroturn CNC. 224

P.R. HANNAH:
A Personal Prospective on the Quest for Higher Precision Diamond Turning Lathes....................................237

SESSION 3 - "MEASUREMENT TECHNIQUES AND EQUIPMENTS"
Chair: D.J. Whitehouse, U.K.

A. FRANKS:
Nanotechnology at the National Physical Laboratory........255

R.S. POLVANI and C.J. EVANS:
Micro-Indentation Assessment of Near Surface Materials Properties and Sub-Surface Damage in Precision Machining..271

K. CREATH and J.C. WYANT:
Measurement of Ultraprecision Components Using Non-Contact Interferometry Based Instrumentation..........287

J.H. RAKELS:
In-Process Surface Finish Measurement of High Quality Components..303

G. HARBERS and M. STRUIS:
Measurement of Steep Aspheric Surfaces Using a Radial Shear Interferometer....................................317

H. TIZIANI, B. PACKROSS and G. SCHMIDT:
Testing of Aspheric Surfaces with Computer Generated Holograms...335

T. PFEIFER, P. HANTEL and R. TUTSCH
Synthetic Holograms for the Testing of Aspheric Surfaces..343

P. LANGENBECK and D. MÜLLER:
Experimental Measurement of Optimum Depth of Cut..........356

Session 1

Manufacturing Technology

Experiences in the Precision Machining of Brittle Materials

W.J. WILLS-MOREN and R.F.J. READ

Cranfield Precision Engineering Ltd,
Cranfield, Bedford. England.

ABSTRACT

Of recent years significant advances have been made in achieving higher accuracies and surface finish in machining. These have been as a result of parallel developments of the machine tool, the cutting tool and the cutting technology. With the recent advances in the development and use of advanced materials, such as ceramics, it is necessary to re-examine the equipment and techniques required to achieve equivalent advances in the machining of these materials.

Work being carried out at Cranfield in the machining of some of these materials using advanced, high precision tools are described. Initial results are presented and some indications of the requirements for further machine development are given.

INTRODUCTION

Remarkable developments in machining accuracy by an increasing variety of means have been achieved within the working lifetime of current practising engineers and scientists (1). These achievements have frequently been brought about as a result of intense research and development efforts to enable emerging technologies to be commercially realised. For example in the field of computers and their peripherals, where not only have the products achieved remarkable improvements in performance, but also manufacturing costs have been reduced by several orders of magnitude to enable systems of immense power to be available on all our desks and even in portable form. In many cases these advances have only been possible by the parallel developments of the manufacturing equipment (machine tools, cutting tools etc.) with the process in question. This has enabled the technology to move from small scale manufacture in the laboratory to high volume production with the attendant need for highly predictable and consistent results on a continuous basis.

In his CIRP Keynote paper, "The Role of Precision Engineering in Manufacturing of the Future" presented last year, McKeown (2) described the need for high precision in manufacture and gave specific examples of fields in which impressive developments have been made and further improvements are needed. Figs 1,2,3,4 and 5 are reproduced from this paper.

Fig. 1 shows past and predicted improvements in the efficiency of civil aircraft propulsion units due to higher precision of aerofoil profiles of rotor and stator blades. From this it will be seen that by improving profile accuracy by a factor of 5 and surface finish by a factor of 2.5, dramatic improvements can be achieved in compressor efficiency.

Fig.2 shows the past and predicted reduction of pattern dimensions for integrated circuits between 1972 and 1996. Whilst these reductions are in general brought about by developments in lithography or pattern generation in various forms they also demand greater precision and predictability in the manufacture of the substrates which necessitates machining in one form or another.

Fig.3 shows the past and predicted reduction in flying heights for read/write heads of commercially available magnetic disc computer storage systems. In these hard disc type systems the flying head is supported only on a self generated air bearing, as shown in fig.4, during disc rotation. Under these conditions the gap between head and disc must be maintained within defined limits and of course it must not "touch down". The gap is produced by air bearing pressure distribution and this is directly controlled by the geometry of the air bearing surfaces. However it can be seen that the achievement in the greater reduction of gap and dimensions over the years to increase disc storage has only been gained by the manufacture of the critical surfaces, to a much higher order of accuracy.

Fig.5 shows the past and predicted increase in storage density of magnetic disc systems, which indicates a factor of 5 improvement every 5 years thanks largely to the developments described above (5). The above examples have related in the main to improvements brought about by the achievement of closer manufacturing tolerances. However in many cases these improvements in performance can be further enhanced by the use of advanced materials which are now becoming commercially available on a wider scale. Foremost among these are the modern fine ceramic materials where advantage can be taken of one or more of their physical properties - mechanical, thermal, electrical, magnetic and optical.

The application of these materials is severely handicapped by two important and related factors:-

- our ability to fabricate the materials to the required shape and geometry, frequently complex, economically and to the necessary accuracy, surface finish and surface integrity;

- our knowledge of how the material will perform in operation, and how components fabricated in the various ways, will behave.

This latter point is of great importance to the successful use of these materials, as the performance of components manufactured from ceramics can be affected by the manufacturing method to a much greater extent than with most more conventional materials. It is not sufficient to consider merely the published material properties when designing for use of these materials (3).

Sub-surface damage is a factor which has great bearing on the performance of ceramic components and it is common practise to lap and/or polish the surfaces with loose abrasives after grinding with a fixed abrasive wheel to overcome this problem. This in general has many disadvantages including:-

- for surfaces other than flat or spherical a loss of form accuracy can be expected;
- the process can be unpredictable and the dimensional relationship between features is difficult to maintain - this is true even when lapping simple flat surfaces;
- the process is costly and time consuming.

The full potential of advanced ceramic materials will only be realised when the limitations described above are overcome and complex shapes can be economically machined with adequate and predictable accuracy, surface finish and acceptable sub-surface damage. Work is proceeding in many centres to this end.

Cranfield Unit for Precision Engineering (CUPE) has been responsible for the development of a number of advanced machines aimed at ultra high precision machining and on which the machining of some advanced ceramics has been carried out. This paper describes some of this work, gives initial results and makes recommendations for further development of equipment and techniques.

SINGLE POINT DIAMOND MACHINING

Single point diamond machining of a wide range of amenable materials is routinely carried out in many branches of industry at the present time and a number of researchers (4) have explored the possibility of employing this technique for the machining of glass and glass ceramics. It is well known that whilst most glass machining takes place by brittle fracture, permanent flow can be induced under certain conditions. It has been stated that permanent flow can exist at the cutting edge of an abrasive grain during grinding (6) and a number of researchers have studied how this can be applied to single point machining. At one extreme a low cost, simple pyramidal shaped tool was used to carry out what was termed "crushing" (7). This process has the great virtue of simplicity and can generate surfaces with remarkably low tool wear. However significant sub-surface damage is always present. Other workers have turned glass at elevated temperatures (8) but it is understood that this is not suitable to high quality optics.

Work at Cranfield has been carried out on a specially developed facing lathe (Fig.6) which has previously been used to obtain high quality surfaces on conventional diamond turnable materials. This machine is equipped with an air bearing spindle as shown in Fig. 7 which combines high axial stiffness of 618 N/um with axial error motion less than 20nm. It is also equipped with a toolholder having a direct piezo electric (PZT) actuator to apply feed as shown in Fig.8. This enables cuts to be applied in increments of 1nm. Drive to the cross traverse is by means of a high torque friction drive coupled to the table in such a way as to provide a highly stiff, smooth drive to the plain bearing guideway. This arrangement provides a combination of uniform feed velocity with a highly damped plain bearing guideway which, together with the PZT actuator for infeed, is well suited to obtaining results of practical value. Various cutting tests are being carried out on a range of optical glasses with a variety of cutting tool geometries and cutting conditions.

It will be some time before this work will be sufficiently well collated for full publication but the results of one series of experiments is worth reporting here (9). In this work specimens were mounted with their front faces normal to the axis of rotation and a spiral cut as shown of 0.4 mm pitch taken. In the course of these cuts the tool was caused to reciprocate with an amplitude of up to 400 nm in a direction normal to the specimen surface using the PZT actuator. This produced a spiral groove of varying depth in the workpiece. By careful control of the nominal depth of cut and the reciprocation amplitude it was found to be possible to produce a groove in which the cutting mechanism alternated between brittle fracture and ductile cutting along its length.

Details of the test conditions are as follows:-

Test 1 Material BK7 Glass

Max. depth of cut (nominal)	400 nm
Min depth of cut (nominal)	0
Tool type	Natural Diamond
Tool geometry	
Radius	0.1mm
Top rake	0 degrees
Clearance	5 degrees
Coolant	Air mist white spirit
Surface speed	25 m/sec max.

Test 2 Material Zerodur

Max. depth of cut (nominal)	400 nm
Min. depth of cut (nominal)	0
Tool type	Natural Diamond
Tool geometry	
Radius	2mm
Top rake	0 degrees
Clearance	5 degrees
Coolant	Air mist White spirit
Surface speed	25 m/sec max.

A transition from brittle fracture mode of cutting to a ductile mode was seen in both specimens. This occurred at a depth of cut less that 100nm for BK7 and between 100nm and 200nm for Zerodur. Fig. 9a shows this effect for BK7 and Fig. 9b shows the ductile type chips adhering to the surface of the specimen.

A large amount of tool wear was observed to occur under the above conditions of greatly varying depth of cut. For a measure of tool wear versus depth of cut to be obtained it is necessary to be able to set and maintain cuts to an accuracy of a few nanometres. For this the PZT toolholder has been modified to incorporate a capacitive feedback system to enable the distance to the workpiece vacuum faceplate to be set and maintained to a resolution in the nanometre region. This system is currently being commissioned to enable further work to be carried out at a range of closely controlled depths of cut.

FIXED ABRASIVE GRINDING

Work carried out at Cranfield has employed a surface grinding machine specially developed for fine grinding (or lap grinding) and embodying many unique features aimed at eliminating or reducing problems usually associated with this type of equipment. The machine employs a spindle similar in principle to that shown in Fig. 10 mounted vertically for use with a cup type grinding wheel. Coolant and a cooling medium are separately fed to the grinding spindle as shown in Fig.11 both fluids being temperature controlled to 0.1^{o}C at entry. The temperature controlled medium circulates round critical zones of the spindle to minimise axial growth so that a warm up period is unnecessary. Grinding coolant is fed to the wheel as shown in the diagram so that it reaches the cutting zone on the inside of the grinding wheel an is fed through by centrifugal force.

The table of the machine is carried in air bearings and its drive is by means of dc motor driven re-circulating ballscrew. To avoid transmission of unwanted vibration (rumble) the nut of the above system is decoupled from the table by an air bearing thrust member (paddle) as shown in Fig. 12 thus the worktable is not connected in any way to the machine base.

Feed rates can be varied between 0.001 and 10.0mm/sec with the feed servo optimised to give high uniformity of feed velocity.

The infeed is controlled to 100nm resolution with position held under grinding conditions by a high performance servo system. The machine is of closed loop bridge type construction with the spindle mounted on the highly stiff bridge. The bed is of Granitan and mounted on pneumatic vibration isolators.

Test machining has been carried out on various ceramic materials of the ferrite, titanium carbide, aluminium oxide and zirconia types. From this it has been shown that given suitable grinding conditions surface textures of the same quality as produced by precision lapping can be achieved as shown in Fig.13 and 14. These traces are from Titanium Carbide Aluminium Oxide samples and are typical of results obtained for a range of materials. In general it can be shown that the surface texture for the lap ground specimens are as regular as lapped surfaces.

It has also been observed that edge chipping of the lap ground surfaces can be better than a lapped surface of equivalent surface texture.

The best results were obtained using resinoid bonded diamond wheels with great care taken in truing and dressing. Truing was carried out by means of grinding wheel mounted on a precision air bearing spindle located on the worktable as shown in Fig. 15. Dressing was carried out by use of a conventional dressing stick but using the precise control of the machine to perform the operation. This was found necessary to achieve consistent results.

CONCLUSIONS

The machining of brittle materials to surface finishes

normally associated with lapping and polishing is a practical proposition. Low sub-surface damage can also be achieved, however the equipment used must conform to stringent requirements.

The machine tool must have:-

- extremely high loop stiffness between the tool and the workpiece;

- particularly low vibration levels;

- the ability to apply and consistently maintain very small depths of cut (as low as a few nanometres in some cases);

- low thermal drift and/or some means of compensating for the resulting errors;

- the ability to operate at very uniform feed rates over a wide range.

In addition to these characteristics the control system for the machine should be sufficiently versatile to enable the optimum machining procedures to be established and then subsequently maintained. In the case of grinding machines the method of truing and dressing should be considered to be part of the machine and the machine designed accordingly. Other aspects of the operation of the machine which must be considered at the design stage include:-

- method of coolant application, filtration and temperature control;

- workholding methods.

Further work is needed to be urgently carried out in a number of other areas however to enable full commercial advantage to be taken of their capabilities. These include:-

- means of grinding wheel conditioning should be re-examined in the light of the machine developments described;

- means of accurately measuring sub surface damage in a wide range of materials;

- the ability to detect initial contact between the tool and workpiece to the above resolution or have a knowledge of their respective positions to this level prior to making contact.

There is no doubt that machine tools to meet the requirements outlined above are emerging albeit from a limited number of sources. Given the parallel developments in grinding wheel and workpiece monitoring technology the needs outlined in the opening paragraphs of this paper will be met.

REFERENCES

1. Tanaguchi N, Current status in and future trends of ultra-precision machining and ultra-fine materials processing. Annals of CIRP 32/2/1983

2. McKeown PA, The Role of Precision Engineering in Manufacturing of the Future. Annals of CIRP vol. 36/2/1987.

3. Richardson DW, Modern Ceramic Engineering, Marcel Dekker Inc. USA.

4. Sanger CM. Baker PC, Feasibility of Glass processing by precision machine technology, SPIE vol. 183 Space Optics 1979.

5. Talke FE, Precision Engineering issues in magnetic

recording technology. 4th International Precision Engineering Seminar May 1985, Precision Engineering 1987, 171-178

6. Huerta M, Malkin S, Grinding of Glass: The Mechanics of the Process, ASME Trans May 1976 98.

7. Bryan JB et el, Single point diamond crushing of glass, Precision Machining Workshop, SME, Berkeley California USA, June 5 1984.

8. Brehm R et al, Transparent single-point turning of optical glass, Precision Engineering 1979, 207-213.

9. Horiuchi O, Internal report CUPE.

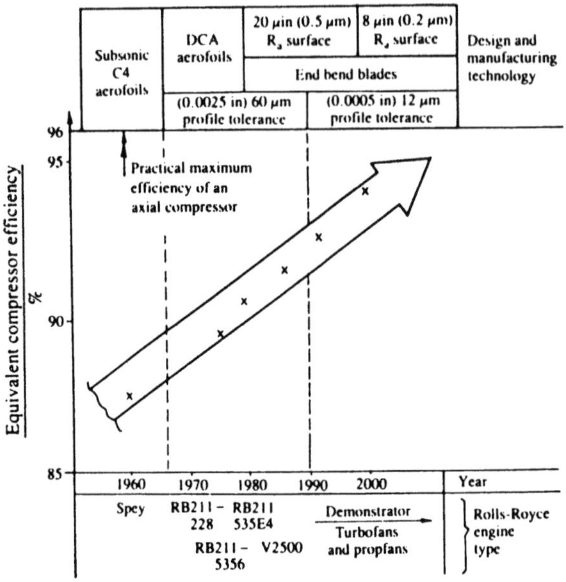

FIG.1 Improvement in compressor efficiency with higher precision aerofoil profile. (courtesy Rolls-Royce Limited)

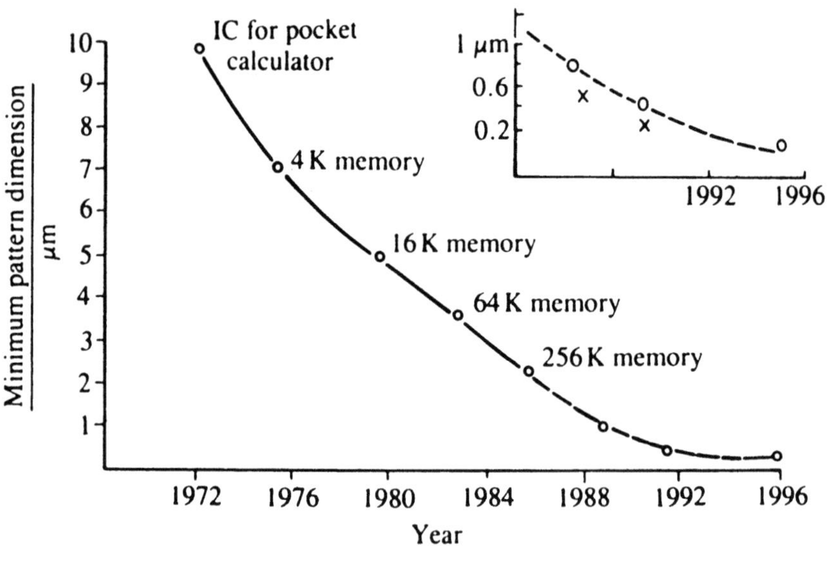

FIG.2 Reduction in pattern dimensions of integrated circuits. ________ volume manufacture. ---------- prototyping

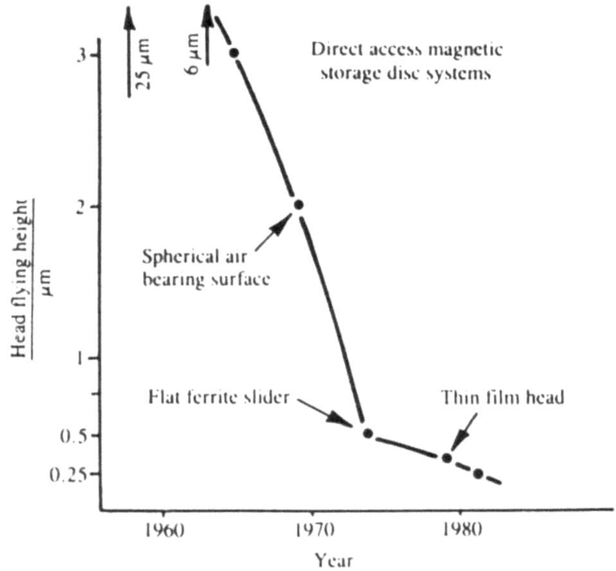

FIG.3 Slider flying height versus year. (commercial systems)

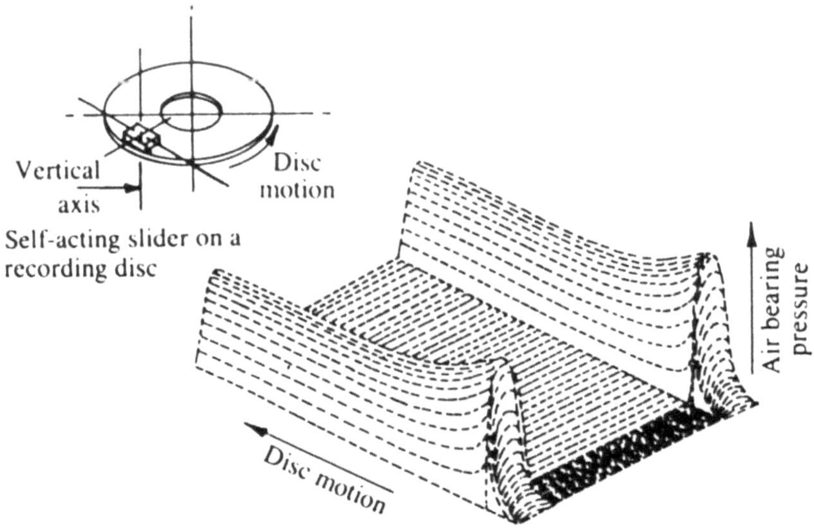

FIG.4 Typical air bearing pressure distribution of flying head

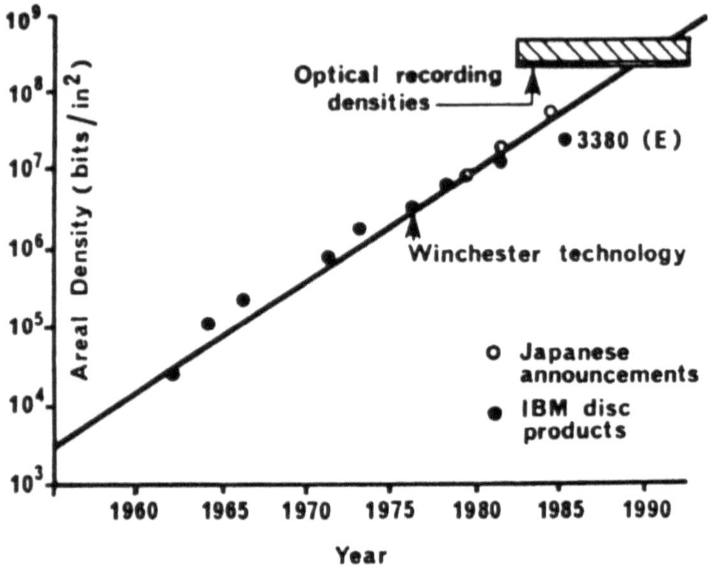

FIG.5 Development of magnetic disc storage density (reproduced from (5))

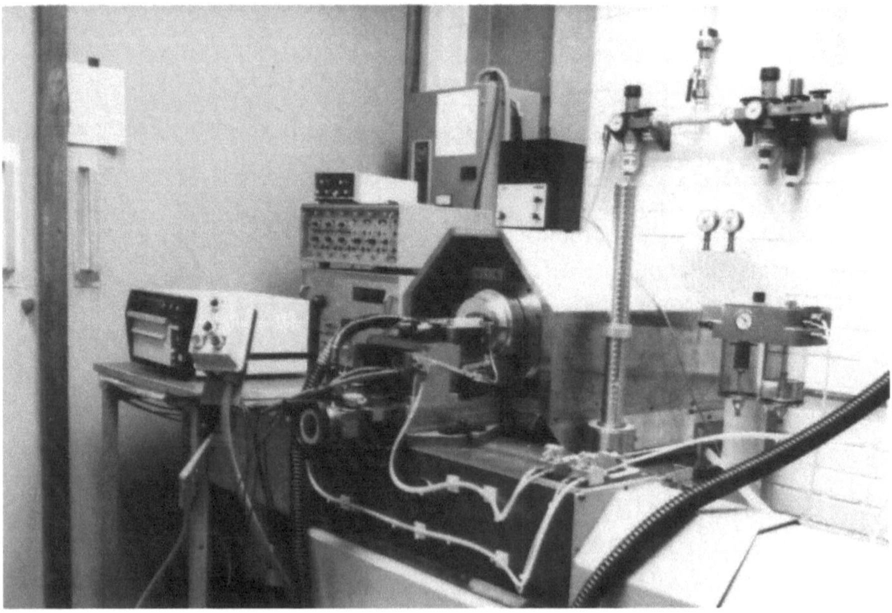

FIG.6 Diamond facing lathe

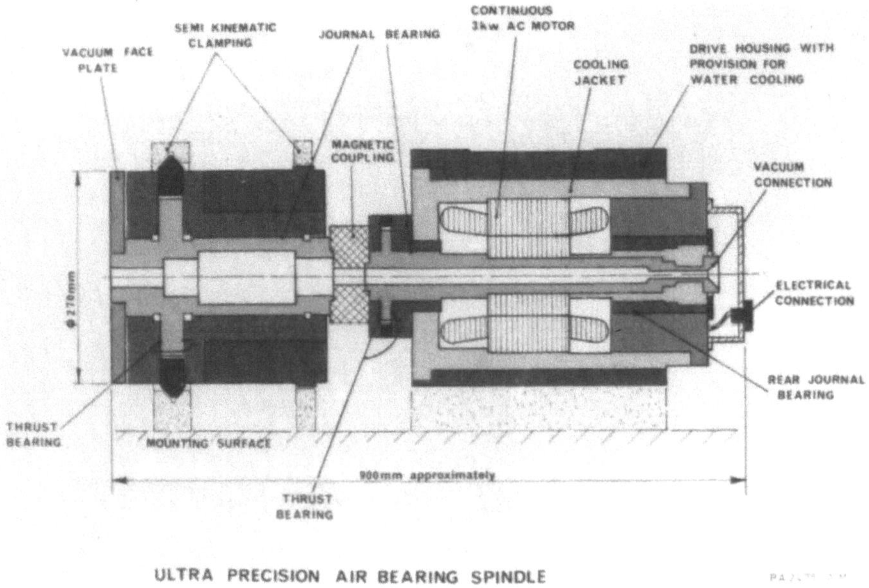

FIG.7 High precision air bearing spindle
with non-influencing drive

FIG.8 Piezo-electric toolholder

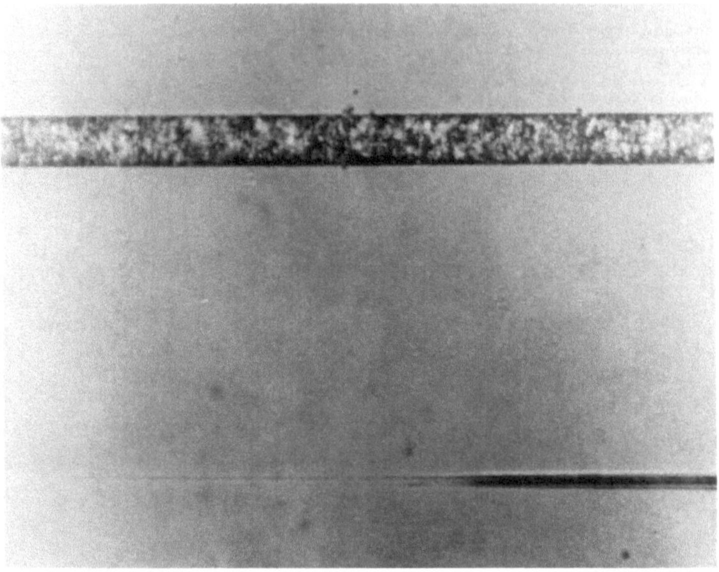

FIG.9a BK7 glass - transition

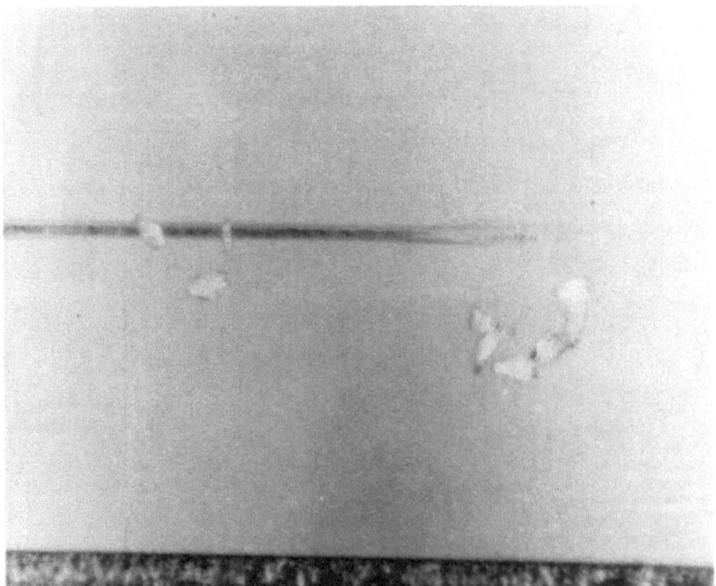

FIG.9b BK7 glass - transition

FIG.10 Ultra precision grinding machine

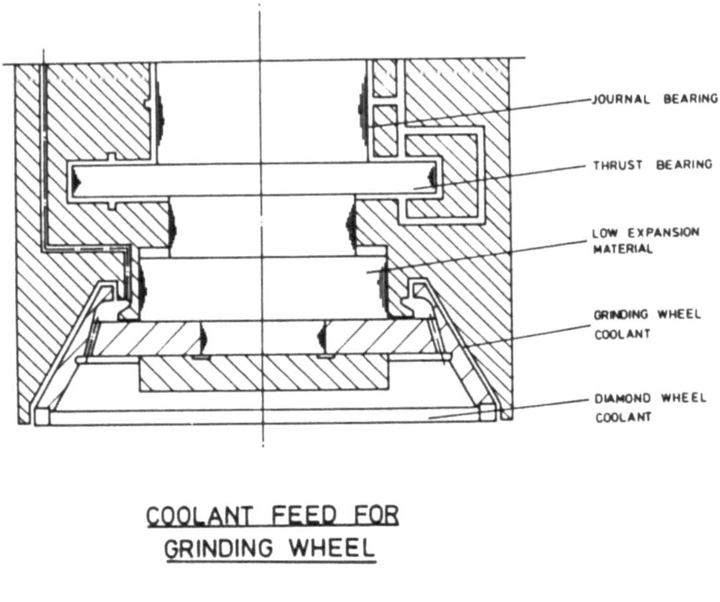

FIG.11

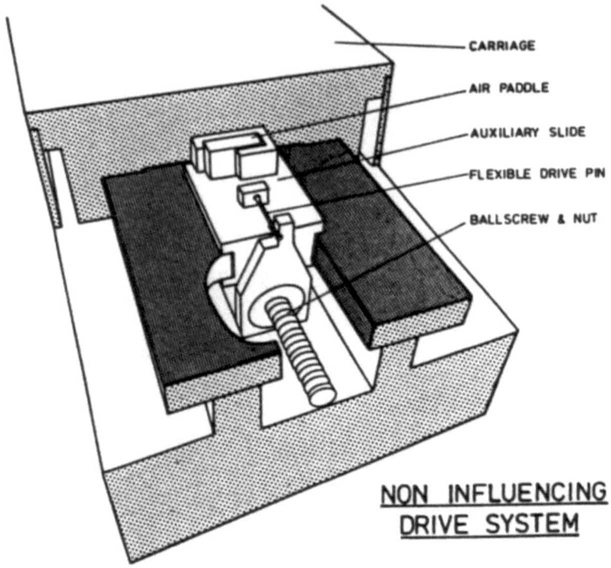

FIG.12

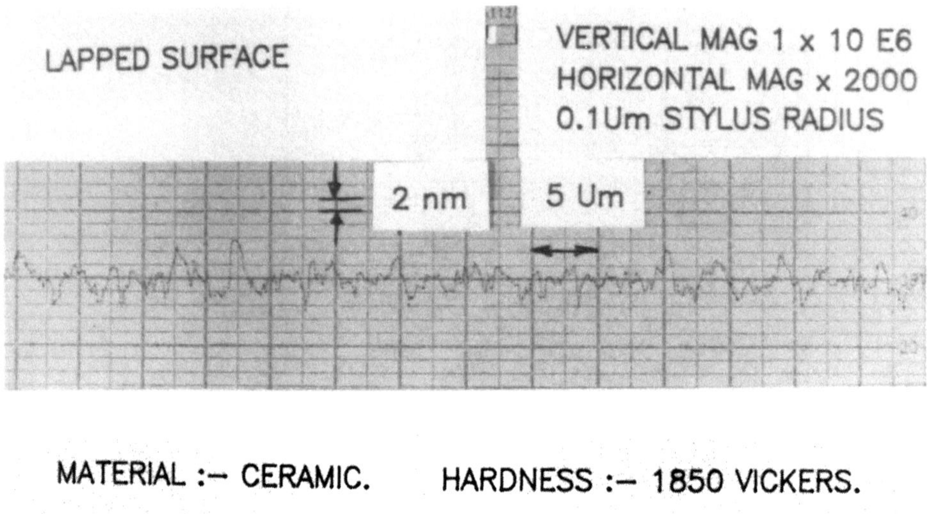

FIG.13 Talystep trace of precision lapped surface

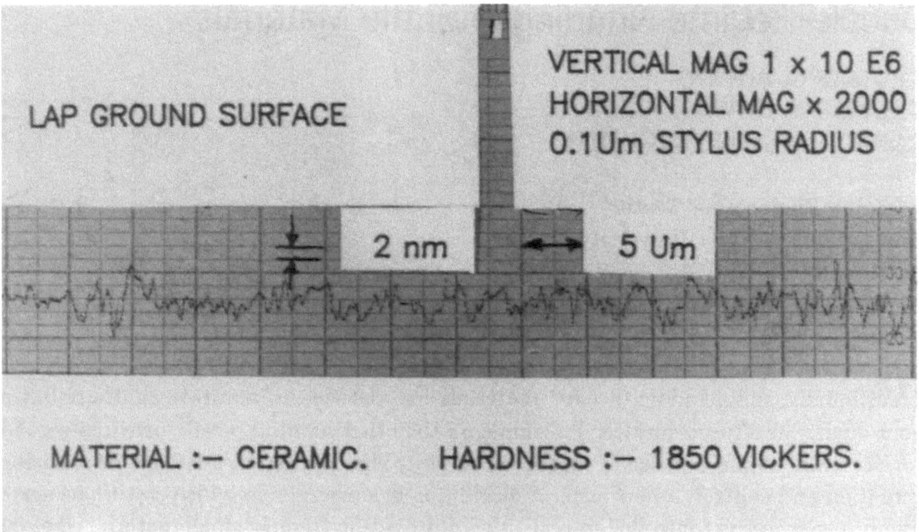

FIG.14 Talystep trace of precision ground surface

FIG.15 Truing spindle

Ductile-Regime Grinding of Brittle Materials

THOMAS G. BIFANO
THOMAS A. DOW
RONALD O. SCATTERGOOD

Precision Engineering Center
North Carolina State University
Raleigh, NC 27695-7918

Traditionally, the grinding of hard materials has resulted in accurate contours but a poor quality, fractured surface. Polishing, on the other hand, typically provides a good surface finish at the expense of contour accuracy. Through recent advances in precision actuation and control, a new form of machining has emerged in which a stiff, accurate grinding machine is controlled in real-time to provide extremely small material removal rates, similar to those encountered in polishing. Under these conditions, the material removal mechanism changes from brittle fracture to plastic deformation, resulting in a much improved surface finish. The micro-plasticity of otherwise brittle materials has been modeled in terms of the mechanical properties of the material surface, and quantitative techniques have been developed for the measurement of surface ductility. Using an experimental apparatus transitions from brittle to ductile material removal, based on in-feed rate, have been demonstrated for a broad range of materials, including single crystals, amorphous glasses, and advanced ceramics.

INTRODUCTION

Progress in fabrication techniques for the production of high precision components is essential for many areas of high technology. Optical systems, bearings, computer memory components, and microelectronic devices all share a direct dependence on the field of precision fabrication for their manufacture. For many advanced technology systems, demands for high fabrication precision are complicated by the use of brittle materials. The past decade has seen a tremendous resurgence in the use of ceramics in structural applictions. The excellent thermal, chemical, and wear resistance of these materials can finally be realized, because of recent improvements in the overall strength and uniformity of advanced ceramics. Magnetic heads for computers, automobile diesel injectors, semiconductor components, and optical lens molds are typical examples of components that require precision fabrication of brittle materials. Advances in fabrication technology are essential to the economic production of these parts.

Impeding the use of these brittle materials in precision engineering applications is the fact that they can not currently be machined in a way that ensures high precision in both form and surface finish. When these materials are ground, the material removal proceeds through a process of controlled, localized surface fracture. The rough surface left by such a grinding operation is unacceptable for many applications. Therefore, a post-grinding polishing operation is typically employed to improve the surface finish. Unlike grinding, polishing proceeds by a predominantly ductile, or plastic material removal process. This plasticity is responsible for the uniform, smooth surfaces generated by polishing. Unfortunately, in the case of polishing, this improvement in surface finish is accompanied by a decrease in the contour accuracy of the workpiece. Also, polishing is a more expensive process than grinding, because it demands skilled labor and is difficult to automate.

The mechanics of grinding and polishing are fundamentally similar: both rely on randomly oriented abrasive particles for material removal. The basic difference between the processes is that grinding uses a fixed abrasive, while polishing uses a loose abrasive. The difference in material removal mechanisms for grinding and polishing (i.e. brittle vs. ductile) is intriguing, especially in view of the similarities between the two processes. The similarity between grinding and polishing brings up an important questions: can the material removal mechanism be changed from one fracture to one of plastic deformation in the grinding of brittle materials? This possibility of a brittle-ductile transition in material removal is the central concern of the microgrinding research described in this paper.

The primary goal of this research [1] has been to investigate microgrinding as a technique for machining brittle materials in a ductile grinding regime. Achieving this goal has required progress along two frontiers of research: determination of the machining parameters relevant to ductile-regime grinding and isolating the material properties that control the transition from brittle to ductile material removal. Through experimentation and analysis along these two fronts, a framework has been constructed to characterize the variables controlling ductile-regime grinding, and to describe the relationship between machining conditions, material properties, and grinding ductility.

GRINDING APPARATUS DESIGN

Microgrinding has been defined as a machining process combining the contour accuracy achievable by fixed abrasive grinding with the fine surface finish achievable by loose abrasive polishing. By utilizing an ultraprecision mechanical system, real-time feedback motion control, and state-of-the-art truing techniques it is possible to merge the high rigidity of a precision grinder with the low machining forces necessary for ductile material removal. Using these principles, a versatile plunge-mode microgrinding apparatus has been designed, constructed, and structurally evaluated. A drawing of this apparatus is shown in Figure 1. The relevant features include:

- Rigid Structural Loop ($50MN/m$)
- Piezoelectric Infeed Actuation ($10\mu m$ range)
- Air Bearing Wheel Spindle
- Closed Loop Infeed Control ($1.5nm/rev$ to $150nm/rev$)
- Vibration Isolation from External Disturbances and Motor Driver
- Precision Wheel Trueing ($125nm$ axial runout)

The material removal rates achievable with this apparatus coincide with the "micro-grinding gap" between conventional grinding and polishing [2]. That is, the apparatus is capable of grinding with smaller controlled cutting depths than are available on any commercial grinder. This has proven to be the most important requirement for ductile-regime grinding of brittle materials.

VARIABLES INFLUENCING GRINDING DUCTILITY

Infeed Rate

A basic hypothesis has been postulated for ductile-regime grinding: all materials, regardless of their hardness or brittleness, will undergo a transition from a brittle machining regime to a ductile machining regime *if* the grinding infeed rate is made small enough[3][4][5]. Below this threshold depth-of-cut, the energy required to propagate cracks is larger than the energy required for plastic yielding, so plasticity becomes the predominant grinding mechanism.

The material property characterizing resistance to plastic flow is the hardness, H. The energy (E_p) required to plastically deform a specified volume of material (V_p) can be written as:

$$E_p = HV_p \tag{1}$$

The material property characterizing resistance to ideal brittle fracture is the crack surface energy per unit area, γ. The energy (E_f) required for fracture is a function of the *area* (A_f) of new surface created by crack propagation. Thus:

$$E_f = \gamma A_f \tag{2}$$

For a machining depth d, it is reasonable to assume that the order of magnitude of both V_p and A_f are determined by d. That is:

$$V_p \propto d^3 \tag{3}$$

$$A_f \propto d^2 \tag{4}$$

Consequently, as a the scale of machining decreases, plastic flow becomes an energetically more favorable material removal mechanism, ie, assuming that H and γ are fixed, E_p/E_f will decrease as d decreases. The specific depth at which a brittle-ductile transition occurs is a function of the intrinsic material properties governing plasticity and fracture.

For a hard material, the predominant material removal mechanism will be brittle fracture for large grinding infeed rates. The ductile-regime grinding hypothesis predicts that there will be a transition to a predominantly ductile material removal mechanism at lower infeed rates [6][7]. The existence of this transitional infeed rate was demonstrated on the PEGASUS apparatus through a series of grinding tests on fused silica.

Using scanning electron microscopy for post-grinding surface analysis, the ground material surfaces were examined for evidence of surface fracture, which would be indicative of the grinding ductility. A grid counting technique was devised to quantify the areal percentage of surface fracture. This technique involved overlaying a grid on the photomicrograph and finding the number of intersections which occurred on damaged regions. By applying this measure of grinding ductility to fused silica samples that were ground using a range of infeed rates, the transitional grinding infeed rate was determined for this material. Figure 2 illustrates this brittle-ductile transition as a function of grinding infeed rate for fused silica.

It was found that a change in grinding infeed rate from $1.5 nm/rev$ to $75 nm/rev$ resulted in a transition from 5% surface fracture to 99% surface fracture. This result is important for three reasons. First, it verifies that the design principles used in the development of PEGASUS were appropriate for ductile-regime grinding. In addition, the microgrinding range of material removal rates is shown to be an important, unexplored region of machining, with significant potential for improving the machinability of glasses and ceramics. Finally, from this series of grinding tests the basic hypothesis of ductile-regime grinding (ie, brittle-ductile transition for a reduced infeed rate) is validated for fused silica. Fused silica is a relevant material for this study because it is particularly brittle and thus difficult to machine in a ductile regime. In spite of this complication, fused silica is an important engineering material, resistant to high temperatures and severe thermal shock. Because of these properties, it is often used in precision molding applications, for which contour accuracy and surface finish are both critical. Thus, both the challenge and the potential benefits of microgrinding are demonstrated by the use of fused silica as a model material.

Critical-Depth-of-Cut Model To investigate the influence of material properties on the brittle-ductile transition rate, a broad range of amorphous glasses, single crystals, and advanced ceramics were chosen. By comparing the grinding ductility of these materials to their intrinsic properties, a critical-depth-of-cut model [8][9] has been established for microgrinding. This model relates the measured critical grinding infeed rate for 10% surface fracture to a calculated critical-depth-of-cut based on the material properties. The basic model consists of the properties of the material that characterize its fracture toughness (K_c), its plastic deformation resistance (H) and its elastic stiffness (E).

The basic form of the model can be simply rationalized in the following manner. The elastic strain energy (E_s) available in the plastic indentation zone is:

$$E_s \propto \left(\frac{H^2}{E}\right) d^3 \tag{5}$$

The energy required to form the crack surface is:

$$E_f \propto \gamma d^2 \propto \frac{K_c^2}{E} d^2 \tag{6}$$

when $K_c = \sqrt{2 \propto E}$ for ideal brittle fracture. A necessary condition for fracture initiation is that $E_s \geq E_f$, which then gives the critical zone size:

$$d_c \propto \left(\frac{K_c}{H}\right)^2 \tag{7}$$

A more detailed analysis [8] [9] leads to the final model equation:

$$d_c \propto \left(\frac{E}{H}\right) \left(\frac{K_c}{H}\right)^2 \tag{8}$$

The relevant properties were measured for each material using microindentation techniques. While the measurement of hardness by indentation is a standard procedure, determination of K_c and E by indentation is a developing area of research.

The properties of the material surface vary with the indentation depth at which they are measured. This size scale variability is especially troubling for the measurement of K_c. Size-scale effects lead to a dependence of K_c on crack size (R-curve behavior), which can be a large effect in certain materials. Such material-related property variations complicate the extrapolation of properties from the scale of indentation ($> 10\mu m$) to the scale of microgrinding ($< 1\mu m$). In spite of these problems calculating the material properties, a reasonable correlation was obtained between the calculated critical-depth-of-cut and the measured critical grinding infeed rate (i.e. the grinding infeed that will produce 10% surface fracture) This correlation is illustrated by the graph in Figure 3.

After establishing a material model for grinding ductility, an investigation was begun to isolate machining parameters other than infeed rate that might influence the grinding ductility. The variables tested were: grinding speed, wheel bond material, and the grinding fluid chemistry. While by no means exhaustive the goal of this series of tests was to discriminate between those variables that affect the grinding ductility and those that do not.

Grinding Speed

The influence of the grinding wheel speed on the grinding ductility was measured for fused silica, zerodur, and silicon. The wheel speeds tested covered a range from $1.5m/s$ to $12.5m/s$, and all other machine parameters were maintained at constant values (including the workpiece infeed rate per wheel revolution) For all three of the materials, no significant change was found in the grinding ductility as a function of wheel speed over this range. To a first order approximation, then, the grinding wheel speed has no effect on the ductility of the material removal process. To some extent, this speed independence implies a minimal influence of the grinding temperature on ductility, since the contact temperature is linked to the grinding speed. If thermal effects in grinding were influential in determining grinding ductility, it would be expected that this influence would be made apparent by a large change in the grinding speed.

Wheel Bond Material

While the contour accuracy inherent in this grinding apparatus is directly related to its large structural stiffness, this quality makes it difficult to achieve the low cutting forces required for ductile material removal. A way to decrease the abrasive grain forces without measurably altering the rigidity of the structural loop is to use a more complaint abrasive grain bonding matrix in the grinding wheel. The primary effect of this less rigid matrix is to more evenly distribute the load among the many diamonds in the contact regions. In addition, this local compliance reduces the forces that accompany small undesired motions (e.g. runout) of the wheel with respect to the workpiece. This theory was tested by grinding fused silica and zerodur using both metal bonded ($E = 110GPa$) and polymer resin ($E = 7GPa$) bonded grinding wheels. The range of infeed rates chosen ($1.5nm/rev$ to $37.5nm/rev$) for these tests encompasses the brittle-ductile transition range for both these materials, so the effect of the bond matrix on grinding ductility could be assessed. Figure 4 illustrates the results of these tests.

As expected, the more resilient bond provided enhanced ductility for both materials, increasing the critical grinding infeed rate (corresponding to 10% surface fracture) by a factor of 2. This significant influence of local grain compliance can be exploited by designing grinding wheels specifically for ductile-regime grinding. The ideal bond matrix would have a low elastic modulus, a strong bonding affinity for diamond, and a good resistance to wear.

Grinding Fluid Chemistry

The influence of the grinding fluid is perhaps the most difficult variable to characterize in microgrinding. A series of experiments on fused silica, quartz, and silicon using both water and a 1:1 ethanol-methanol mixture has provided some insight into the potential effect of the grinding fluid on the ductility of the material removal process. For both fused silica and quartz, a dramatic increase in the brittleness of the material removal process accompanied a change in grinding fluid from water to the alcohol mixture. For single crystal silicon, a slight decrease in the brittleness of the grinding process accompanied the same change in grinding fluid. This result points toward a chemical effect: the chemical compositions of amorphous fused silica and crystalline quartz are virtually identical (silicates) and different from that of crystalline silicon. If the effect could be attributed to differences between the thermal or lubrication properties of the two fluids, it would be expected to influence the grinding ductility of the three materials in the same way, if not by the same amount. Chemical interactions between the grinding fluid and the material workpiece can harden or soften the surface layer, and advance or retard crack growth. As such, these effects must be individually categorized for each fluid/workpiece pair to determine the optimum fluid for a given material workpiece.

IN-PROCESS MEASUREMENT OF GRINDING DUCTILITY

The hypothesis relating fracture damage to the infeed rate appears to be a reasonable model for the transition from the ductile to the brittle regime. However, accilinary variables such as wheel material and coolant can greatly influence the location of the transition zone. Therefore, a method of monitoring grinding ductility in real-time is necessary and in the following section two candidate techniques are discussed.

Acoustic Emission

Acoustic Emission (AE) monitoring is a promising sensor for real-time control of grinding ductility. AE signals result primarily from fracture events in the material workpiece, and thus a direct measure of the grinding ductility can be made. Measurements of the grinding ductility (i.e. surface fracture) versus the peak amplitude of the AE power spectrum for both germanium and soda line glass are plotted in Figure 5. In each case the AE power is a strong function of the percentage of surface fracture.

As a measure of grinding ductility the peak amplitude of the AE power spectrum is remarkably sensitive (more than a hundredfold increase in amplitude for a 20% change in grinding ductility for germanium). In addition, for these two materials, the relationship between the AE peak and percent surface fracture is linear over the ranges measured. The only complication to this technique is that both the sensitivity and the amplitude of the AE signals are intrinsically linked to the specific workpiece material. To use such a measure of grinding ductility in process, then, would require individual calibration for each material type that is machined.

Grinding Energy

An alternative technique for in-process measurement of grinding ductility is obtained by monitoring the specific grinding energy during the machining process. This variable, which is uniquely determined by the tangential grinding force and the material removal rate, represents the energy required to remove a unit volume of material. Because plastic deformation requires more energy per unit volume of material removal than fracture, monitoring the specific grinding energy can allow a sensitive measure of the grinding regime. In particular, the *ratio* of specific grinding energies for two different grinding infeed rates is a reliable measure of the *change in ductility* over that particular range of grinding infeed rates. If there is no change in grinding ductility, the specific grinding energy ratio (R) will be ~ 1. For a change toward more brittle grinding R will be less than one, and in-fact will tend toward zero. This unique inverse relationship between the specific energy ratio and the ductile-brittle transition has been demonstrated to be largely independent of the material properties for all of the hard materials tested. Figure 6 is a plot of the specific grinding energy ratio versus the percent of brittle-ductile transition for nine different materials, illustrating the inverse relationship between these variables.

Though it requires real-time measurement of both grinding infeed and grinding tangential force, this material independence makes the use of specific grinding energy ratio a promising candidate for in-process control of grinding ductility.

DISCUSSION

Several important variables have been left unexplored for their influence on grinding ductility. Notably, the size and concentration of diamonds in the grinding wheel will undoubtedly influence the material removal process. The traditional philosophy of superfinishing has been that smaller abrasive grains produce a better surface finish, but this is not necessarily true in ductile-regime grinding. Recent efforts by Miyashita have challenged this idea by demonstrating that large grain wheels can be effectively utilized for superfinishing if the exposed diamond grains are first machined flat. In this way, each flattened diamond grain appears to the workpiece as a large, sharp, zero degree rake cutting tool. One direction for future research, then, is to compare this type of wheel to a fine grain wheel with random diamond orientation, to see which is more effective in ductile-regime grinding. In any case, a study of the role of the diamond grain characteristics in material removal would contribute significantly to the overall understanding of the micromechanics of ultraprecision grinding.

This research was aimed at a fundamental understanding of the brittle-ductile transition in grinding. Extension of the results of this work to the development of a prototype grinding station is the next step. In this case wheel trueing, motion control, and machine stiffness must all be addressed in two or three dimensions, instead of just one dimension. Such an extension is a vital step in the research of ductile-regime grinding. Indeed, an extension to two dimensions might *simplify* characterization of grinding ductility; flat surfaces could be measured using profilometry techniques to assess both their form and surface finish. This measurement could provide information about the ground surface quality that is easier to obtain and quantitatively more repeatable than the SEM "grid-counting" technique.

A promising area for research is the pursuit of a closed-loop control of grinding ductility. The potential for monitoring either acoustic emission or specific grinding energy ratios as an in-process measure of grinding ductility has been clearly identified. The utility of such real-time, in-process control is that it would allow material removal to progress in stages: by a brittle process for larger rates of stock removal of material and by a ductile process for finishing. The principles and guidelines established by this study of ductile-regime grinding make this type of sophisticated control of the grinding process a realistic research goal.

References

[1] Bifano, T.G. Ductile Regime Grinding of Brittle Materials, PhD Thesis, North Carolina State University, Raleigh, North Carolina, January, 1988.

[2] Miyashita, M., 1st Annual Precision Engineering Conference, North Carolina State University, Raleigh, North Carolina, November,1985.

[3] King, R. F., and Tabor, D., Proc. of the Roy. Soc. London, A223, p.225, 1954.

[4] Huerta, M., Malkin, S., J. Eng. for Industry, pp.459-467, May 1976.

[5] Chandrasekar, S., Sathyanarayanan, G., 15th North American Manufacturing Research Conference Proceedings v.2, Manufacturing Technology Review, pp. 499-505, 1987.

[6] Yoshioka, J., Koizumi, K., Shimizu, M., Yoshikawa, H., Miyashita, M., Kanai, A., SME Technical Paper MR82-930, 1982.

[7] Yoshioka, J., Koizumi, K., Shimizu, M., Yoshikawa, H., Miyashita, M., Kanai, A., Milton C. Shaw Grinding Symposium, R. Komanduri, D. Maas, eds., ASME Production Engineering Division, v.16, pp.209-227, 1985.

[8] Toh, S. B. and McPherson R., 2nd Int. Conf. Science of Hard Materials, Rhodes, 1986.

[9] Marshall D. B., and Lawn, B. R., Microindentation Techniques in Matls. Sci. and engrg., p.26, 1985.

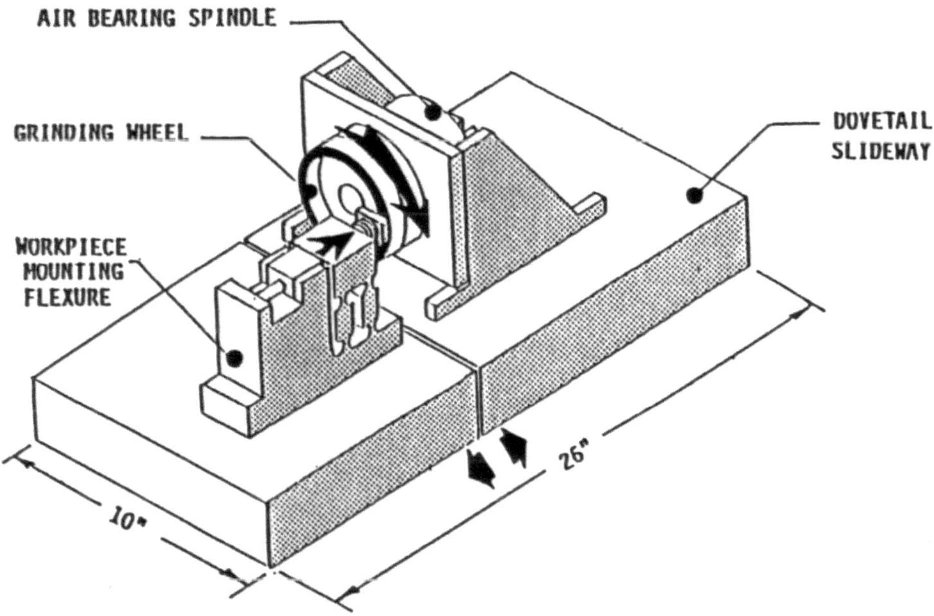

Figure 1: Drawing of the PEGASUS microgrinding apparatus.

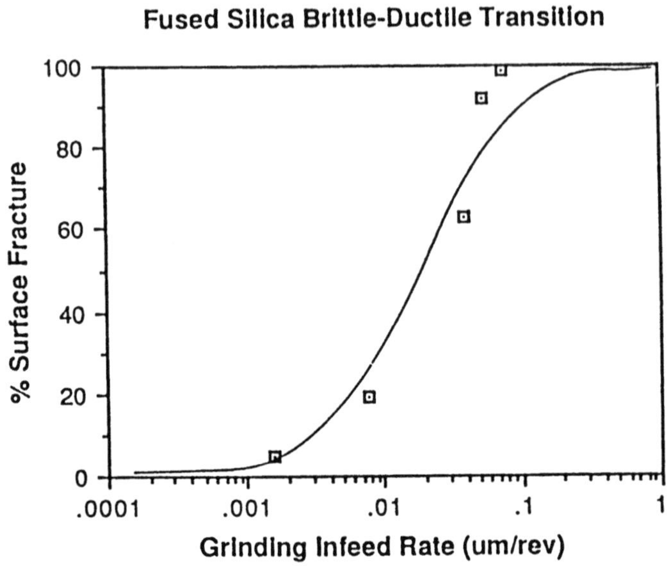

Figure 2: Brittle-ductile transition as a function of grinding infeed rate for fused silica.

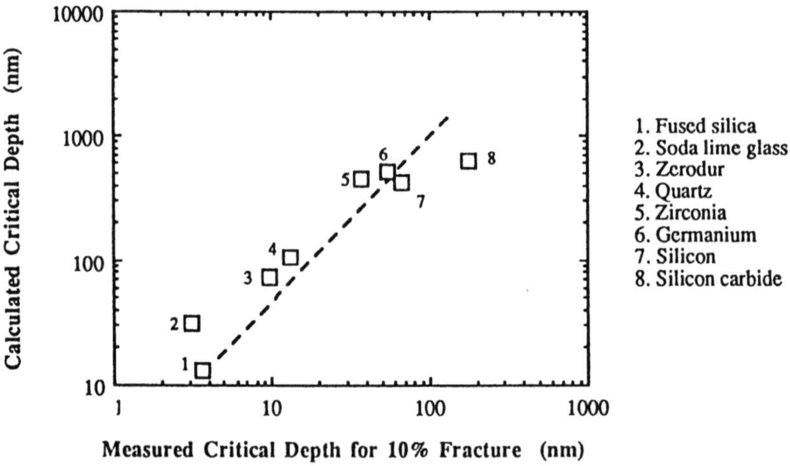

Figure 3: Measured critical grinding infeed rate for 10% surface fracture versus the calculated critical-depth-of-cut

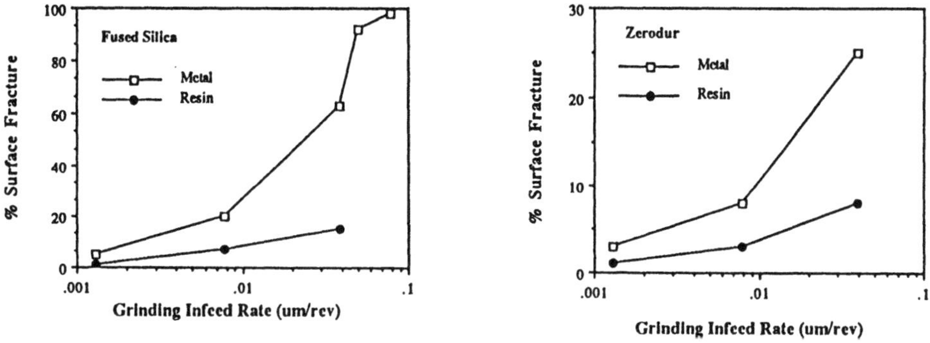

Figure 4: Effect of wheel bonding matrix on grinding ductility. Right: Fused Silica. Left: Zerodur.

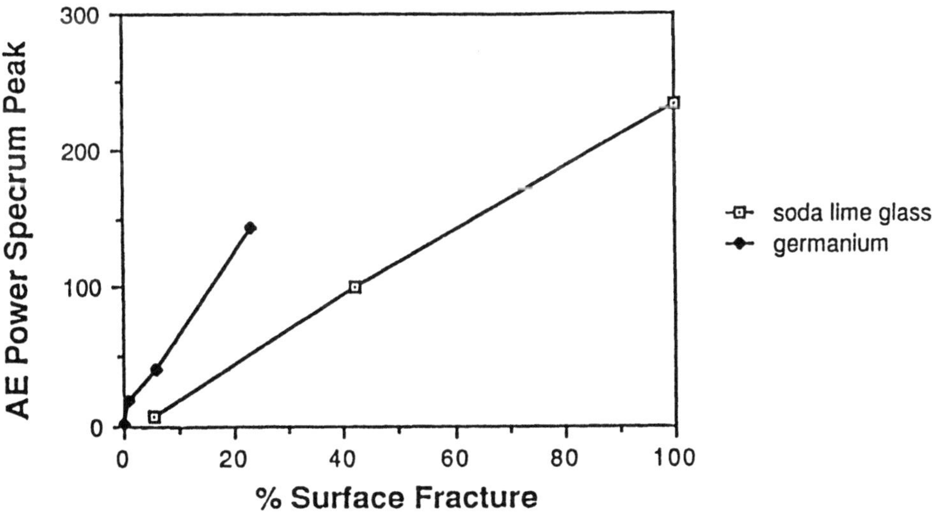

Figure 5: AE power spectrum peak level versus grinding ductility

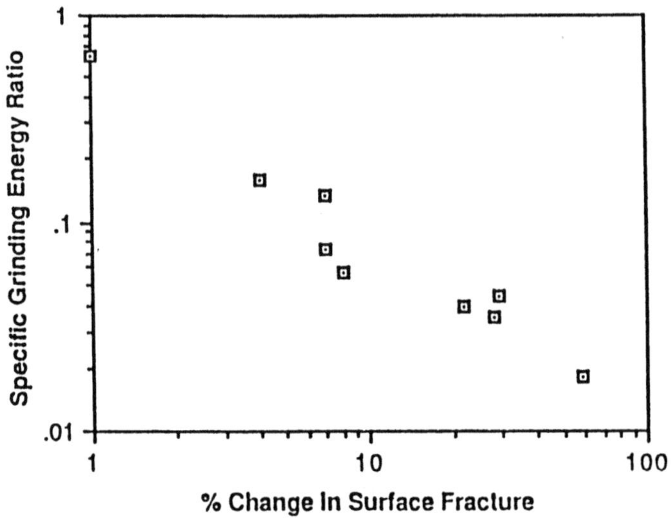

Figure 6: Specific grinding energy ratio versus % ductile-brittle transition for nine brittle materials, measured over a range of infeed rates from $1.5nm/rev$ to $37.5nm/rev$.

Review on Ultraprecision Grinding Technology

M. MIYASHITA

Mechanical Engineering Department
Ashikaga Institute of Technology
268 Omaecho, Ashikaga, Tochigi, 326 Japan

Abstract
Today's art of grinding technology is based on the brittle mode application of grinding wheel, which results the poor motion copying capability on the grinding wheel in the trueing process, difficulties of controlling the wheel wear characteristics and a lack of predictability in grinding processes. For overcoming these problems for realisation of ultraprecision grinding, the microgrinding technology is proposed, basing on the ductile mode application of grinding wheel in both trueing and grinding processes, which means the crack free and precise surface generation on the grinding wheel and strategies for improving the wheel wear characteristics. And several test results and the expectation in future are shown.

1. Can the today's art of grinding technology achieve ultraprecision machining

The survey on industrial problems in grinding by R. Kegg [1] has pointed a critical lack of predictability and resulted difficulties of synthesizing reasonable grinding processes. This shows that the grinding phenomena are characteristic of statistical variation and it is difficult to derive formulas as causality. The survey gives us industrial problems for "today's grinding accuracy". Can the today's grinding technology apply to ultraprecision machining aiming at "ultraprecision"? N. Taniguchi[2] has insisted grinding processes not to be applicable to ultraprecision machining. It is requested to realize very fine material removal process for ultraprecision machining. And at the depth of cut in the range of submicrometer specific cutting energy will be extraordinary increased due to low dislocation density of work materials and resulted increased resistance of cutting grains accelerates wheel wear. This is the reason why not achievable "ultraprecision grinding".

R. McClure[3] expresses that the grinding process is a kind of the mutual wear phenomena of both the workpiece and the grinding wheel, which arises critical difficulties of achieving ultraprecision grinding.
The followings are typical problems of the today's art of grinding technology in the viewpoint of ultraprecision machining.

2. Problems of the today's art of grinding technology
- brittle mode application of grinding wheel -

In the today's art of grinding technology, single or multipoint diamond dresser generates specified surface on grinding wheel. Some of the grinding characteristics commonly resulted from such application method of grinding wheel are shown here.
Fig.1[4] is an example of shape error of regulating wheel trued with single point diamond dresser. The runout of 14 um and the generator profile error of 60 um/150 mm are very poor in comparison with the spindle runout of 0.1 um and the truer carriage straightness of 1.0 um/150 mm correspondingly, which means a poor motion copying. Fig.2 is the wheel wear characterstics, which is composed of the hard initial wear region and the semi-stationery region. The former is very dependent of residual cracks and chips caused in the trueing process and the grinding ratio in the latter region is dependent of the applied grinding load. Fig.3[5] is the variation of surface roughness in grinding. The transient roughness variation and the semi-stationary roughness variation dependent on the grinding load show a good correspondence to the wheel wear characteristics.
Generally, failure modes of brittle materials are separated into two modes, a brittle mode and a ductile mode. The transition is decided by the size of plastically deformed layer caused by the interference between the material and the tool[6]. Fig.4[7] is the ground surface finishes of quartz crystal in brittle mode and ductile mode. By reducing the depth of cut per pass from 1.7 um to 0.2 um, the brittle mode failure changes to the ductile mode failure in grinding.
In a normal trueing process with single point dresser, the

dresser tip is about 0.2 mm to 0.4 mm in spherical radius and the depth of cut is about 0.01 mm to 0.1 mm. Such material removal process will arise a brittle mode failure on the wheel comparing with the grinding results in Fig.4. As a result, it is easy to understand that the motion copying errors are greater by more than one order comparing with the relative motion errors between the workpiece and the tool. The today's grinding theory is based on the principle of self-sharpening of grinding wheel, which means the brittle mode failure of grinding wheel.
Mechanical fractures of brittle materials are understood to be complicated and discontinuous brittle failure phenomena like the mechanical strength being expressed its variation with Weibull parameter m. Similarly, in the today's grinding technology using brittle mode failure of grinding wheel, simulation of grinding process, estimation of parameters etc. are not easy, and repeatability of grinding performance and predictability of grinding phenomena are poor, all of which are considered to be caused by the brittle mode failure of grinding wheel. And such grinding technology characteristic of essential variation is very difficult to achieve ultraprecision machining.
In addition, in the surface integrity viewpoint for ultraprecision machining, such problems as remained stress, hardened layer and so on are remained.

3. Microgrinding technology using ductile mode failure of grinding wheel-new application of grinding wheel

As shown in Fig4, the ductile mode grinding is superior to the brittle mode grinding in respect of both geometrical accuracies and surface integrity. From this point of view, trueing the grinding wheel by ductile mode grinding is necessary to achieve the wheel surface with small shape errors and remained cracks sufficient for ultraprecision grinding. Trued wheel surface with single point diamond dresser and ground wheel surface at the depth of cut of 0.2 um are shown in Fig.5. In (a), grains and bond material fracture in brittle mode, and so the distribution of cutting points is coarse. In (b), they are ground in ductile mode uniform and almost no crack and chip are observed. Trueing by ductile mode grinding is called "microtrueing". Fig.6[4] shows the shape errors of microtrued regulating wheel. Comparing with Fig.1, shape errors are reduced by more than one order. Fig.7[7] (a) is the shape error of microtrued grinding wheel. These results say that shape errors of microtrued wheel are small and motion copying accuracy is high. Fig.7(b) is the generator profile of grinding wheel after continuous through-feed centerless grinding for 150 min. Fig.8[7] shows the wheel wear characteristics for 150 min. and the worn cutting grains. Fig.9[7] is the variation of surface roughness during this period. The common characteristics to these test results is that there is no initial or transient region in both the wheel wear and the surface roughness, and the wheel wear rate and the surface roughness are kept almost constant during continuous grinding for 150 min. This shows cutting point distribution being kept constant during the period. As Fig.8(b) shows ductile mode wear of cutting grains, the wheel wear is also ductile mode wear and so continuous phenomenon.

Grinding process in the range, in which grinding wheel wears continuously in ductile mode is called "microgrinding".In microgrinding, mechanical strength of cutting grain is estimated to be stronger due to ductile mode failure than brittle mode failure strength, and finer material removal rate and reduced wear rate of cutting grains due to the maximum cutting point distribution density and the uniform height of cutting points are expected. Such possibilities are shown in Fig.10 [8].

4. Several test results in microgrinding

The followings are the photographs of the shapes of cutting grains, the debris removed by them and the grinding streaks on the workpiece with a common mognified scale, when the hardened steel being microground with SiC wheel. Fig.11 is the cutting grains at the cumulative removal of Vw= 640 mm^3/mm after microtrued. It is characteristic of the uniform cutting height of grains. The sizes of cutting grains are about 20 um to 30 um. Fig.12 is the SEM microphotograph of debris removed with these cutting edges. The sizes of debris are about 2 um to 3 um in width and the continuous ribbon-like debris are observed. The resulted ground surface from removing these debris is shown in Fig. 13. The grinding streaks are observed being about 2 um to 3 um in width, which correspond to the size of debris. The followings are the list of the root mean square measure of the spacial wavelength λ_q and the mean spacing of adjacent local peaks S for 6 microground samples.

Sample No.	1	2	3	4	5	6
λ_q (um)	3.72	10.38	3.91	9.91	10.64	3.87
S (um)	1.33	5.20	1.38	4.95	5.33	1.36

From the above, the width of grinding streaks is approximately equal to the λ_q or the integer multiplification of it.
Surface integrity of microground surfaces is also tested, and Figs.14 and 15 (10) are examples of the resulted residual stress and surface hardening from microgrinding. Here, it is remarkable that the residual stress is compressive and damaged layer thickness in stress and hardness is reduced to less than one tenth, 5 um to 6 um.

5. Possibilities of microgrinding technology for ultraprecision machining

For reducing motion copying error and machined surface damage to the minimum, ultrafine material removal process has to be achieved. Fig.16 shows the ranges of normal grinding and polishing in material removal rate Z'. For ultraprecision and massive production technology, material removal process based on the principles of motion copying and capable of achieving optical accuracy by the today's art of polishing has to be developed. Microgrinding is expected to expand the controllable range of Z' up to the range of polishing $Z' \lesssim 10^{-4}$ mm^3/mm.s, ultrafine material removal process.

6. Overview from microgrinding to nanogrinding

For meeting the frontier of machining for nanometer accuracy (11) the present microgrinding has to be improved to a nanogrinding, which will be compared on Table 1.

Table 1. Comparison of microgrinding with nanogrinding

	Microgrinding		Nanogrinding	
Spec. of G/M				
Motion error	100 nm,	1 urad	10 nm,	0.2 urad
Resolution of positioning	100 nm		10 nm	
Work/wheel support				
stiffness	order of 10 kgf/ m		order of kgf/nm	
Spec. of G.W.				
Shape error	100 nm		10 nm	
Height uniformity	100 nm		10 nm	
of cutting grtains				
Target of grinding accuracy				
Size errror	100 nm		10 nm	
Shape error	100 nm		10 nm	
Roughness	10 nm		order of nm	

References

(1) Kegg, R., Industrial problems in grinding, Annals of CIRP Vol.32/2/1983
(2) Taniguchi, N., Current stauts in, and future trends of ultraprecision CIRP Vol. 32/2/1983.
(3) McClure, R., Development and research trend in ultraprecision machining, ONRFE SCI. BUL. 10(4), 1985
(4) Hashimoto, F., Kanai, A., Miyashita, M., High precision trueing method of regulating wheel and effect on grinding accuracy, Annalls of CIRP Vol.32/1/1983
(5) Verkerk, J., The influence of the dressing operation on productivity in precision grinding, Annals of CIRP Vol. 28/2/1979.
(6) Puttick, P.E., The correlation of fracture transitions, J.Phys. D: Appl. Phys., 13 (1980)
(7) Yoshioka, J., Hashimoto, F., Miyashita, M., Kanai, A., Abo, T., Daito, M., Ultraprecision grinding technology forfor brittle materials, ASME Winter Ann. Meeting, PED-Vol.16.
(8) Yoshioka, J., Hashimoto, F., Miyashita, M., Kanai, A., Daito, M., Application of grinding wheel to ultraprecisionmachining, Intersociety symp./A Cer S, JSME, AES in Pittburgh, April, 1987.
(9) Yoshioka, J., Hashimoto, F., Koizumi, K., Miyashita, M., Kanai, A., Daito, M., Hasebe, T., Proposals of ultraprecision diamond grinding technology for brittle materials, NAMRC, 1985.
(10) Yoshioka, J., Koizumi, K., Shimizu, M., Yoshikawa, H.,Miyashita, M., Kanai, A., Surface grinding with a newly developed ultraprecision grinding machine, NAMRC, Mfg. Eng. Transaction (1983)
(11) Franks, A., Nanotechnology, to be published in J. Phys.E.,

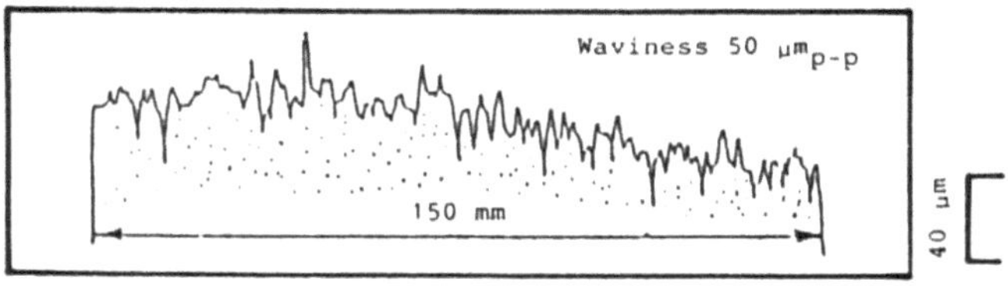

(b) Generator profile

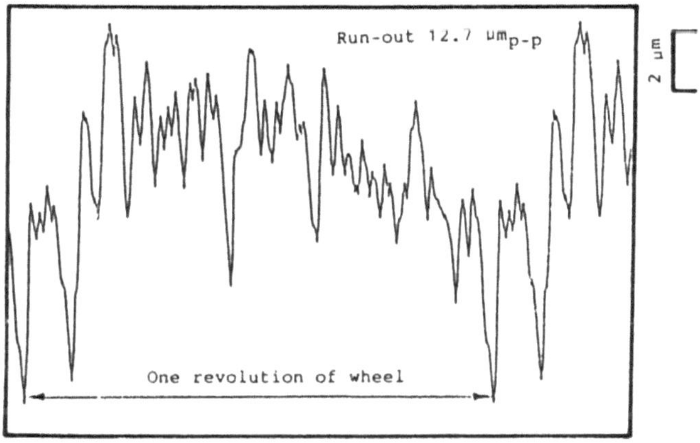

(a) Run-out of regulating wheel periphery at work contact point

Fig.1. Trueing accuracies of regulating wheel with single point diamond dresser

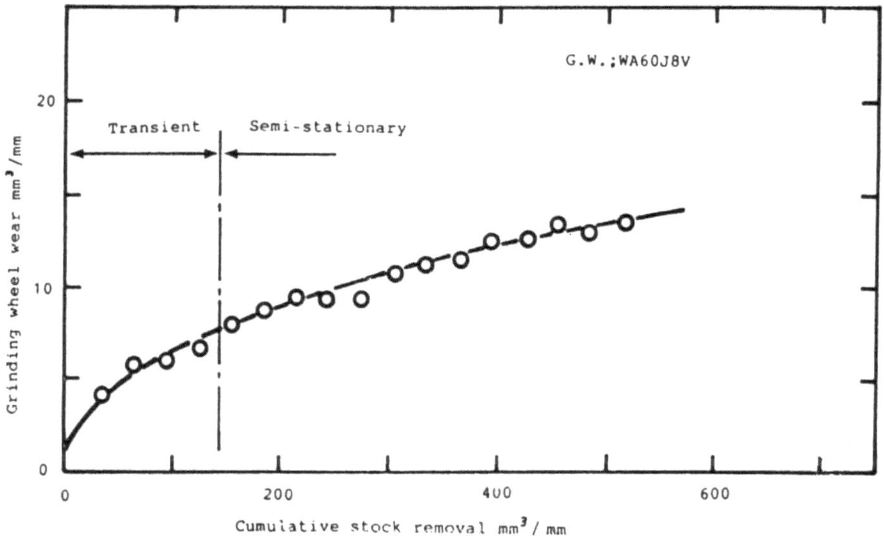

Fig.2. Wear characteristics of grinding wheel

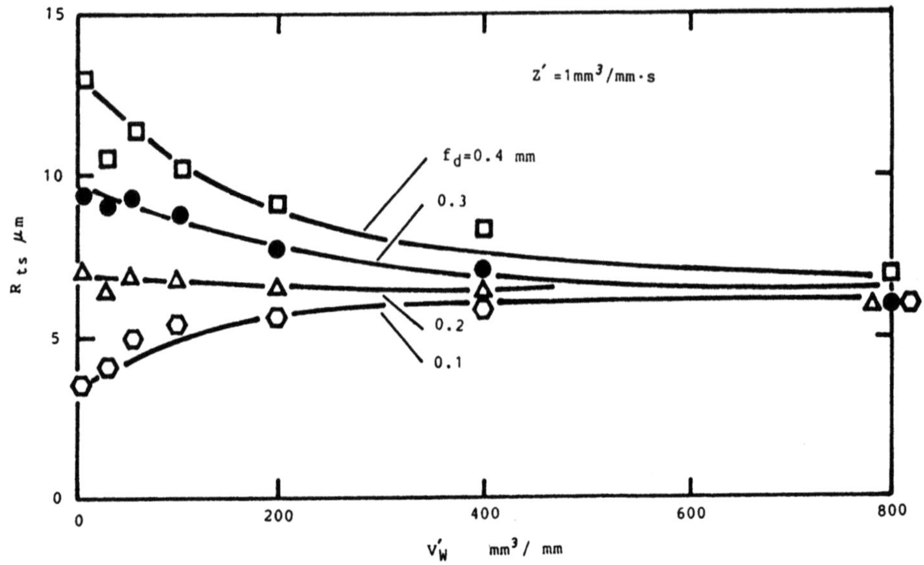

Fig.3. Variation of surface roughness in grinding process

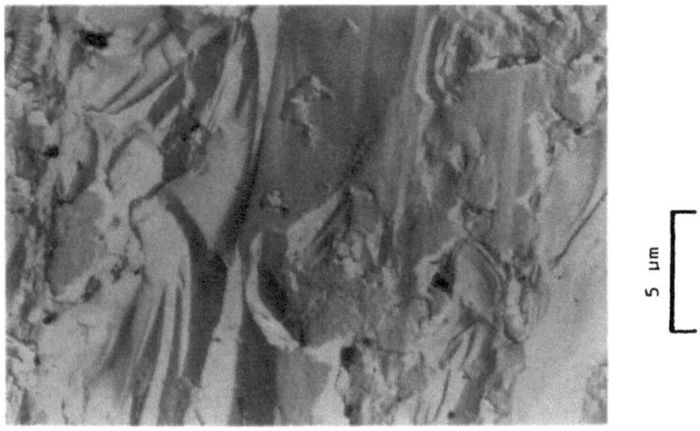

(a) Brittle mode grinding

(b) Ductile mode grinding

Fig.4. Grinding of quartz cryatal in brittle mode and ductile mode

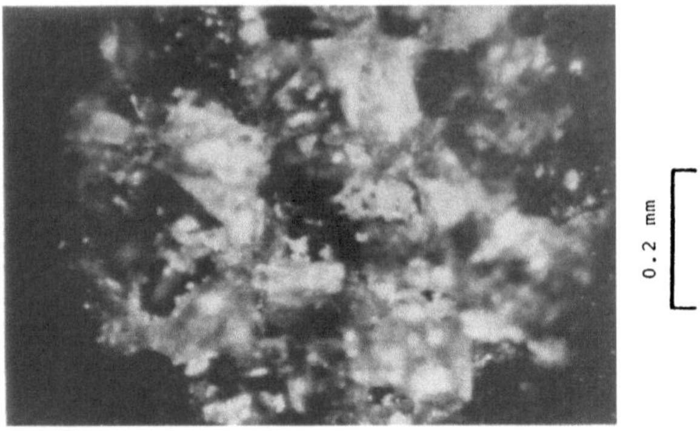

(a) Brittle mode trueing

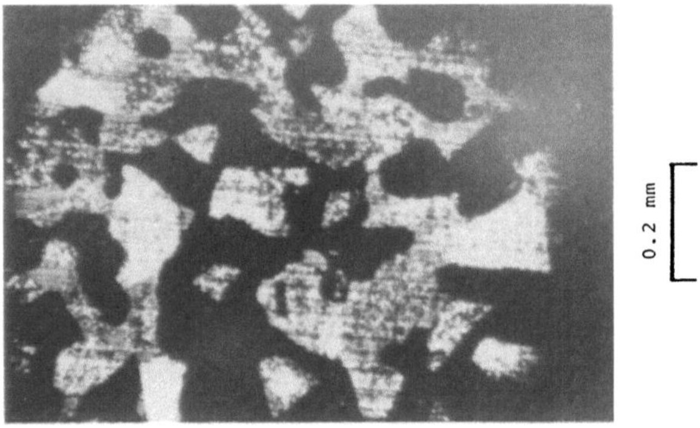

(b) Ductile mode trueing

Fig.5. Brittle mode and ductile mode trueing of grinding wheel

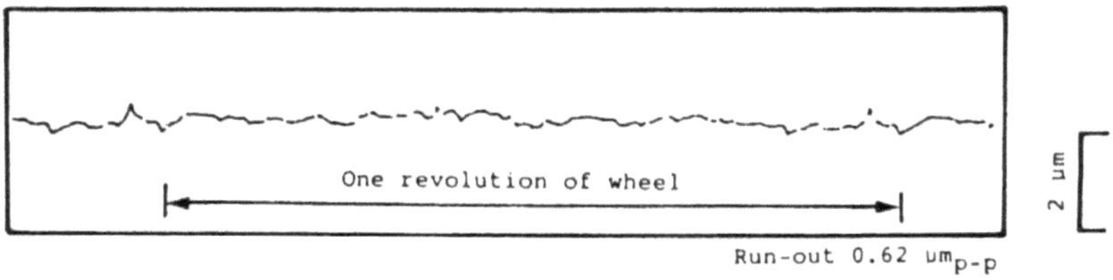

(a) Run-out of regulating wheel periphery at work contact point

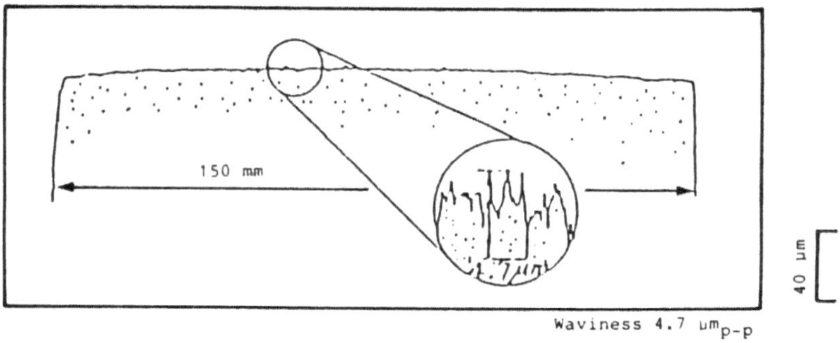

(b) Generator profile

Fig.6. Shape accuracy of microtrued regulating wheel
-trueing with grinding wheel

Grinding wheel : $350^{Dia} \times 150^{W}$
Workpiece : SUJ2, $10^{Dia} \times 10^{W}$
Chip removal : 5 µm/pass

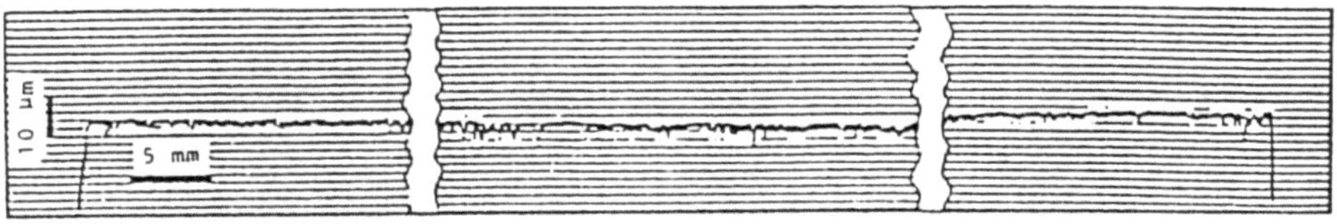

(a) Generator profile of microtrued grinding wheel.

(b) Generator profile of microtrued grinding wheel after continuation of grinding for 150min..

Fig.7. Microtrued wheel accuracies as trued and after continuous through-feed grinding for 150 min.

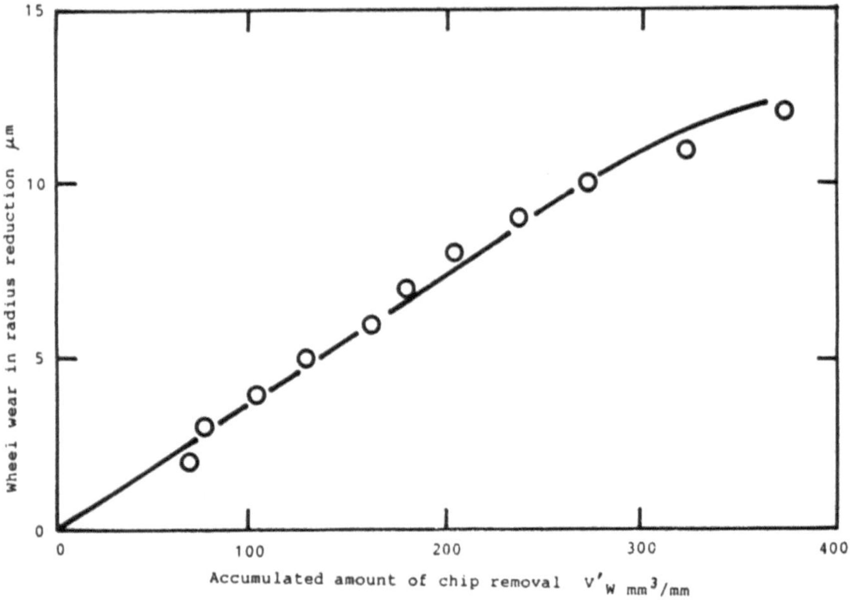

Fig.8(a). Wear characteristics of microtrued grinding wheel

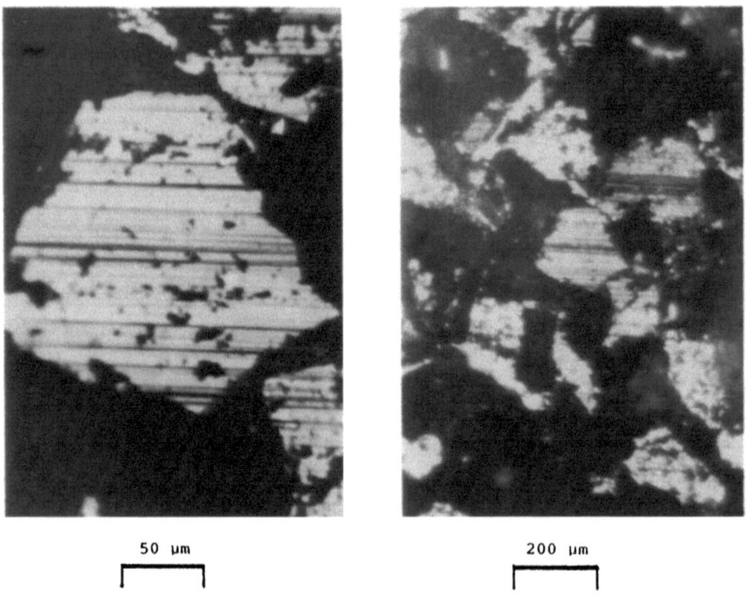

Fig.8(b). Microphotograph of microtrued grinding wheel surface after continuation of grinding for 150 min.

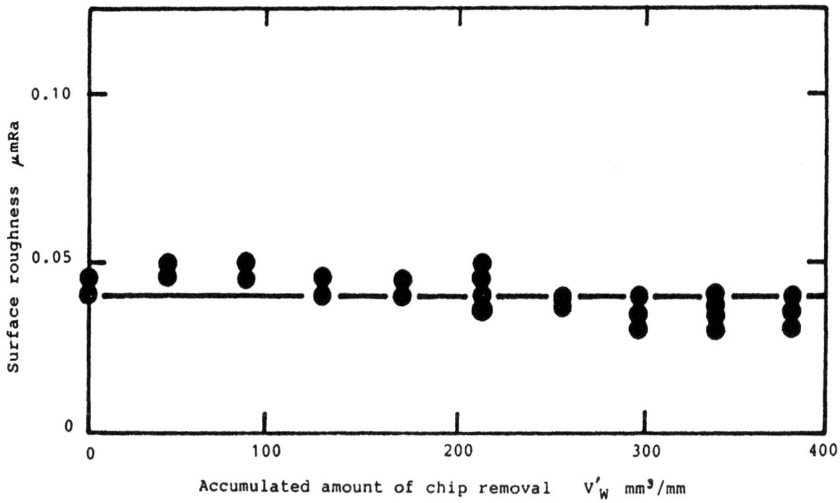

Fig.9. Variation of surface roughness in microtrueing process

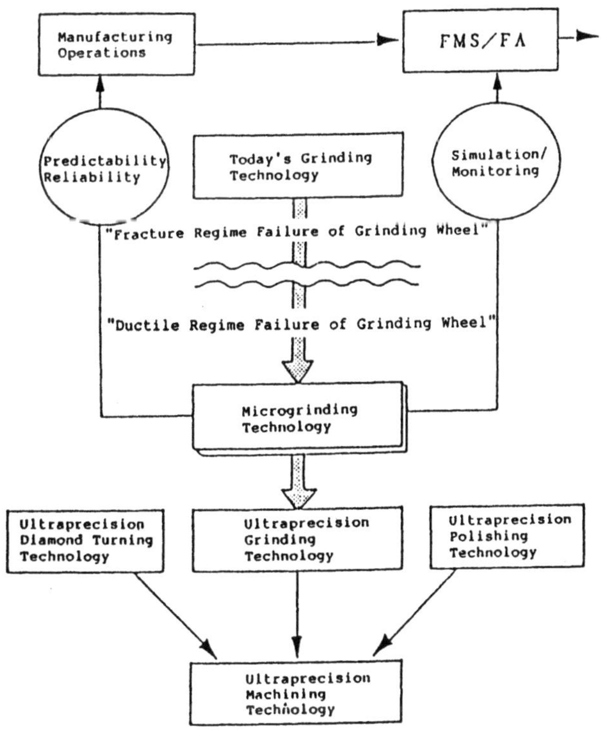

Fig.10. Possibilities of microgrinding technology

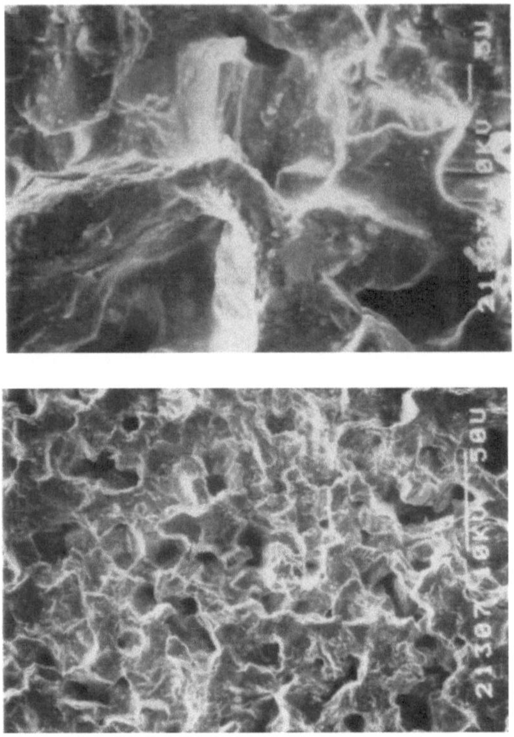

G.W. : GC800MV

V_w : 640 mm^3/mm

Fig.11. Microphotos of microtrued cutting points of grinding

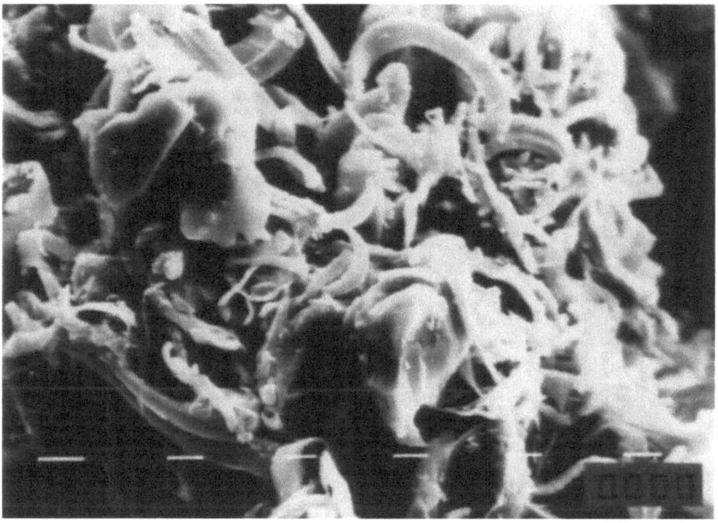

Z' : 0.2 $mm^3/mm.s$
W.P. : Hardened steel

Fig.12. Microphotos of debris in microgrinding

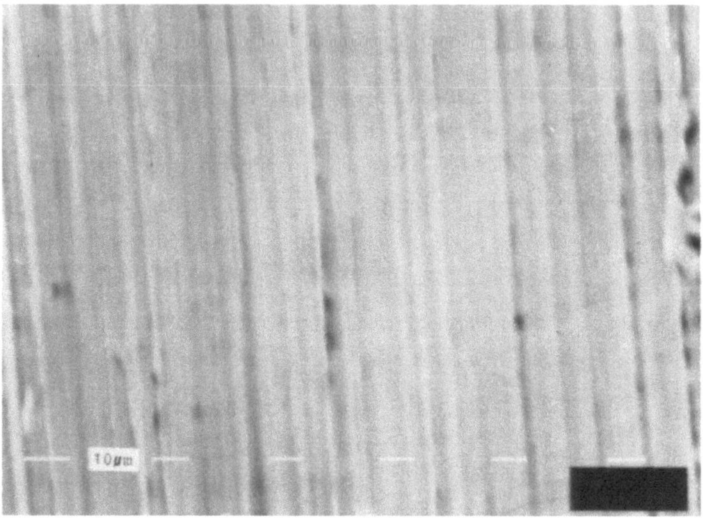

Fig.13. Microphotograph of microground surface

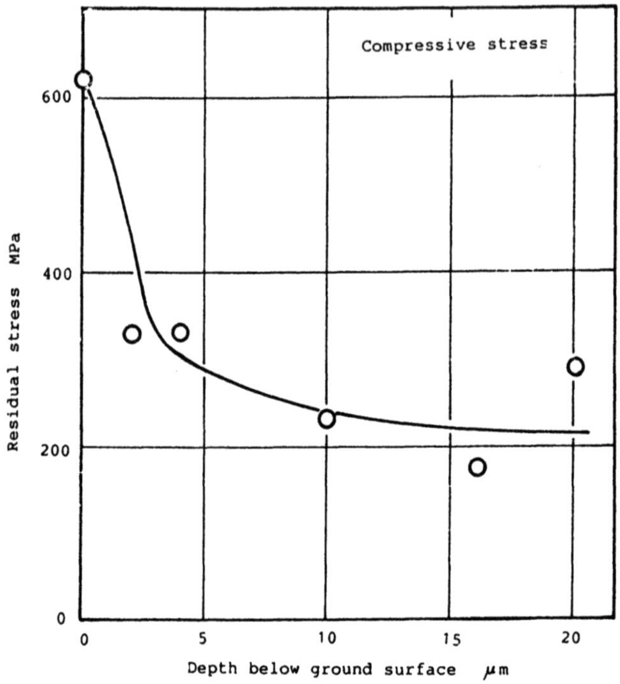

Fig.14. Resulted residual stress from microgrinding

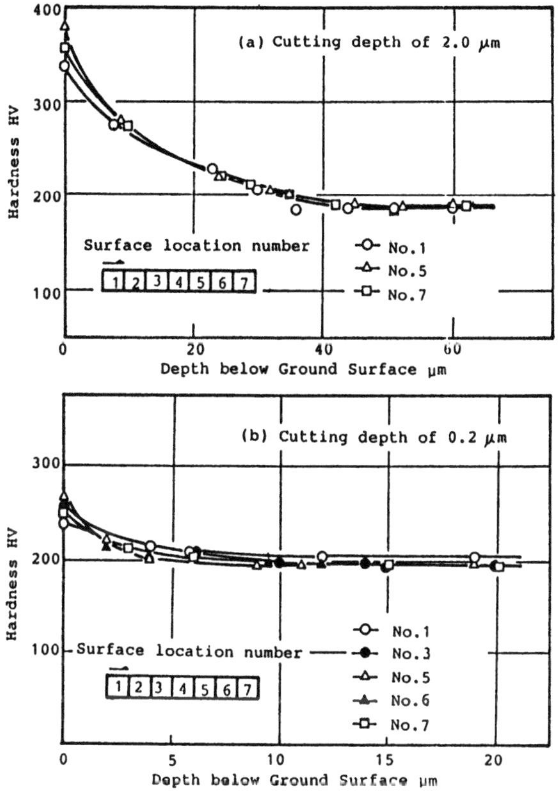

Fig.15. Resulted hardening of austenitic stainless steel from microgrinding

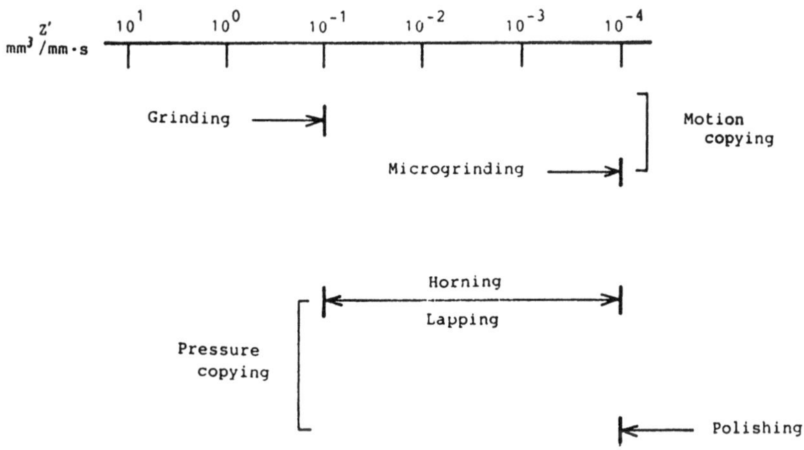

Fig.16. Ranges of material removal rate for different machining process

Precision Machining of Advanced Ceramics – A Challenge in Production Technology

W. KÖNIG; M. POPP

Fraunhofer - Institut für Produktionstechnologie IPT - Aachen

Summary

Whenever component surfaces have to meet highest requirements in terms of hardness and strength, advanced ceramics are the most appropriate choice of material.

The advantageous performance features of these materials, however, are opposed by extremely delicate machining properties. The difficulties of machining even multiply with tightening quality specifications. Consequently precision engineering still is far away from a common use of such materials.

This paper will give a review of the problems related to precision machining of advanced ceramics and present some solutions of a considerable potential in terms of reliable quality and efficiency.

Introduction

High requirements on material properties are the reason for using ceramic materials for many applications in mechanical and process engineering.

Ceramic materials are defined as inorganic, nonmetallic and not easily dissoluble in water. 30% of the ceramic volume have to be crystalline phases, the material properties are achieved by a treatment at a temperature of more than 800°C /1/. This definition is applied to an extremely large group of materials (figure 1). The term of advanced ceramics mainly covers the subordinate groups of nitrides, borides, carbides and oxides.

As compared to metals ceramic materials show superior hardness and compression strength which decreases only slightly with temperatures up to 1000 °C. Other advantageous performance features are the enormous resistance to wear and to corrosion as well as the low specific weight. A disadvantage of ceramics is its high brittleness so that even the smallest defects in the component surface, which are caused by improper machining, can lead to disastrous failure if conditions of load distribution are unfavourable in operation.

Due to the almost complete lack of run-in behaviour of ceramics there is an additional requirement of high precision for the quality of the functioning surfaces of components for instance when surfaces are subject to friction and sliding wear (figure 2). Thus the application of silicon infiltrated silicon carbide (SiSiC) for bearings of pumps which transport corrosive and abrasive liquids requires a roughness value of $R_a \leqslant 0.08$ µm as well as an out-of-flatness value of $f_e \leqslant 0.3$ µm. Circon oxide (ZrO_2) or silicon nitride (Si_3N_4) as materials for rocker arm inserts and rams with a roughness value R_a smaller than 0.01 µm are to be mentioned as further examples.

With regard to the advantages of using ceramics for such applications the question is why these materials are not employed more often and above all on larger scale production.

The reason for this is - apart from the brittleness mentioned above - the extremely difficult machinability of ceramics. This problem becomes even worse when increasing requirements are posed in terms of component quality (figure 3).

Thus only a limited range of possible machining methods can be employed for finish machining. As for industrial practice only the techniques of cutting with geometrically undefined cutting edges such as grinding, honing, lapping and polishing can today be applied for precision machining.

Due to the extreme hardness of ceramics diamonds have generally to be used as cutting materials; the combination of high hardness and strength, however, causes considerable wear on the diamond itself or on the tools. On the one hand this results in higher machining costs, on the other hand the resulting high machining forces can lead to problems with regard to the accuracy to shape and dimension to be achieved. Last but not least the low material removal rates achievable result in long machining times thus impeding an economical process.

Concerning these problems this report will offer some considerations on the development of process parameters appropriate for machining ceramics with high precision. To this end examples of plane and cylindrical machining will be chosen. Furthermore, the potential of each technology in terms of achievable surface and shape quality will be examined.

Polishing plane surfaces

A traditional machining technology to produce plane surfaces is flat polishing.

Here, the workpieces to be machined are held in workpiece carriers between two polishing wheels which rotate in opposing direction. Due to the planetary motion of the carriers and the loose guiding of the workpieces this method performs advantageous machining without traces in a preferential direction (figure 4).

As contrasted to grinding and honing polishing is carried out with loose diamond grain which is held in a suspension and reaches the contact area between polishing wheel and workpiece surface through boreholes in the upper polishing wheel.

Depending on the amount of perpendicular force aplied and on the material properties of the polishing wheel employed the diamond grains will be temporarily embedded in the polishing wheel during machining. Removal type and, above all machining quality are affected by this.

If a hard polishing wheel is used the diamond grains with predominantly move in a sliding and rolling way and are forced into the brittle workpiece material due to the perpendicular force. Minute cracks are then induced into the surface of the workpiece so that material particles break away leading to material removal. In the case of employing a soft polishing wheel, however, diamonds can be embedded deeply into this wheel, thus producing a cutting material removal because of the relative motion between polishing wheel and workpiece surface.

In terms of achieving optimum surface quality the latter method of removing material is to be preferred, as here it is possible to completely remove the roughness peaks which are left from the previous rough machining process.

Flat machining of a silicon carbide workpiece shows the effects the use of different polishing wheel materials can have on the surface quality achievable. Roughness, which is characterized by the average roughness height R_a, is reduced from Ra = 48 nm to Ra = 7 nm, when aluminium or polyurethane-impregnated polyester is used instead of cast iron (figure 5).

The pliability of polyester provides an additional possibility for the diamond grains to recede back. Thus the generation of scratches by projecting grains can be avoided.

The pores in the polishing wheel are rather large as compared to the diameter of the diamonds used and provide a spacing large enough to hold diamonds which have got loose. They furthermore prove to be advantageous with regard to chip removal as well as to a constant supply of cooling liquid and new diamond grains. In this case there is no need for grooves as is necessary with non-porous polishing wheels.

The soft polishing wheels, however are detrimental with regard to the achievable shape accuracy. The smooth material can easily be distorted under the influence of the perpendicular force applied to the polishing wheel. Convex workpiece surfaces with more or less strongly rounded edges are a result.

When soft polishing wheels are used the advantage of extremely high surface qualities has to be considered against a better degree of shape accuracy. As illustrated in the example the out-of-flatness of f_e = 0.5 µm referring to the workpiece diagonal of 70 mm increased to f_e = 1.2 µm.

A compromise between achievable surface quality and shape accuracy can be realised with the use of aluminium as polishing tool. Due to the low hardness of aluminium the diamonds can fix themselves deeply in the polishing wheel which is a preconditon for generating a minute chip thickness. At the same time, the homogeneity of the material and its compression resistance, which is relatively high as compared to polyurethane impregnated polyester, avoid a deformation of the polishing wheel with the resulting disadvantages for the shape accuracy of workpieces to be machined.

In total the illustrated results impressingly show the potential of this technology. The surface qualities achieved even meet the requirements on plane mirrors for material processing lasers.

In spite of the long machining time caused by the low effective velocity this machining technology holds considerable possibilities as far as economy is concerned. For example, a number of workpieces can be machined on both sides at the same time. This proves to be an advantage, especially, when the problem of plane parallelism to be produced, e.g. when machining bearing rings, is taken into consideration. Moreover, the fact that this machining technology does not require any tool conditioning is an important simplification for carrying out the process.

Face grinding with cup wheels

Another technology for machining of plane surfaces is face grinding with diamond cup wheels (figure 6).

The workpieces are clamped onto a rotating table thus being passed along the ring-shaped diamond layer of the cup grinding wheel. Depending on the arrangement of the workpieces on the table infeed is either continuous or stepwise per table rotation.

An important feature of grinding in terms of precision machining is the contact length which is determined by the engagement conditions between tool and workpiece. It is longer for surface grinding with cup wheels than for peripheral grinding where it is limited to few millimeters.

The resulting high overlap factor is an advantage. This means that the workpiece surface is machined with a larger number of cutting edges within the period of contact. This considerably improves the roughness of the surfaces ground. This advantage can be further intensified by lowering the rotational motion v_w of the workpiece or by increasing the peripheral speed of the wheel and by enlarging the active grinding layer width. Such grinding conditions result in a small chip thickness and, as far as is known at present, diminish the risk of surface damage by inducing microcracks /2/.

Another advantage of face grinding can be seen in the twofold engagement of each workpiece crossing the ring-shaped grinding layer. Here, however, the material removal rate is not equally distributed to the outside and the inner side of the grinding layer. Most of the cutting is carried out rather by the outside which engages at first, whereas the layer inside only works with smallest chip thicknesses. Beyond these advantages in quality a unidirectional orientation of the grinding traces and thus the formation of a surface roughness dependent on machining direction cannot occur because of the remaining cross-shaped surface finish.

The generation of highest quality in surface roughness is achieved in spark-out. The deformation of the machine caused by high machining forces does relax during this period. Increasing spark-out time results in a diminishing cutting depth. Minute chip thickness is the cause for the material no longer chipping but being removed by plastic chip formation.

In the illustrated example the process of spark-out results in an average roughness height of less than 0.035 µm for silicon carbide (figure 7). These values could be even reached by use of a relatively rough tool (D91 C50) in one machining step.

The exact accuracy-to-shape achievable (f_e = 1.2 µm, referring to a measuring length of 55 mm) of the workpieces machined is a proof of the kinematic advantages of this technology. The generating motion in which the workpiece shape is produced by a straight line (projection of tool diameter), is the more effective the larger the workpiece surface to be machined is. This leads to a considerable improvement of the productivity of this technique, as cup wheels can cover a large area thus making a silmutaneous machining of several workpieces possible.

Important criterion for the efficiency of a machining technology is the cost for tool conditioning. As for this process, there is no need for profiling because of its generating kinematics. Wheel sharpening can be carried out, for example, with a sharpening stone. It is more effective, however, to keep the tool in a self-sharpening condition which can be achieved by adapting tool specification, process parameters and machining task. With the use of cup grinding wheels it is possible to support certain desired changes in the layer condition by choice of a special layer geometry and by application of different diamond layer specifications on the same tool /3/.

Cylindrical grinding with diamond belt

In terms of machining cylindrical workpieces external cylindrical plunge grinding is unfavourable as far as process kinematics is concerned. Due to engagement conditions there is only an extremely short contact length between workpiece surface and tool, leading to a small number of active cutting edges and a low overlap factor.

In contrast to face grinding the average roughness height can only partly be influenced by choice of parameters but more by grain size of the grinding wheel employed.

This makes a multi-step process strategy necessary, featuring decreasing grain sizes. This, however, requires a considerable effort for tool changes. Such a method is possible but not efficient.

A multi-step process strategy of belt grinding, however, proves to be far easier, as here a fast tool change is carried out by inserting another type of grinding belt. Problems, as are known with conventional grinding wheel changes, such as concentricity errors or difficult balancing out, do not occur.

The use of non-endless diamond belts is positive for the finishing in many respects (figure 8) /4/:

For instance, problems of non-accuracy caused by the joints of endless belts here can be excluded. As the tool is used just once, the belt is lead over the contact trolley at a relatively low speed; the cutting speed required for this machining process is largely generated by workpiece rotation. As a result the diamonds remain in the contact zone for a rather long time, thus being exposed to a strong wear. In addition, the free space between the diamond grains will be quickly clogged with removed particles of the workpiece surface. It is advantageous, that new diamond grains are continuously added and an optimum cutting tool is available at any time. There is no need for additional dressing and sharpening.

In order to improve the machining quality it is possible to move the belt in an oscillating motion. Thus higher cutting edge overlap can be realized leading to a better quality in surface finish. At the same time the copying of the contact trolley contour into the workpiece surface can be avoided.

The influence of the oscillation on the final quality of the machined workpiece can be clearly seen in the illustrated example (figure 9). The average roughness height R_a of 0.28 µm can be reduced to R_a = 0.12 µm when a belt with diamond grain size of 9 µm is used. On the whole roughness values R_a of 0.04 µm can be achieved through increasing machining periods and above all through using the extremely fine diamond grain of D 0.5 µm.

However, a disadvantage has to be mentioned here: With this type of machining principle it is not possible to improve accuracy-to-shape and, perhaps, even a deterioration has to be accepted. Therefore the accuracy-to-shape quality has to be generated in the previous machining steps, for instance by grinding with grinding wheels.

Summary

The production of ceramic components with surfaces of highest precision is still accompanied by a lot of problems. For this reason a commercial use in large-scale products has hardly been realized yet.

Inspite of the limited range of available machining technologies and the complicated machinability of advanced ceramics, there are ways to improve the efficiency of the machining process as well as the quality of the resulting workpiece finish.

Chosing the right polishing tool material as for example aluminium it is possible to produce extreme quality in terms of surface roughness. The deficiency of the polishing process concerning workpiece shape accuracy can almost be eliminated.

Adapting the process parameters in grinding with cup grinding wheels finishing becomes possible by use of this technology. In this way the rough and finish machining of plane surfaces using one tool can be realized. Consequently face grinding can be developed as an excellent machining process for the finishing with high efficiency in the future.

Under certain circumstances, however, chosing a different machining technology may be the appropriate alternative to provide a machining process with higher efficiency. In peripheral machining of circular workpieces, for example, grinding with diamond belts can help to make a multi-step machining process economically competitive by its easy and quick tool change.

References

1. Salmang, H.; Scholze, H.: Keramik. Springer Verlag, Berlin, Heidelberg, Tokio 1983.

2. Wemhöner, J.: Werkstoffgerechte Prozeßstellwerte - eine Voraussetzung für die wirtschaftliche Schleifbearbeitung von heißgepreßtem Siliziumnitrid. Industrieanzeiger 109(1987)81, 36-37.

3. Kleinevoss, R.: Spiegelglanz an keramischen Bauteilen. Industrieanzeiger 97(1987), 11-13.

4. Böckling, H-A.; Quambusch, A.: Neue Perspektiven zur Erzielung feinster Oberflächengüten unter Einsatz von Mikro Schleifmitteln. die maschine 12(1986), 33-34.

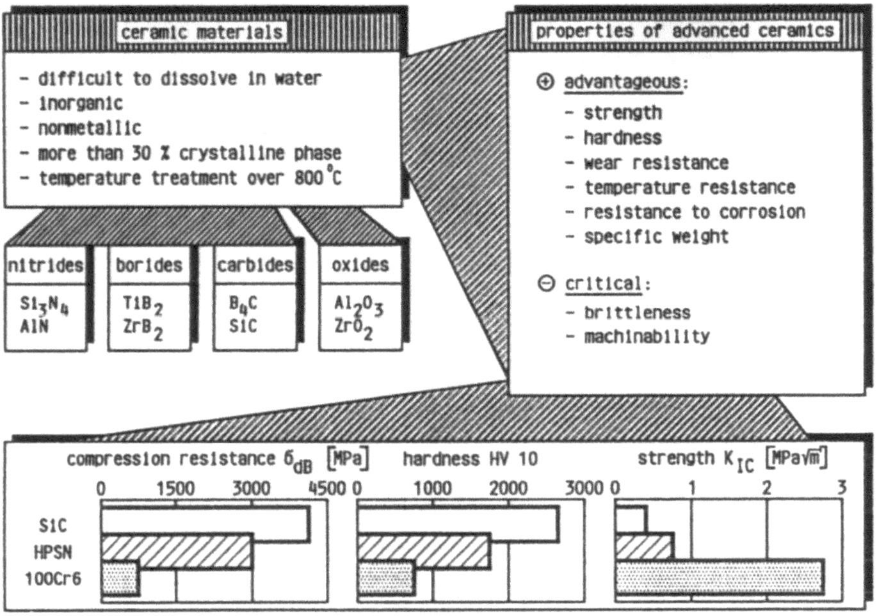

Fig.1. Ceramic materials - definition, division and properties in comparison to metals

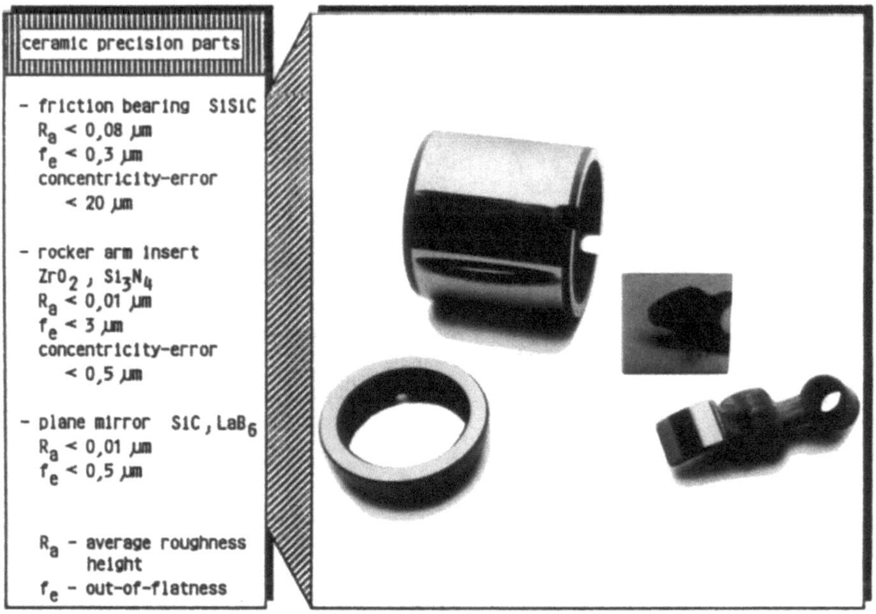

Fig.2. Ceramic precision parts

difficulties in precision machining of advanced ceramics

- small range of machining methods
- application of superabrasives necessary
- high tool wear
- small material removal rates
  - long machining time
- high machining forces
  - problems in accuracy of shape
- potential surface damage

Fig.3. Difficulties in machining ceramics

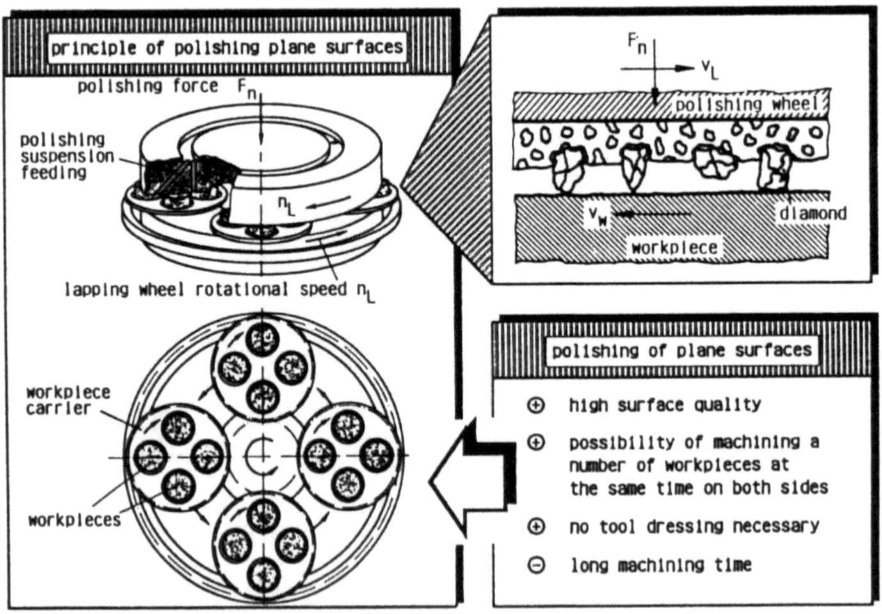

Fig.4. Principle of polishing plane surfaces

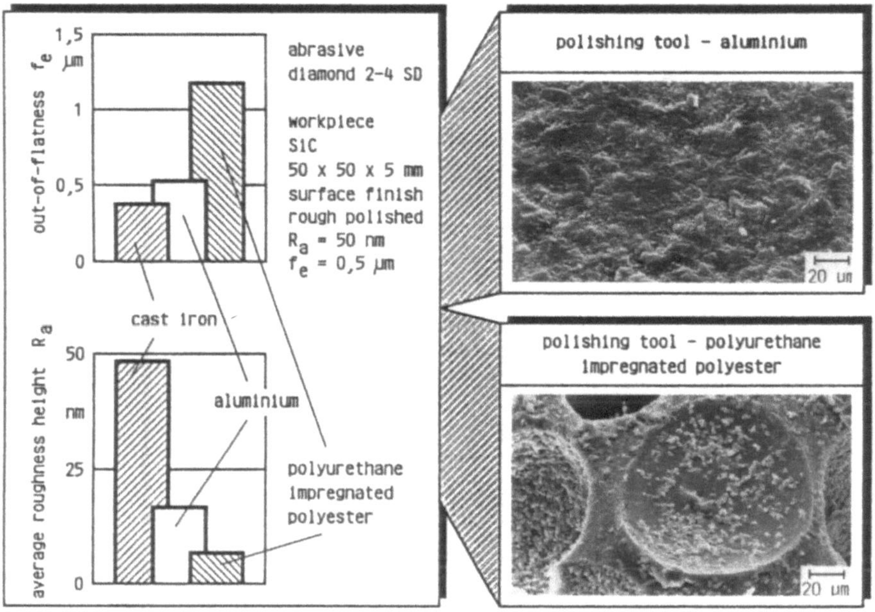

Fig.5. Better surface finish but lost of accuracy by using soft polishing pads

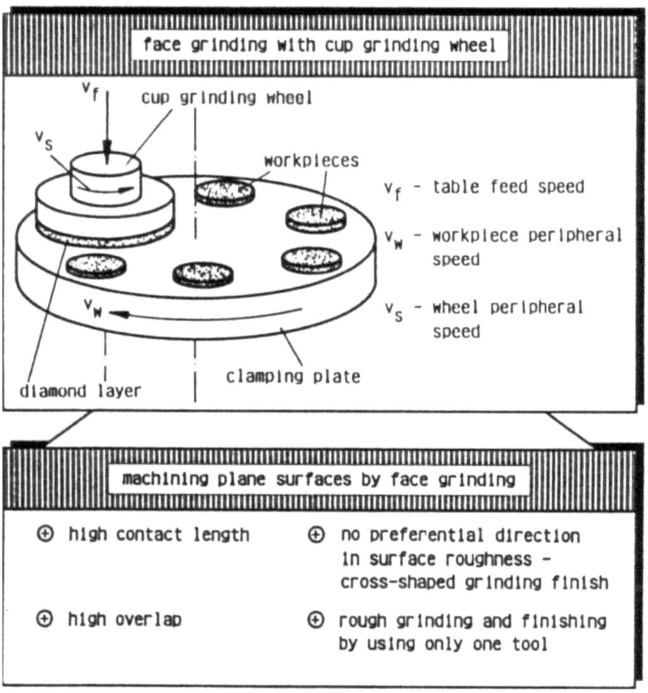

Fig.6. Face grinding with cup grinding wheels

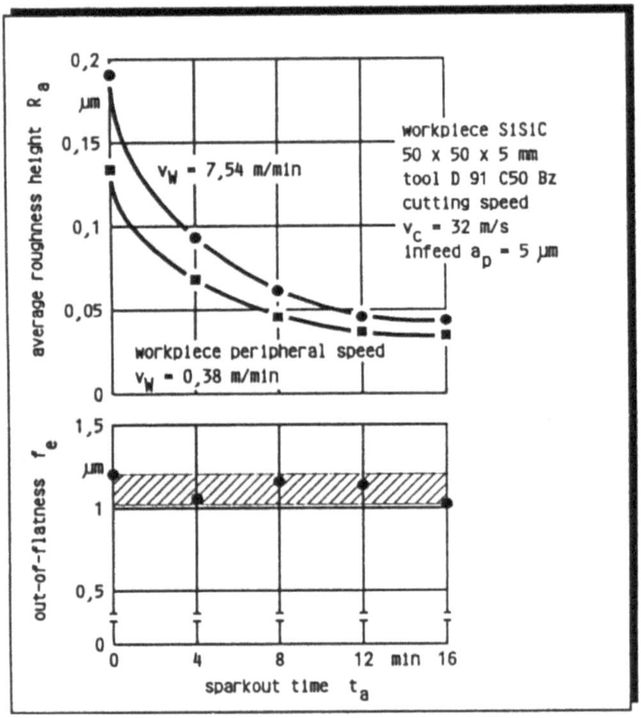

Fig.7. Sparking out leads to highest surface quality

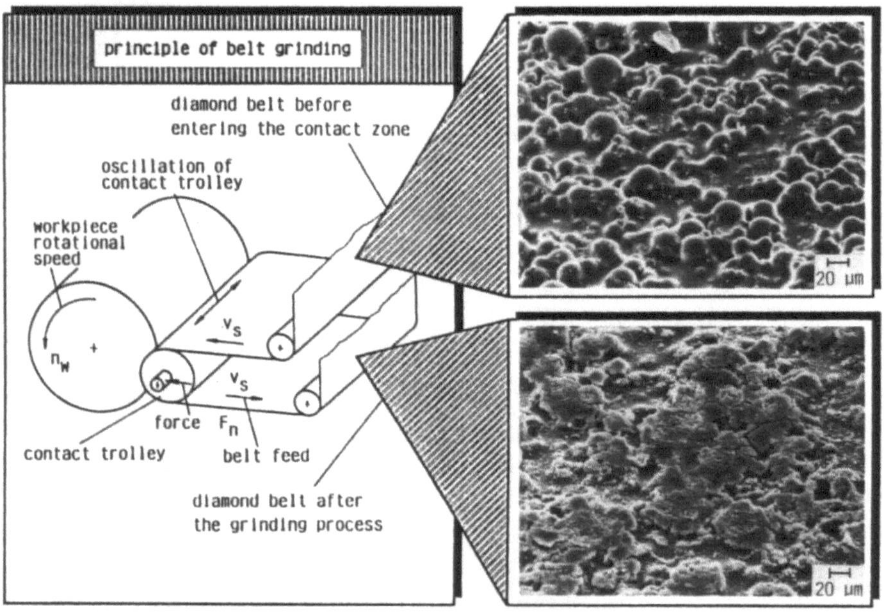

Fig.8. Principle of belt grinding process

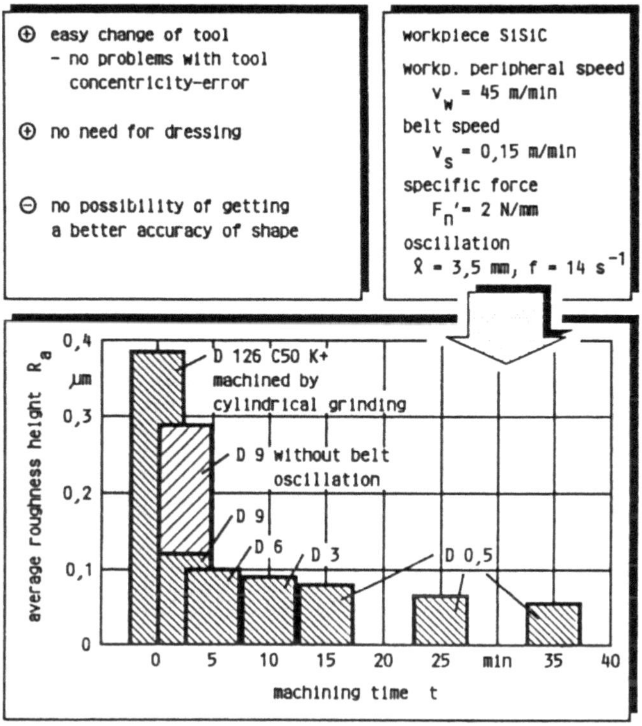

Fig.9. Small grain size and high overlap lead to high surface quality in belt grinding

Precision Grinding of Brittle Materials

K.H. Fiedler

Carl Zeiss, Oberkochen (FRG)

Abstract

The ALG 120 has been used successfully for about 2 years for the manufacture of aspheric precision optics. It has been proved that optimization of the machine component characteristics and modernization of the process parameters permits the viscoplastic grinding of optical glass, which decisively increases the economic efficiency of aspheric manufacturing.

Introduction

The objective of all technologists concerned with designing and manufacturing ultra-precision machines is to produce processing machines which guarantee sub-micrometer quality with respect to accuracy and repeatability. The prerequisite for the repeatability is an appropriately rigid and effectively controlled tool/workpiece machine unit with special consideration of thermal stability and tool wear. Repeatability and accuracy regarding slope errors and surface finish are largely dependent on detrimental vibrations being kept to a minimum, on optimum tool conditioning and optimum material-dependent process parameters.

Aspherizing Grinding Machines in Production of Precision Optics

At Carl Zeiss, Oberkochen, precision optics are manufactured on several different aspherizing grinding machines. The feature common to all these machines is that all spindles and carriages decisive for the grinding result are mounted on air bearings or hydrostatic bearings. Cutting is performed with the external contours of rotationally symmetrical tools with diameters of between 150 and 10 mm. In order to ensure functional cutting speeds for all tool diameters, different drive systems are used: from brushless d.c. motors with maxiumum speeds of >7000 rpm to air-actuated systems with maximum speeds of >40.000 rpm.

Fig. 1 shows the operating principle of the last Aspheric Lens Generator (ALG) 120 to be installed, with which convex workpieces with radii ranging from 25 to 160 mm and diameters of up to 120 mm can be processed:

The grinding machine was mounted on an anti-vibration base. As shown in Fig. 1, the workpiece carriage adjustment D is used to define the center thickness of the workpiece. The tool spindle unit is moved horizontally in two steps by so-called coarse and fine carriages. Here, the position of the coarse carriage G defines the so-called best-fit circle radius RP of the workpiece, taking into consideration the tool radius RW. Aspherization of the workpiece is performed by the fine carriage feed F, which is a function of the swivel angle.

The machine bed consists of a casting construction to provide optimum machine stability. The aim of the box-like design was to produce the best rigidity possible. The machine bed construction permits the mixture of grinding coolant and glass slurry to be drained from the working area.

All guideways, bearings, drives and measuring systems of the machine were enclosed from the environment and the work site. The work site itself was partitioned off by elastic curtains to avoid microphonic effects. The mist produced during grinding is removed upwards from the work area by an exhaust system mounted on the base unit.

The machine is controlled by a PDP 11/23 from the company Digital Equipment. A torque motor and tachogenerator control are provided for the air-bearing mounted swivel and rotary motions of the workpiece. Both the axial and the radial runouts of the workpiece cementing plate barely exceed 1μm. The coarse carriage runs in a dovetail guideway on the machine bed direct. It has a setting range of 150 mm and is moved via a threaded spindle by a torque motor. During the grinding process the air supply of the bearings is cut off and the carriage therefore lowered. The fine carriage rests on air bearings on the coarse carriage. A d.c. motor with a lever-type gear with setting speeds of approx. 40 ms provides practically stick-and-slip-free travel of the fine carriage over the dynamic range of 1000 μm. The measuring system used is a laser interferometer. The tool spindle is driven continuously over the range of 0 -7200 rpm by an integrated brush-less d.c. motor. This permits cutting speeds of up to about 60 m/s. Tool wear is monitored by an impact pressure meter, the dressed tool spindle has a residual imbalance of <300 mg . mm/kg and the runout of the grinding wheel is approx. 0.3 μm. To maintain an optimum machine temperature, part of the grinding coolant is conducted through the spindle housing. Under grinding conditions, the elasticity of the machine in the tool/workpiece contact area is >100 N/μm in the X direction and >18 N/μm in the Y direction.

Asphericizing Technology at Carl Zeiss Today

When selecting the process control in ultra-precision machining, viscoplastic abrasion is the mode also preferred for brittle materials such as optical glass. In the production of aspheric image-forming precision optics made of glass, final polishing has remained an indispensable process to this very day, but viscoplastically shaped glass surfaces have two essential advantages over surfaces generated by conventional brittle fracture:

(1) The ground surface can be checked laser-interferometrically directly after processing, permitting correction of the machine settings within seconds, if necessary.

(2) The surface ground in a suitable viscoplastical mode permits the material-dependent reduction of the required polishing rate to fractions of a micrometer.

Both characteristics are prerequisites for effective and cost-efficient series production of aspheric glass precision optics.

Unlike the grinding of plane and spheric surfaces [1], aspherizing does not involve extended tool/workpiece contact, but only a sort of point contact with all the problems associated with single point diamond turning. In addition, aspherizing necessitates more moving parts such as carriages and spindles, reducing machine rigidity. Experience, however, has shown processing to be successful if the load penetration tests [2] and diamond scratching experiments [2,3] performed with Vickers diamonds are taken into account. Thus the machine operates at the highest possible cutting speeds with vibrations kept to a minimum in order to guarantee adiabtical-viscoplastic processing of as many glass types as possible. This avoids incorrectly shaped abrasive products being generated, which is a critical factor with multi-diamond tools. The best material feed rates, which are dependent on the glass type, are determined empirically.

Processing results have now been obtained on numerous types of optical glass such as fused silica (HERASIL), BK 7, LaK 8, SK 2, SF 5, SF 10, LaSF N15 and ceramic glass (ZERODUR). In the more easily worked glass types the slope error is about 10 arc sec. With optimum process control the roughness values longitudinal and transverse to the cutting direction are 1.5 nm and 3.2 nm rms respectively, requiring a polishing rate of only 0.2 to 0.3 μm. Fig. 2 shows a LaK 8 sphere which was ground with an active aspherizing carriage and on which a polished lens was optically contacted: this sample produced in 1986 still shows clear rotationally symmetrical brittle fraction structures and a central circle which suggests unacceptable tool wear.

Acknowledgement

Development and production of the above ALG 120 was promoted by the "Bundesministerium für Forschung and Technologie (BMFT)" (West German Federal Ministry for Research and Technology).

References

1. Bifano, T.G. et al., SPIE Proc. Vol 803 (1987) 12
2. Schinker, M.G. et al., SPIE Proc. Vol 802 (1987) 70
3. Schinker, M.G. et al., Industrie Diamanten Rundschau IDR (1984) No. 4

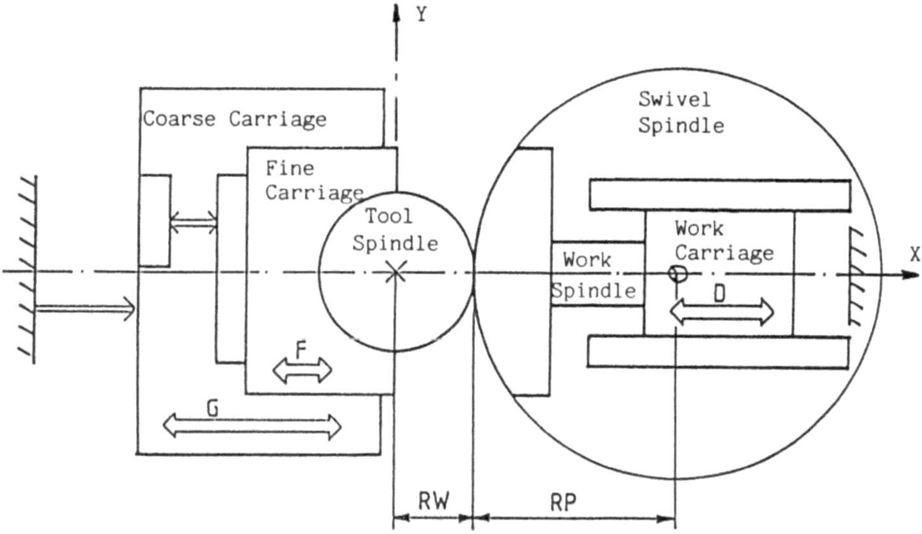

Fig.1. Illustration of the process related machine functions

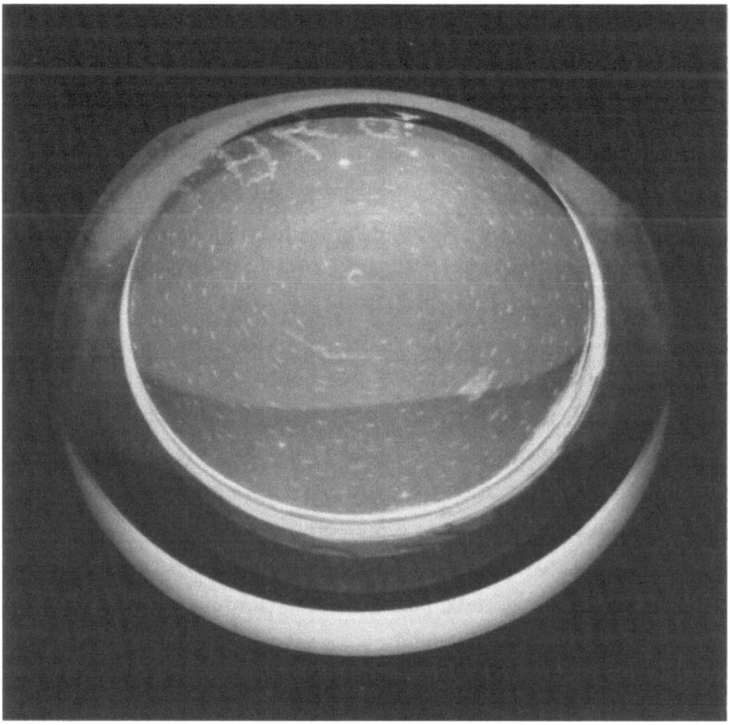

Fig.2. Example of a precision ground glass surface optically contacted to a polished lens surface

Form Accuracy and Surface Integrity in Plane Lapping

Prof. Dr.-Ing. Dr. h.c. Dr.-Ing. E.h. G. Spur,
Dipl.-Ing. D. Simpfendörfer,
Dipl.-Ing. I. Sabotka

Institut für Werkzeugmaschinen und Fertigungstechnik (IWF)
Technische Universität Berlin

Summary
Deriving from an analysis of the lapping process, a kinematic-based model is presented that enables attainment of optimal workpiece form accuracies. It is exemplified that, in addition to kinematics, particularly the lapping-wheel wear behaviour is of causal importance to the quality of the working result achievable. Furthermore, investigations concerning the results of a lapping operation on the development of residual stresses in the workpiece surface layer are also discussed in this paper.

1 Introduction

More than ever before, modern industrial production technology is becoming a factor which penetrates the functional structures and mechanisms of society. Social changes in the sense of a development of the quality of life, prosperity and security can take place only so far as it becomes possible to exploit hitherto undiscovered or unused production resources in the fields of energy technology, material technology and information technology for industrial purposes.

Manufacturing technology is therefore faced with the task of fulfilling the growing demands for accuracy, complexity and profitability of machining in ever expanding fields of application. Manufacturing technology can meet these challenges both by means of constant adaptation and evolution of conventional machining operations and through the development and utilization of entirely new ones /1/. In this context one must refer to the numbers of more recent manufacturing processes in eroding and coating, which already rank highly in the actual industrial practice /2/.

An analysis of the development in the field of conventional cutting shows above all the efforts towards higher accuracies. According to *Taniguchi* /3/ the attainable manufacturing accuracy rises on the average to a power of ten within a range of about 20 years (*fig. 1*). The limits of accuracy in measurement expand in the same way /4/. A continuation of this development is ensured by the application of ultra-precision machines. These are mainly turning or milling machines whose specific design in combination with the use of mostly mono-crystal diamond tools makes it possible to attain manufacturing accuracies up to an order of about 5 nm under defined ambient conditions. The reason for the development of such machines was the necessity of having to finish optical components from soft metals such as aluminium or copper for applications in high-performance lasers /3, 5/. Moreover, ultra-precision grinding and polishing machines have recently emerged which allow similarly high accuracies in the machining of hard and brittle materials /6 to 9/.

Furthermore, new problems for manufacturing technology result from changing material requirements in such special key industries as the automobile industry, the aviation and space industry, the power industry and the electronic industry. According to the growing practical importance of *New Materials* such as high-performance polymers, fibre-reinforced plastics, nickel-based alloys, amorphous metals, ceramics or ceramic-metal compounds, manufacturing technology is confronted with the question of the treatment of these materials. Especially ceramic materials, which are gaining a considerable importance, must be classified as very difficult to machine due to their extreme hardness and resistance to wear combined with their remarkable brittleness; only very few manufacturing processes appear to be suitable at all for these materials.

Apart from laser cutting and electro-erosion in some special cases /10/, only the fine-finishing procedures grinding, honing and lapping, including ultrasonic lapping /11/, allow a machining of ceramics in the hard state /12/. Due to their special properties, ceramic materials could be used profitably for various technical reasons. Nevertheless, the number of applications is still unsatisfactory mainly because of an insufficient reproducibility of ceramic-machining which has been very difficult to realize until now /13/.

Against this background a high demand for abrasive machining can be forecast for the future. Lapping will participate in this development and will especially offer advantages in reaching high form accuracies and surface qualities. It will also remain indispensable for the machining of extremely thin, non-chuckable, fragile parts such as wafers or quartz crystals in the foreseeable future /14 to 25/. Despite high yield in many cases /26/, certain economic disadvantages are inherent to the process of lapping which must be reduced by a gradual increase of the degree of automation. There is potential for automation especially with regard to:

- in-process measuring of the active partners,
- dressing of the lapping-wheels,
- cleaning of the workpieces and
- loading and unloading of the machine.

In general, it must be considered that only a reliable control of the process creates the option for an automated production. Since the lapping process is especially complex, due to the high demand for accuracy among other factors, it has been easily perceivable until now that an unmanned production could be realized only in a few cases /25, 27 to 30/.

2 Analysis of the Lapping Process

Lapping Medium

Lapping grains are of block- or slate-like shape (*fig. 2*). Our own sedimentation analyses of special SiC-lapping compounds indicate with a good reproducibility the grain size variations illustrated in *fig. 3 A*. According to DIN 66 141, the grain size is interpreted as the equivalent diameter of a ball whose cross-section surface corresponds to the projection surface of the lapping grain which is positioned orthogonally to the direction of the incident light (*fig. 3 B*).

The motional behaviour of lapping grains can be characterized as inhomogeneous; it is dependent on the specific workpiece material and its structure. The numerous grain indentations (*fig. 4 A*) on metallic surfaces suggest a predominant rolling of the lapping grains, but there are also occasional sliding tracks detectable (*fig. 4 B*) whose number rises with increasing hardness of the workpiece material.

In the lapping of metals, the stock removal process caused by the rolling of lapping grains can be interpreted according to the model shown in *fig. 5*, regardless of the material structure condition. For this model formation, a section is drawn through a very small surface layer area orthogonally to the surface, and the material continuum is replaced by a number of small beams, oriented parallel to the surface. Assuming that two lapping grains with a certain distance to each other penetrate into the surface simultaneously in a wedge-like manner, a microplastic shear deformation of the material would result which is recognizable by the shortening especially of the outer beams. In deformation processes the volume of the deformed material is approximately constant, so that the material volume displaced by the grains can only be pushed up. This phenomenon is abstracted by an increase in the number of beams. Transferred to real surfaces, this effect means that, starting with an incipient increase in the thickness of the workpiece, stock removal can be expected only after a certain time of lapping /31/. If further lapping grains penetrate into the deformed surface layer, this results predominantly in an alternate material displacement parallel to the surface, and subsequently there is a possibility that this material can be pushed into already existing craters.

The shear deformation processes are a consequence of dislocation sliding motions. If dislocations are not slidable any more due to lattice deformations or defects and no further dislocations are created, the critical material shear strength is surpassed which results in incipient cracks and following clod-shaped material removal. In this way a specific surface structure appears whose ideal form is a sequence of craters of different depths and sizes and of levels of different heights (fig. 5). The peculiarity of material overlapping at the edges of craters, as it can be seen in the model, is often found in real surfaces.

The thickness of the surface layer area represented by the beams, the so-called separation layer, depends on the grain size, the number of active grains, i.e. actually indented grains at a given moment, the material structure condition and on the engaging pressures specific to the lapping grains (*fig. 6*). According to assumptions by *Martin* /32/, lapping grains reach a depth of penetration into metallic materials of 5 to 10 % of their size depending on the structure of the material and the load. These data have been confirmed by our own findings to a large extent. For instance, the surfaces of strain-relief annealed workpieces of steels 100 Cr 6 and Ck 45 showed penetration depths of about 3 to 5 µm after a machining with the lapping compound PWS 30 (fig. 3) and an engaging pressure of $5 \cdot 10^{-2}$ Nmm^{-2}.

However, the course of the integral width of the X-ray diffraction line IB as shown in *fig. 7* documents hardening effects combined with plastic deformations up to a depth of about 30 µm. Even up to a depth of approximately 50 µm, which corresponds to the tenfold width of the separation layer, compressive residual stresses are extended, induced by grain indentation within the surface layer. In the case under consideration here, the residual stresses at the workpiece surface reach a maximum of -400 Nmm^{-2}.

The high pressure peaks at the projection surfaces A_{kw_i} and A_{kl_i} of active lapping grains created by scattering of the engaging pressure can be assumed to be nearly constant due to the low penetration depths. These pressure peaks come about in such a way that there appears a balance of forces at each lapping grain orthogonal to the surface (fig. 6). Too high specific grain engaging pressures can lead to an early destruction of active grains (*fig. 8 A*) and to an irreversible embedding of grain fragments into the workpiece surface /33 to 35/; see *figures 8 B and 9*. This means that a larger number of smaller or originally unfavourably positioned so-called passive lapping grains become involved in the force transmission. It should also be mentioned that due to the relative motion between the active partners, a hydrodynamic pressure p_{ef} builds up in the lapping film which reduces the load on the active lapping grains.

The machining of a totally different type of materials, the brittle sintered ceramic materials, is in principle similar to the machining of metals as regards to the grain motion behaviour /36/. However, the rolling and sliding motion results in a quite different kind of stock removal because of the completely different structure of such materials. Due to the inability of ceramics to plastically deform, in the course of the rolling motion of the lapping grains, the stock removal mechanisms, as shown in *fig. 10*, can occur:

- disengagement of complete ceramic grains or grain clusters along the bond,
- disengagement of ceramic grain fragments and
- inducement of microcracks in the ceramic grain as the preliminary stage of material disengagement.

Moreover, the machining of hard materials involves the danger of a temporary anchoring of lapping grains in the more ductile lapping-wheel material which normally results in undesirable scratching, comparable to the grooves resulting from machining with bound grains /37/. *Figures 11 A and B* illustrate the various appearances of surfaces for Si-SiC after a lapping operation using B_4C lapping compound.

The fact that the structure of lapped surfaces depends on the material and even more that the shape and size of the lapping compound are specific to the lapping grain, leads to the conclusion that the influence exerted by the lapping compound can be seen as stochastic and everywhere the same on the average. Taking as a basis a constant engaging pressure within the effective area, a most general description of the lapping process can be made by way of the kinematics.

Kinematics

The determination of the movement of a workpiece can only be achieved by dividing its face into a theoretically infinitely large number of infinitesimal segments degenerated into points whose path curves (motion forms) can be exactly defined either absolute or relative to the lapping-wheel (*fig. 12*).

Although for plane lapping with constrained workpieces up to three independent drives are used, the relative motions of the active partners basically possess only one degree of freedom (DF). However, this is valid only if no self-rotation of parts relative to their holders occurs, as is the case when the faces of prismatic parts are being lapped.

As far as types of motion with one degree of freedom are concerned, there are as a rule either epi- or hypocycloidal and, in certain cases, circular path curves. Cycloidal motion forms can be further subdivided according to *fig. 13* into lengthened, normal and shortened curves. The type of curve which arises for any particular case is dependent on the factor

$$K := \left| \frac{\omega_h - \Omega}{\omega_z - \Omega} \right| \qquad (1)$$

the position (e_h) of the workpiece-point under consideration and the dimensions (a_z) of the machine /39/.

Compared to this situation, the additional degree of freedom, brought about by the self-rotation of workpieces within the holder cut-outs, causes a so-called modulated cycloidal type of motion (fig. 13). Curve forms of this kind cannot be called purely lengthened, normal or shortened any more. Rather, depending on the self-rotation of workpieces occuring with a constant angular velocity, modulated cycloidal path curves successively pass through quasi-lengthened, quasi-normal and quasi-shortened stages within certain repeated periods /40/.

For the assessment of the stock removal rate at an observed workpiece-point, which is important also for the achievable form accuracy of the workpiece, the parameters path velocity and path length can be considered. *Path velocity* is defined as that velocity v, with which any particular point P of the workpiece moves along its path curve on the lapping-wheel; the stretch of path covered on the lapping-wheel is termed as *path length* s in the case of a permanent contact and as *effective path length* s_{eff} in the case of temporary free-wheeling at the edges of the lapping-wheel (*fig. 14*).

Starting from the coordinates of point P, as given in fig. 12, the following equation for the path velocity may be derived:

$$v\,(t, e_h(t)) = \sqrt{A_1 + A_2 \cdot \cos\,(\omega'_w - \omega'_h) \cdot t + A_3 \cdot \cos\,(\omega'_h - \omega'_z) \cdot t + A_4 \cdot \cos\,(\omega'_z - \omega'_w) \cdot t} \qquad (2)$$

where

$$A_1 := a_z^2 \cdot \omega'^2_z + c^2 \cdot \omega'^2_h + a_w^2 \cdot \omega'^2_w$$

$$A_2 := 2 \cdot c \cdot a_w \cdot \omega'_w \cdot \omega'_h \qquad (3)$$

$$A_3 := 2 \cdot a_z \cdot c \cdot \omega'_h \cdot \omega'_z$$

and

$$A_4 := 2 \cdot a_w \cdot a_z \cdot \omega'_z \cdot \omega'_w$$

as well as

$$\omega'_w = \omega_w - \Omega \;;\quad \omega'_h = \omega_h - \Omega \;;\quad \omega'_z = \omega_z - \Omega \;. \qquad (4)$$

The path length is determined by means of the time-based integral

$$s = \int_0^{t'} v\,(t, e_h(t))\;dt \quad . \tag{5}$$

The influence of self-rotation on the amount of stock removal to be expected becomes evident if the gradients of path length g_s and mean-velocity g_m possessing the same value are defined for workpieces with holder-invariant motion behaviour by means of the cycle time

$$T_Z = \frac{2 \cdot \pi}{\left|\omega_z - \omega_h\right|} \tag{6}$$

under the assumption of a total surface covering as

$$g_s := \left[\frac{s\,(e_h)}{s\,(c)}\right]_0^{T_z} \quad \text{and} \quad g_m := \frac{v_m\,(e_h)}{v_m\,(c)} = \frac{\int_0^{T_z} v\,(t, e_h)\;dt}{\int_0^{T_z} v\,(t, c)\;dt} \quad \text{respectively.} \tag{7}$$

Fig. 15 A shows the gradients belonging to a holder-invariant motion of workpieces with rotational-symmetric faces possessing the radius r_W. It is significant that the gradients become larger by an increase of the K-factor as well as by a decrease of the occupation radius c; both of these effects lead to worse lapping. An optimum is given by the choice of such kinematic setting parameters that K becomes zero. In this case the path length becomes independent of the holder eccentricity.

However, if the workpieces considered in fig. 15 A perform a self-rotation relative to the holder around their mid-points with an angular velocity in the most favourable case corresponding to half a rotation per lapping time T_L, the considerably smaller and thus more favourable gradients shown in *fig. 15 B* result. Higher angular velocities of the workpiece self-rotation in principle lead to the same characteristics, but on the whole they cause higher gradients of the outer workpiece points.

Thus one can speak of an optimal choice of kinematic setting parameters corresponding to $K = 0$ if at the same time a self-rotation of the workpieces relative to the holder is prevented, e.g. by means of a form-closing. All kinematic constellations deviating from $K = 0$ require a self-rotation of the workpieces relative to the holder with half a rotation per lapping time in the most favourable case. The direction of rotation of the workpieces is of peripheral importance. Parts with a prismatic face should be given the possibility of self-rotation in the case of $K \neq 0$ by means of putting them into rotatable holder fixtures.

3 Form Accuracy

General Overview

Form accuracy of parts to be machined by double-sided plane lapping includes the evenness of the faces as well as their plane parallelism. The following influences on the workpiece form accuracy are of importance:

- the lapping kinematics; with given dimensions of the machine these are determined by
  - the combination of kinematic setting parameters and
  - the number of degrees of freedom of the relative motion between the active partners,
- the specific arrangement of the workpieces within the holders; this determines, among other factors, whether during the machining there occurs
  - a total surface covering or
  - a temporary workpiece free-wheeling, as well as
- the wear state of the lapping-wheels.

All experiments had in common the combination of the kinematic setting parameters in such a way that the motion between the upper lapping-wheel and the workpieces corresponded to an optimal K-factor of zero and a time-constant path velocity of 20 m/min. Moreover, both lapping-wheels were dressed before performing the experiments in such a manner that they were in an ideal flat initial state, which could be recorded, however, only on the lower lapping-wheel. The wear was measured with a computer-aided measuring device developed at the Institute of Machine Tools and Manufacturing Engineering Berlin /41/.

The wear state of the upper lapping-wheel was inferred from the lower wheel on the basis of the hypothesis, confirmed by practical experience, that the profile of one lapping-wheel generally is the exact negative copy of the profile of the other one /42, 43/.

Influences of Kinematics and Holder Occupation

To start with, between the workpieces and the lower lapping-wheel motion conditions of an epicycloidal kind corresponding to an unfavourable K-factor of 4 were set. Due to the local and temporal dependence of the path velocity given in this case, the temporal-arithmetic mean value of the path velocity at the mid-point of the workpieces was chosen and calculated as 8.3 m/min.

First, the holders were occupied in such a way that the workpieces were secured against self-rotation (1 DF) and that there was a total surface covering between the active partners (*fig. 16*). After a lapping time of 35 min and a load collective consisting of initial load, main load and final load, the lapping-wheels still showed no significant wear so that their form accuracy could be disregarded as an influence parameter in this series of experiments.

Whereas the workpiece faces engaged with the upper lapping-wheel showed an unevenness which could not be resolved anymore by means of the used plano-spherical lens (< 67 nm), the face lapped with $K = 4$ possessed a slightly crowned shape (< 0.5 μm) on the average. Furthermore, a mean parallelism deviation of the workpiece thickness of about 2.5 μm appeared on the lower face (fig. 16). The reason for this is the slight progressive rise of the gradients g_s and g_m connected with growing holder eccentricity (fig. 15 A).

The model tests carried out in part 2 yielded the result that a suitable method of optimizing the motion between the active partners in many cases consists of enabling the workpieces to perform a self-rotation relative to their holders. The resulting modulated motion forms (fig. 13) cause significantly smaller and thus more favourable gradients g_s and g_m (fig. 15 B) than motion forms with one degree of freedom (fig. 15 A). The results of the experimental verification of this phenomenon are represented in *fig. 17*.

The parts lapped under otherwise equal conditions have indeed a considerably better form accuracy. For instance, the mean parallelism deviation of the workpiece thickness, which appeared on the lower face, was reduced by about 80 % to about 0.5 μm. The approximately wedge-shaped sectional form of the workpieces (fig. 16) gave way to a slightly crown-shaped one (fig. 17) according to the course of the gradients g_s and g_m. It is furthermore notable that the evenness of the workpiece face previously corresponding with the upper lapping-wheel slightly deteriorated to values of less than 0.3 μm, as could be expected, since self-rotations in the basically optimal case $K = 0$ are accompanied by a small loss of accuracy.

If kinematic setting parameters are changed in such a way that the motions between the lower lapping-wheel and the workpieces can also be described by $K = 0$, a necessary, but not sufficient criterion is hereby given as regards the optimization of the achievable workpiece form accuracy. This is the case because special attention has to be paid to keep the lapping-wheels flat. Model investigations permit the conclusion that lapping-wheels show fast deviations from the ideal flat surface profile especially in such cases where there is a total surface covering between the workpieces and the lapping-wheels (also see fig. 21). However, if there is a partial free-wheeling of the workpieces at the edges of the lapping-wheel, the latter can be kept flat for an essentially longer period of time. This is due to the fact that the engaging pressure $p_e(t)$ overcomes the pressure p_v preset at the machine as workpiece free-wheeling increases (*fig. 18*). This results in load peaks at the edges of the lapping-wheel which counteract the comparatively low wear strain in these areas caused by the kinematics. Thereby it was possible in the present case to keep the lapping-wheels almost ideally flat for a period of more than 5 hours (*fig. 19*), so that the assumptions for the achievement of an optimal workpiece form accuracy could be regarded as sufficiently fulfilled.

The form accuracies subsequently attained on the initially non-rotatable workpieces were correspondingly high. Since the thickness inaccuracies of the workpieces were no longer detectable by the use of our mechanical measuring system, there appeared characteristic interference-structures on the faces of the workpieces (fig. 19) which indicated a minimal unevennesses of these parts (< 0.3 µm) resulting from the free-wheeling of workpieces.

In the course of a further test series, the workpieces were permitted to self-rotate under the conditions listed in fig. 19 (*fig. 20*). Self-rotations in the case of $K = 0$ result in minor deviations from the kinematically optimal state due to the gradients g_s and g_m rising towards the workpiece edges (fig. 17), but this disadvantage is apparently just compensated by the workpiece free-wheeling at the lapping-wheel edges. In any case, this can be gathered from the achieved workpiece form accuracies which can be regarded as unsurpassable, especially considering the relatively coarse grain size of the lapping compound PWS 12 (fig. 3). This concerns the evenness values attained which surpassed the evenness of the plano-spherical lens used (< 67 nm) as well as parallelism deviations of the workpiece thickness which could not be mechanically resolved any further (< 1 µm).

Influence of the Lapping-Wheel Wear

If machining occurs without workpiece free-wheeling at the lapping-wheel edges, significant wear with projecting edges arises even after a lapping time of 70 min under the conditions listed in fig. 19 (*fig. 21*). In spite of optimally set motional conditions between the active partners, this kind of wear results in significantly reduced part accuracies (fig. 21). For instance, after a total lapping time of 175 min, workpiece thickness reduced by 5 µm on the average was found in the workpiece areas which were in contact with the lapping-wheel edges. Moreover, the evenness of the workpieces became successively worse with increasing lapping time until no interference occured at all.

4 Residual Stresses

Established Level of Research

It can be regarded as a fact that surface integrity of workpieces, which is characterised by the grain structure condition, the hardness of the material, and the state of stress, is substantially influenced by any kind of mechanical machining /44 to 46/. It is all the more striking that investigations of surface integrity in the field of lapping have been performed only to a small extent.

Early investigations into the influences on the surface layer in lapping by *Letner* and *Snyder* /47/ identified isotropic residual stresses of -700 Nmm^{-2} on the average at the surface of oil-hardened manganese steel after the machining with an Al_2O_3 lapping compound of the grain size F 800 under very small engaging pressures of 0.05 or 0.45 Ncm^{-2}. These compressive residual stresses subsided at an average depth of 5 µm. Interestingly enough, they found out that marginal values decreased with a growing load, and that the depth effects of the residual stresses increased. They attributed this phenomenon to a cutting rather than chiselling stock removal effect of the lapping grains.

Investigations on high-speed steel by *Sridhar et al.* /48/ established increasing compressive residual stresses and a concomitant increase in wear resistance of the machined surfaces with increasing lapping time. *Matalin* /49/ determined compressive residual stresses growing strongly with increasing grain wear in a lapped C-steel of unknown specification. On the other hand, *König* /50/ argues that the lapping process has the advantage of reducing residual stresses.

Kedrow /51/ reports that he succeeded in completely removing the microhardness measured after grinding hardened steel by a subsequent lapping operation. In contrast to this, *Özhan* /52/ found in the soft magnetic material *Mumetall* a clear increase of microhardness as a result of lapping. He determined that the depth of the affected marginal zone and the hardness value at the surface both depend on the grain size and the amount of introduced pressure. In addition, he determined at the surface irregular compressive residual stresses of considerable height [200 Nmm^{-2} ($p_v = 0.5$ Ncm^{-2}; $d'_k = 2$ µm) and 400 Nmm^{-2} ($p_v = 10$ Ncm^{-2}; $d'_k = 30$ µm) respectively]. *Rystsova et al.* /53/ found an increase of microhardness with decreasing pressure in steel materials (40 HRC). From this they drew the conclusion that the lapping grains, predominantly freely moving under low pressure, have a special hardening effect. Lapping grains which are temporarily embedded into the tool surface at higher pressures have a cutting and thus less strengthening effect on the workpiece surface. A statement by *Miller* /54/ is in contrast to this theory. He suggested a decrease of load especially in the case of machining with small grains (3 to 6 µm) in order to reduce the plastically deformed volume percentage.

Chandrasekar et al. /55, 56/ have carried out residual stress investigations on brittle Ni-Zn-ferrites which were machined with bronze lapping-wheels using diamonds (3 µm) as a lapping compound at a pressure of 7 Ncm^{-2}. The residual stresses of -37.5 Nmm^{-2} measured at the surface were considerably higher than those obtained by grinding under comparable conditions. They turned into tensile stresses at a depth of about 4 µm, and finally subsided at 15 µm. Since components made from Ni-Zn-ferrites are mainly used in magnetic storage systems and residual stresses are responsible for the emergence of magnetically inactive surface layers (< 1 µm), ways had to be found to diminish the residual stresses. A clearly reduced residual stress of -24 Nmm^{-2} at the surface could be obtained by combining the lapping process with chemical etching using a hot 7 % solution of orthoposphoric acid. It was also proposed to minimize the deformability of the material by means of deep-freezing of the workpieces before lapping.

Residual Stresses in Steels

Within the scope of our own investigations, measured residual stresses in the upper surface layer in all cases were of compressive nature. Their modulus values as well as the integral width of the X-ray diffraction line IB of (211)-ferrit-reflex at preset engaging pressure under a strain-relief annealed grain structure condition proved to be nearly independent from material and also from path velocity (*table 1 A*). Furthermore, a perfectly isotropic state of compressive residual stress at the surface could be determined (*table 1 B*). This fact reflects the irregular machining as a significant feature of lapping. The amount of compressive residual stresses may be evaluated as an index of the marginal meaning of thermo-effects in lapping. *Chandrasekar et al.* /55/ arrived at the same conclusions.

As shown in *fig. 22*, depth-courses of residual stress achieved by the use of lapping compound PWS 30 (fig. 3) indicate that the mechanical loads introduced during machining are also absorbed by deeper layers of material. The residual stresses decrease rapidly below the surface; however, considering the low engaging pressure of $5 \cdot 10^{-2}$ Nmm^{-2}, the amount of residual stresses directly at the surface is remarkable. The reason for this are the high specific grain engaging pressures (see fig. 6).

The characteristic curves of the integral width of the X-ray diffraction line IB in fig. 22 document for both materials that even at a depth of 12 µm, compared to the basic structure no mentionable value of material solidification has been recorded. Residual stresses beyond this range are only induced indirectly by the plastically deformed marginal zone, and moreover they are almost elastically absorbed by the material. The compressive residual stress condition changes into a tensile one at a depth of about 50 to 60 µm, whereby an asymptotic subsiding of the residual stresses in deeper material layers is assumed; this is the case as far as the sum of all the induced residual stresses must equal zero /57/. For both materials, the approximate congruity of the σ_E- and IB-characteristic curves confirming the small influence of path velocity suggest invariant stock removal processes.

Experiments performed by application of the lapping compound PWS 12 (fig. 3) had the result that the induced compressive residual stresses at the surface of hardened workpieces were approximately twice as high as those which arose at the surface of strain-relief annealed workpieces lapped under identical machining conditions (*fig. 23*). A reason for this are the specific grain engaging pressures which arise with increasing material hardness due to the decrease of the projection surfaces below the lapping grains (fig. 6). In addition to this, there was found an influence deriving from the path velocity which only slightly exceeds the range of accuracy in measurement and which is therefore difficult to estimate. This fact applies to all material/structure combinations investigated here.

Fig. 24 illustrates marginal zone effects resulting from the lapping operation as a function of the material structure condition. The reason for the extremely fast subsiding of the residual stresses in the marginal zone of the workpieces in the hardened material state consists of the lower penetration depth of the lapping grain peaks and their subsequently more rapid fracture. The stress-influence due to the lapping operation already at a depth of approximately 10 µm therefore dies nearly totally away; this is recognizable by the attainment of the tensile residual stress condition resulting from the heat treatment which remains constant by a depth range of several millimeters. The slight decrease of the thermal-caused tensile residual stresses observable towards the surface of the steel Ck 45 may be caused by a certain decarburization in the marginal zone.

5 References

1. Spur, G.: Einführung in die Abtragtechnik. In: Handbuch der Fertigungstechnik; Bd. 4/1: Abtragen, Beschichten. Hrsg. G. Spur, Carl Hanser Verlag, München 1987.

2. N.N.: Handbuch der Fertigungstechnik; Bd. 4/1: Abtragen, Beschichten. Hrsg. G. Spur, Carl Hanser Verlag, München 1987.

3. Taniguchi, N.: Current Status in, and Future Trends of, Ultraprecision Machining and Ultrafine Materials Processing. Annals of the CIRP 32 (1983) 2, S. 573-582.

4. Kunzmann, H.: Grenzen der Meßgenauigkeit. In: Tagungsberichtsband des 5. Int. Feinbearbeitungskolloquiums, Braunschweig 1987.

5. Weck, M.; Hartel, R.: Anforderungen und konstruktive Ausführungen von Ultrapräzisionsmaschinen für die Bearbeitung von Metalloptiken. In: Tagungsberichtsband des 5. Int. Feinbearbeitungskolloquiums, Braunschweig 1987.

6. Namba, Y.; Tsuwa, H.: Ultra-Fine Finishing of Sapphire Single Crystal. Annals of the CIRP 26 (1977) 1, S. 325-329.

7. Yoshioka, J. et al.: Ultraprecision Grinding Technology for Brittle Materials: Application to Surface and Centerless Grinding Processes. In: Proc. of the ASME – M.C. Shaw Grinding Symposium, Miami Beach/USA 1985.

8. Inasaki, I.: Entwicklung von ultrapräzisen Werkzeugmaschinen in Japan. In: Tagungsberichtsband des 5. Int. Feinbearbeitungskolloquiums, Braunschweig 1987.

9. Namba, Y.; Tsuwa, H.; Wada, R.: Ultra-Precision Float Polishing Machine. Annals of the CIRP 36 (1987) 1, S. 211-214.

10. Panten, U.: Funkenerosive Bearbeitung von keramischen Werkstoffen. Ind.-Anz. 108 (1986) 81, S. 48-49.

11. Haas, R.: Mit Hilfe der Ultraschall-Erosion; Keramische Werkstoffe dreidimensional bearbeiten. Ind.-Anz. 108 (1986) 97, S. 10-13.

12. Spur, G. u.a.: Keramische Werkstoffe. 4. Trennende Fertigungsverfahren zur Keramikbearbeitung. In: Neue Werkstoffe; Fachaufsätze zur Materialforschung. Hrsg. Bundesministerium für Forschung und Technologie, Bonn 1986.

13. Spur, G.; Sabotka, I.: Fundamentals in the Lapping of Advanced Engineering Ceramic Materials. In: Proc. of the 3rd Conference of the Irish Manufacturing Committee, Galway 1986.

14. Martin, K.: Das planparallele Läppen von dünnen Halbleiterscheiben und Schwingquarzen bis auf 0,06 mm Dicke. Neuentwicklung einer Speziallläppmaschine. In: Jahrbuch der Oberflächentechnik; Bd. 20. Metall-Verlag, Berlin 1964.

15. Fischer, E.: Wechselbeziehungen zwischen dem Schneiden, Läppen und Ätzen von Germanium bzw. Silizium bei der Herstellung von Halbleiterbauelementen. Wiss. Z. TU Dresden 13 (1964) 3, S. 989-992.

16. Fischer, E.: Zur Technologie feinmechanischer Bauteile aus monokristallinen Werkstoffen für elektrische und mechanische Bauelemente. Wiss. Z. TU Dresden 13 (1964) 1, S. 277-286.

17. Fischer, E.: Zur Kinematik an reibungsgekoppelten umlaufenden Planläppmaschinen. Wiss. Z. TU Dresden 15 (1966) 3, S. 599-610.

18. Caspers, W.: Wirtschaftliches Läppen zwischen zwei Scheiben. Fachberichte für Oberflächentechnik 5 (1967) 5, S. 177-178.

19. Herb, C.O.: Precision Lapping the Hard-to-Work Materials. Abrasive Engineering (1970) August, S. 24-25.

20. Fischer, E.: Zur Kinematik und Käfigbelastung beim doppelseitigen Feinstbearbeiten planparalleler Werkstücke auf Anschlag. Wiss. Z. TH Otto von Guericke Magdeburg 15 (1971) 5, S. 523-530.

21. Bogenschütz, A.F. et al.: Über die Polier- und Läppwirkung einiger Oxide und Carbide bei der Oberflächenbehandlung von Siliciumeinkristallen. feinwerktechnik + micronic 76 (1972) 7, S. 349-354.

22. Martin, K.: Läppen. VDI-Z. 119 (1977) 9, S. 475-476.

23. Blum, G.: Läppen und Polieren von Siliciumwafern. In: Jahrbuch Schleifen, Honen, Läppen und Polieren, 50. Ausgabe. Hrsg. E. Saljé, Vulkan-Verlag, Essen 1981.

24. N.N.: Läpp- und Poliermaschine für Wafer. IDR 17 (1983) 2, S. 114.

25. N.N.: Lapmaster brings superfine finishes to mass production. Metalworking Production (1987) March, S. 62, 66.

26. Miller, P.C.: Lapping machine boosts quality output. Tooling & Production 52 (1986) 3, S. 60-61.

27. Buckley, B.R.A.; Smith, J.D.; Welbourn, D.B.: Eliminating the Human Touch; Paper F2. In: Proc. of the Automation in Production Conference. Belle Vue, Manchester 1974.

28. Kühn, K.-D.; Rohde, G.: Schleif-, Hon- und Läppmaschinen. wt-Z. ind. Fertig. 73 (1983) 8, S. 516-519.

29. Zurawski, W.; Simpfendörfer, D.: 6. EMO Hannover. Stand der Technik in der Feinbearbeitung. VDI-Z. 127 (1985) 23/24, S. 957-963.

30. Häuser, K.: Beschickungseinrichtungen steigern Produktivität. Feinbearbeitungsverfahren Läppen und Honen gewinnen an Bedeutung. Maschinenmarkt 91 (1985) 92, S. 1930-1932.

31. Martin, K.: Der Werkstoffabtragvorgang beim Feinbearbeitungsverfahren Honen. Maschinenmarkt 82 (1976) 60, S. 1074-1078.

32. Martin, K.: Neue Erkenntnisse über den Werkstoffabtragsvorgang beim Läppen. Fachberichte für Oberflächentechnik 10 (1972) 6, S. 197-202.

33. Grünwald, F.; Schilling, G.: Werkstoffabtrag und Rauheit beim Planläppen metallischer Werkstoffe. Feingerätetechnik 28 (1979) 8, S. 345-348.

34. Sulies, P.: Untersuchungen über das Einlaufläppen von breitenballigen Stirn- und Kegelradgetrieben. Dr.-Ing.-Diss. TH Aachen 1970.

35. Barylski, A.; Feld, M.: Zusetzen der Oberfläche sphärolitischen Gußeisens mit Mikro-Schleifkörnern nach dem Läppen. Werkst. u. Betr. 119 (1986) 12, S. 1025-1026.

36. Spur, G.; Sabotka, I.: Planparallelläppen keramischer Werkstoffe - Technologie und Abtrennmechanismus. ZwF 82 (1987) 7, S. 375-380.

37. Paulmann, R.: Vergleich der Feinbearbeitungsverfahren Schleifen, Honen und Läppen. In: Tagungsberichtsband des 5. Int. Feinbearbeitungskolloquiums, Braunschweig 1987.

38. Spur, G.; Sabotka, I.: Läppen technischer Keramik. ZwF 81 (1986) 6, S. 320-323.

39. Simpfendörfer, D.: Kinematik des Planläppens mit zentrischer Wälzanordnung und Zwangsführung. Ind.-Anz. 107 (1985) 44, S. 49-50.

40. Spur, G.; Simpfendörfer, D.: Teileeigenbeweglichkeit beim Planläppen mit Zwangsführung. Fortschr.-Ber. VDI, Reihe 2, Nr. 124. VDI-Verlag, Düsseldorf 1987.

41. Spur, G.; Simpfendörfer, D.; Trilk, H.: Rechnergestützte Werkzeugverschleißmessung beim Planläppen. ZwF 82 (1987) 1, S. 30-34.

42. Grösel, H.: Vorrichtungen auf Zweischeiben-Läppmaschinen. Werkst. u. Betr. 87 (1954) 4, S. 163-166.

43. Orlov, P.N.: Basic Principles of Abrasive Lapping. Machines & Tooling 48 (1977) 1, S. 43-47.

44. Tönshoff, H. K.; Brinksmeier, E.: Eigenspannungen durch Schleifen – wesentliche Einflußgrößen des Prozesses. In: Eigenspannungen; Entstehung - Messung - Bewertung; Bd. 2. Hrsg. E. Macherauch und V. Hauk, Deutsche Gesellschaft für Metallkunde e.V., Oberursel 1979.

45. Degner, W.: Bedeutung der Oberflächenbeschaffenheit für die Erhöhung der Qualität und Zuverlässigkeit der Bauteile. Feingerätetechnik 25 (1976) 2, S. 85-88.

46. Kloos, K. H.; Macherauch, E.: Development of Mechanical Surface Strengthening Processes from the Beginning until Today. In: Shot Peening; Science, Technology, Application. Hrsg. H. Wohlfahrt et al., Deutsche Gesellschaft für Metallkunde e.V., Oberursel 1987.

47. Letner, H. R.; Snyder, A. A.: Grinding and Lapping Stresses in Manganese Oil-Hardening Tool Steel. Trans. ASME 75 (1953), S. 873-882.

48. Sridhar, H. K.; Krishnamurthy, R.; Balasubramanian, G.: On some Aspects of Fine Finishing. In: Proc. of the 8th AIMTDR Conference I.I.T., Bombay/Indien 1978.

49. Matalin, A. A.: Restspannungen beim Schleifen. Werkst. u. Betr. 103 (1970) 9, S. 639-644.

50. König, W.: Fertigungsverfahren; Bd. 2: Schleifen, Honen, Läppen. VDI-Verlag, Düsseldorf 1980.

51. Kedrow, S. M.: Neuzeitliche Läppmaschinen. Maschinenbautechnik 2 (1953) 7, S. 300-306.

52. Özhan, F.: Feinbearbeitung weichmagnetischer Werkstoffe. Produktionstechnik-Berlin, Forschungsberichte für die Praxis, Bd. 36. Hrsg. G. Spur, Carl Hanser Verlag, München 1985.

53. Rystsova, V. S. et al.: Improving the wear resistance of lapped surfaces. Russian Engineering Journal 53 (1973) 6, S. 52-54.

54. Miller, N.E.: Three-Body Abrasive Wear with Small Size Diamond Abrasives. Wear 58 (1980), S. 249-259.

55. Chandrasekar, S.; Shaw, M.C.; Bhushan, B.: Comparison of Grinding and Lapping of Ferrites and Metals. In: Proc. of the ASME - Machining of Ceramic Materials Symposium, Miami Beach/USA 1985.

56. Chandrasekar, S.; Shaw, M.C.; Bhushan, B.: Morphology of Ground and Lapped Surfaces of Ferrite and Metal. In: Proc. of the ASME - Machining of Ceramic Materials Symposium, Miami Beach/USA 1985.

57. Spur, G.; Simpfendörfer, D.: Einflüsse der Kinematik des Planläppens auf die Randzonenausprägung metallischer Werkstoffe. In: Vorträge des Symposiums "Prozeßmodelle von trennenden und umformenden Fertigungsverfahren", Braunschweig 1987.

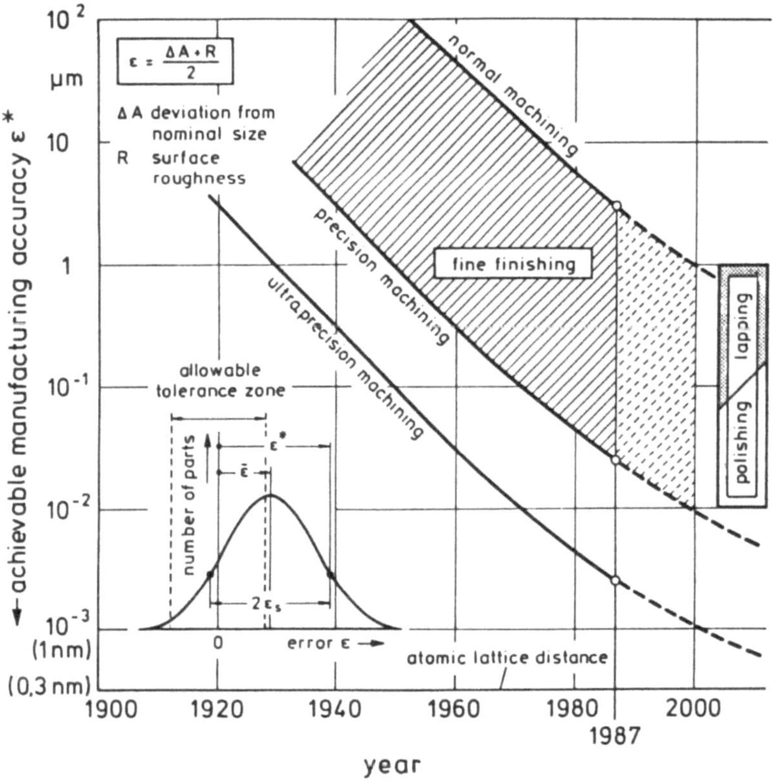

Fig. 1. Increase of the manufacturing accuracy according to *Taniguchi* /3/

Fig. 2. SiC lapping grains

A)

lapping compound SiC			
PWS 12		PWS 30	
share	size	share	size
0,5 %	< 4,5 µm	0,5%	< 20 µm
5 %	< 5,7 µm	5 %	< 25,5 µm
20 %	< 7,5 µm	20 %	< 28 µm
d'_k	= 10,3 µm	d'_k	= 33,8 µm
80 %	< 11,5 µm	80 %	< 35,8 µm
95 %	< 13,1 µm	95 %	< 38 µm
99,5 %	< 15 µm	99,5 %	< 40 µm

B)

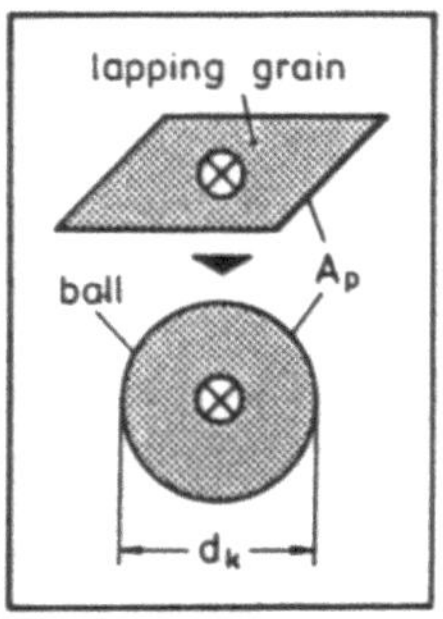

Fig. 3. SiC lapping compounds; lapping grain size variations (A) and definition (B)
d'_k mean equivalent diameter of a lapping grain

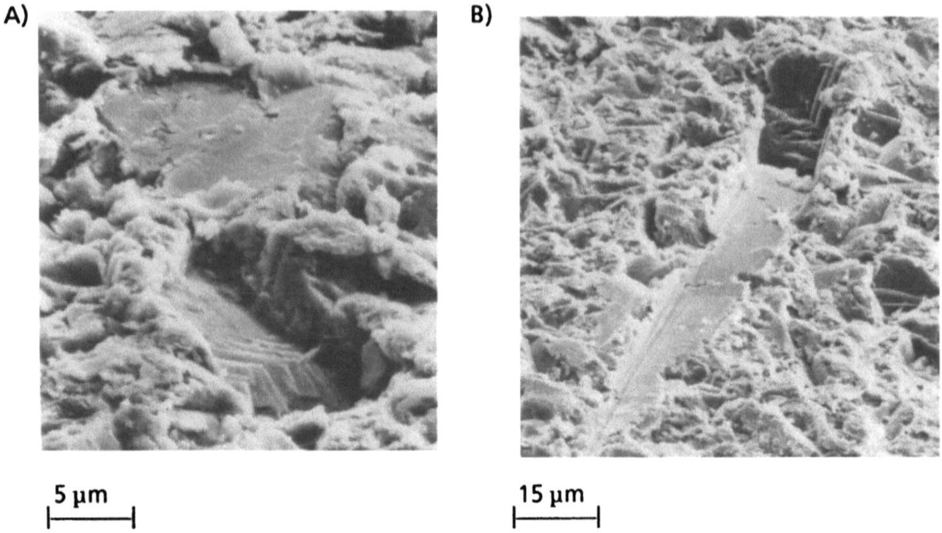

Fig. 4. Indentations (A) and sliding tracks (B) in steel 100 Cr 6 resulting from SiC lapping grains ($d'_k = 34\ \mu m$) .

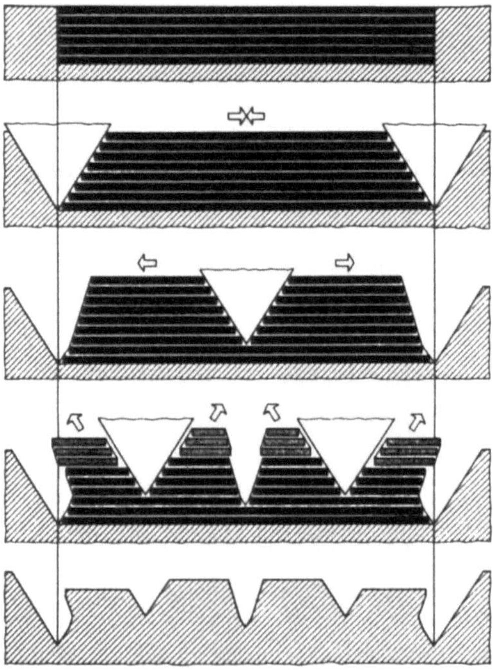

Fig. 5. Model of the mechanisms which occur in the upper surface layer[1]) of steel materials

[1]) ductile (hardened) material condition: high (low) density of dislocations, long (short) dislocation sliding paths resulting from large (small) numbers of defects in the lattice structure —> material easy (difficult) to deform.

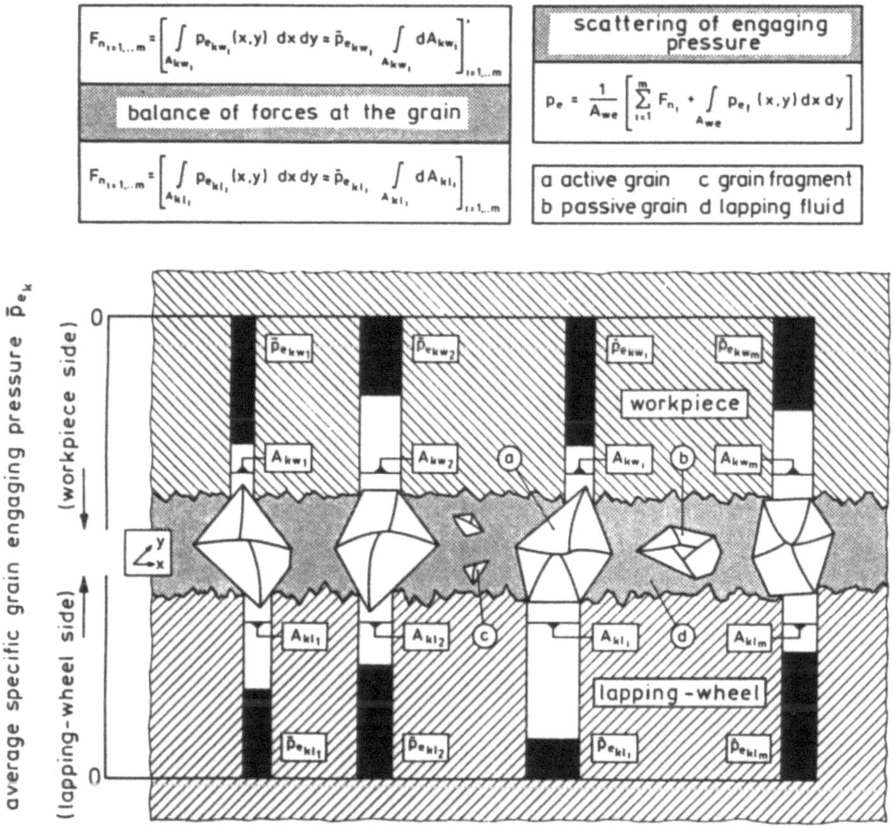

Fig. 6. Specific lapping grain engaging pressures

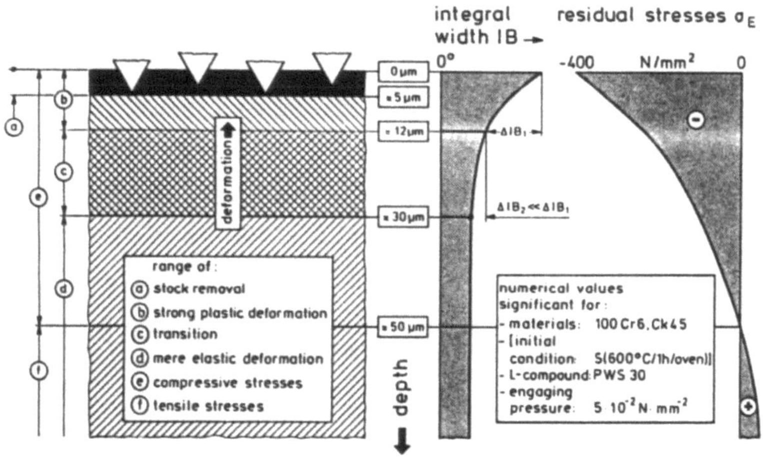

Fig. 7. Model of the different ranges of a surface layer influenced by the lapping process
$\Delta IB > 0$: plastic deformations; $\Delta IB = 0$: mere elastic or no deformations

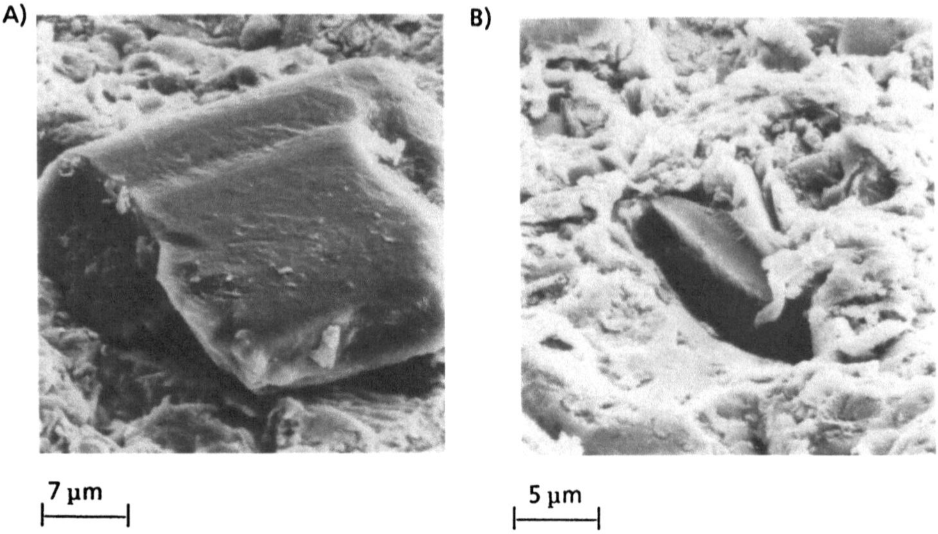

Fig. 8. Lapping compound SiC ($d'_k = 34$ µm); a destroyed lapping grain (A) and a lapping grain fragment embedded irreversibly into the workpiece surface (B)

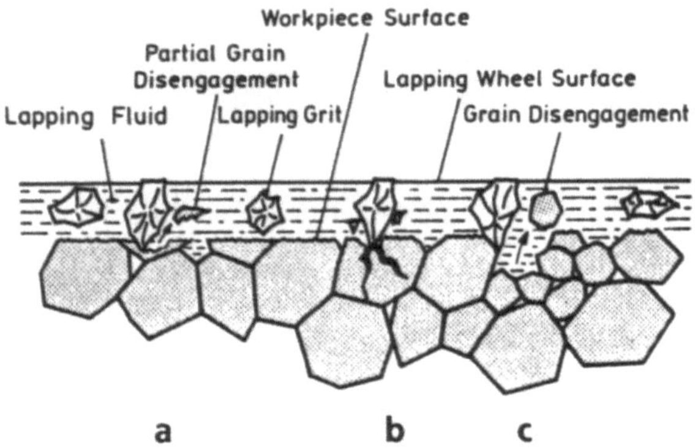

Fig. 10. Mechanisms of stock removal occuring when lapping brittle sintered ceramic materials /36/
a disengagement of ceramic grain fragments, b microcrack inducement in the ceramic grains, c disengagement of complete ceramic grains

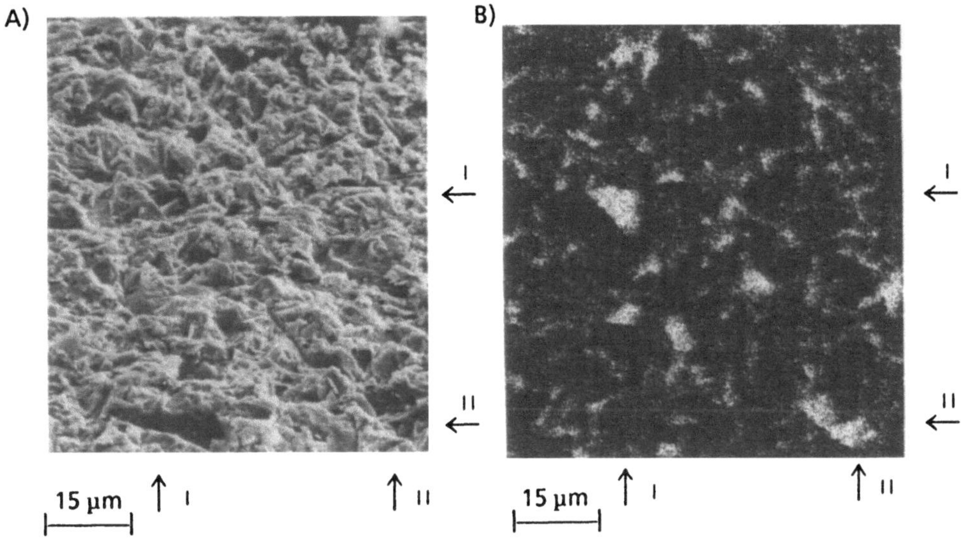

C)

material	Ck 45				100 Cr 6					
grain structure condition	strain-relief annealed 2)		hardened 3)		strain-relief annealed 2)			hardened 4)		
element percentage [Gew. %]	Fe	Si	Fe	Si	Fe	Cr	Si	Fe	Cr	Si
path velocity 5 m/min	88,6	11,4	[89,5]	[10,5]	88,1 [87,6]	3,0 [3,1]	8,9 [9,3]	86,5 (86,2)	2,6 (2,6)	10,9 (11,2)
path velocity 20 m/min	89,6	10,4	[93,0]	[7,0]	87,9	3,0	9,1	89,7	2,7	7,6
path velocity 35 m/min	[91,4]	[8,6]	92,0	8,0	88,4	3,0	8,6	90,1	3,0	6,9

lapping compound: PWS 12
engaging pressure: 5N/cm²
measure area:
55 × 55 µm
[ ] 400 × 400 µm
( ) 750 × 750 µm
2) 600°C / 1h / oven
3) 54 HRC
4) 62,5 HRC

Fig. 9. Detection of lapping grain fragments1) embedded into the workpiece surface by means of the energy-dispersive X-ray distribution analysis (spectral analysis of the stimulated X-ray radiation)
A) lapped surface, B) analysis of the X-ray distribution
[material: 100 Cr 6 (heat treatment condition: 600°C/1h/oven), lapping compound PWS 30], C) element distribution2) in the upper surface layer of the workpieces

1) regions of high pixel concentration correspond to lapping grain fragments embedded into the workpiece surface; pixel clouds of low concentration result from scatter radiation effects.
2) silicon percentage serves the purpose of estimating the extent of the embedding of SiC lapping grain fragments into the workpiece surface (bremsstrahlung eliminated).

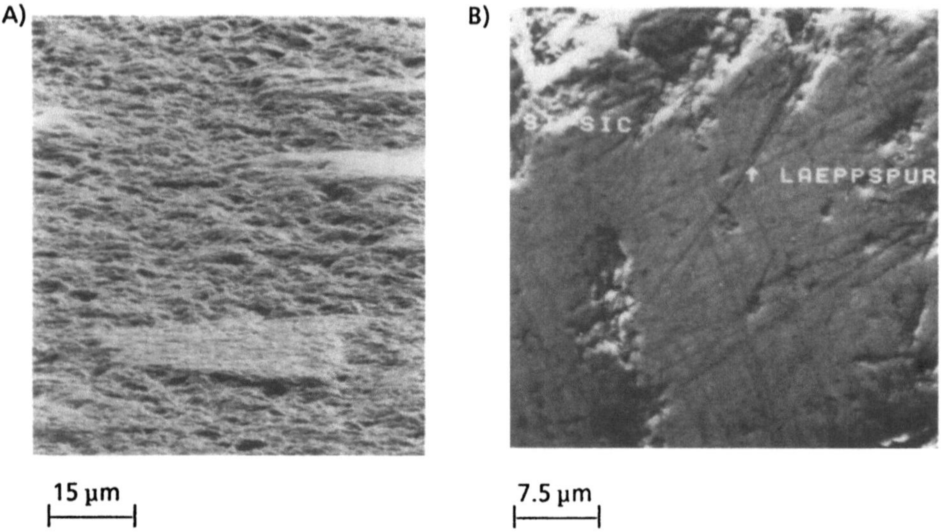

Fig. 11. Surface structure of the non-homogeneous ceramic material Si-SiC after lapping operation /38/
A) projecting SiC primary ceramic grains in a fissured Si-SiC matrix, B) scratches on a SiC primary ceramic grain resulting from sliding lapping grains
lapping compound B_4C (d'_k = 7 µm)

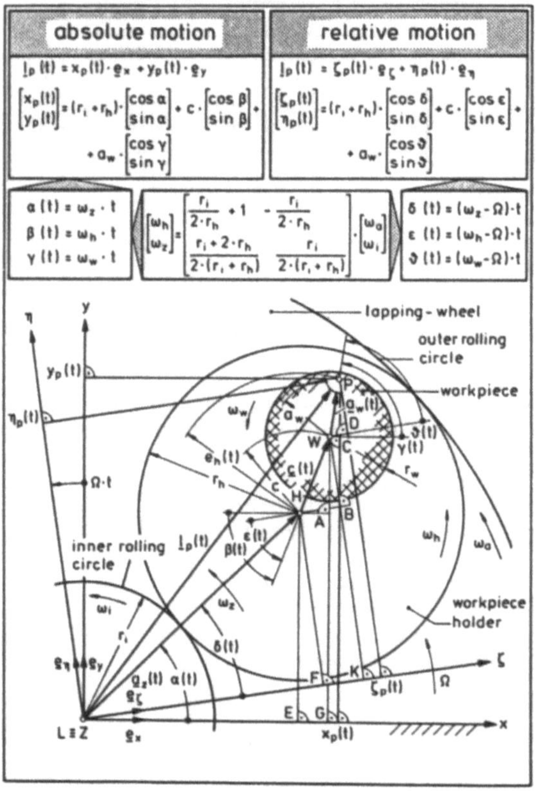

Fig. 12. Description of the absolute and relative motion of individual workpiece points when plane lapping with constrained workpieces
t time, ω_a absolute angular velocity of the outer pin ring, ω_i ... of the inner pin ring, Ω ... of the lapping-wheel, ω_h ... of the workpiece holder rotating around its mid-point H, ω_z ... of the workpiece holder mid-point H rotating around the central axis Z of the machine, ω_w ... of the workpiece rotating around its mid-point W

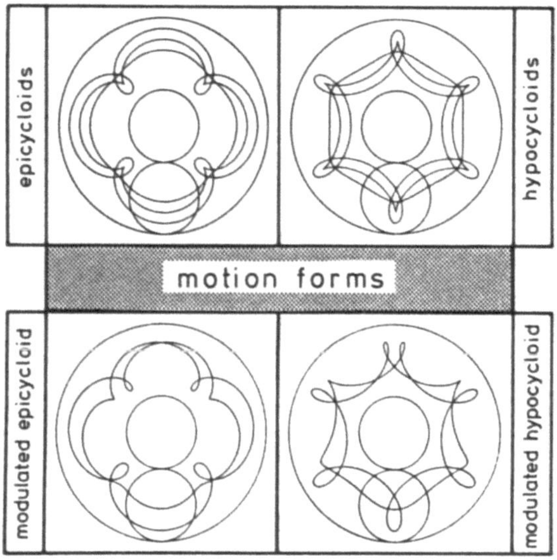

Fig. 13. Motion forms with various degrees of freedom

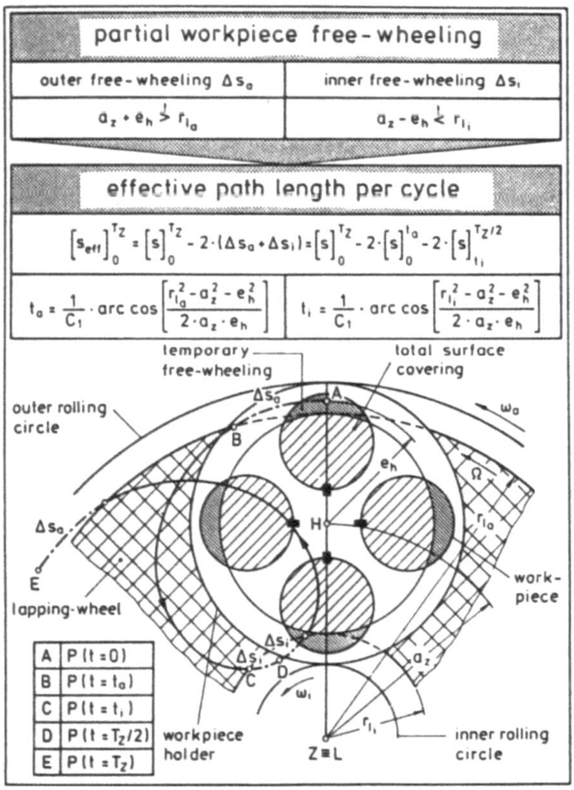

Fig. 14. Partial workpiece free-wheeling at the edges of the lapping-wheel

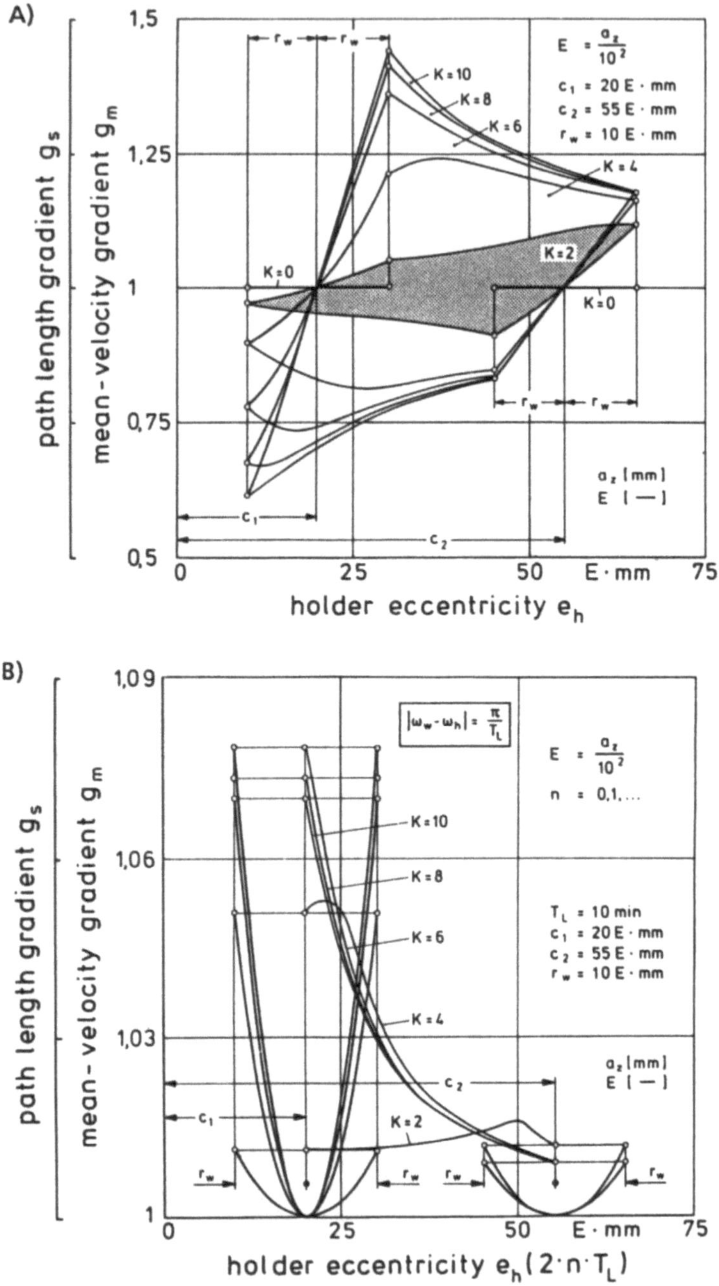

Fig. 15. Gradients of mean-velocity and path length as a function of the holder eccentricity and the K-factor for various machine dimensions; workpiece self-rotation within the holders: A) prevented, B) enabled
c_1, c_2 occupation radii, r_w radius of the workpiece face, T_L lapping time, a_z distance between the mid-point H of the workpiece holder and the central axis Z of the machine

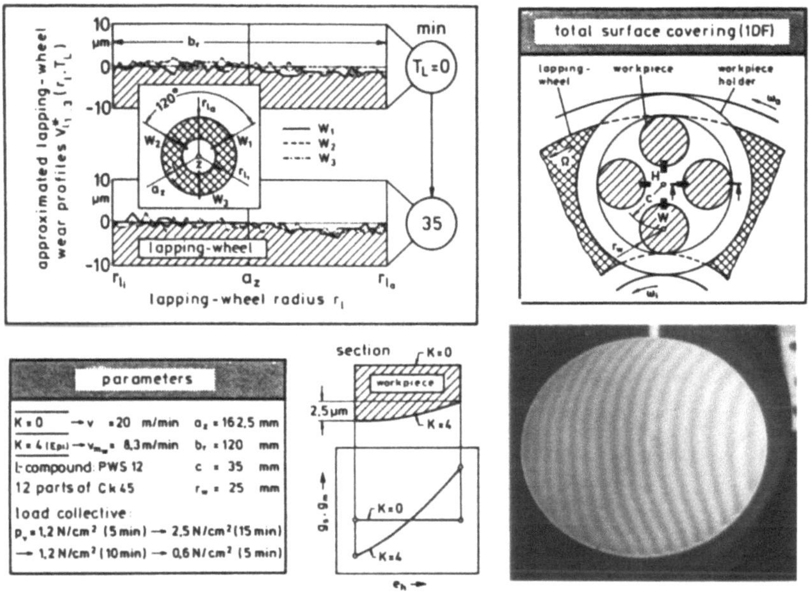

Fig. 16. Achievable workpiece form accuracy in the case of unfavourably chosen kinematic setting parameters and prevented workpiece self-rotation (1 DF); total surface covering
[Interference image: $T_L = 35$ min, $K = 4$]

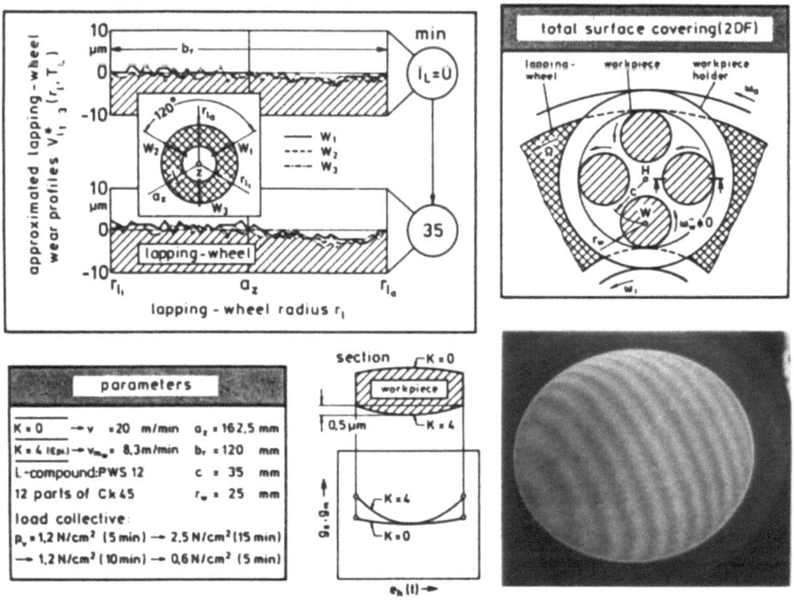

Fig. 17. Achievable workpiece form accuracy in the case of unfavourably chosen kinematic setting parameters and enabled workpiece self-rotation (2 DF); total surface covering
[Interference image: TL = 35 min, K = 4]

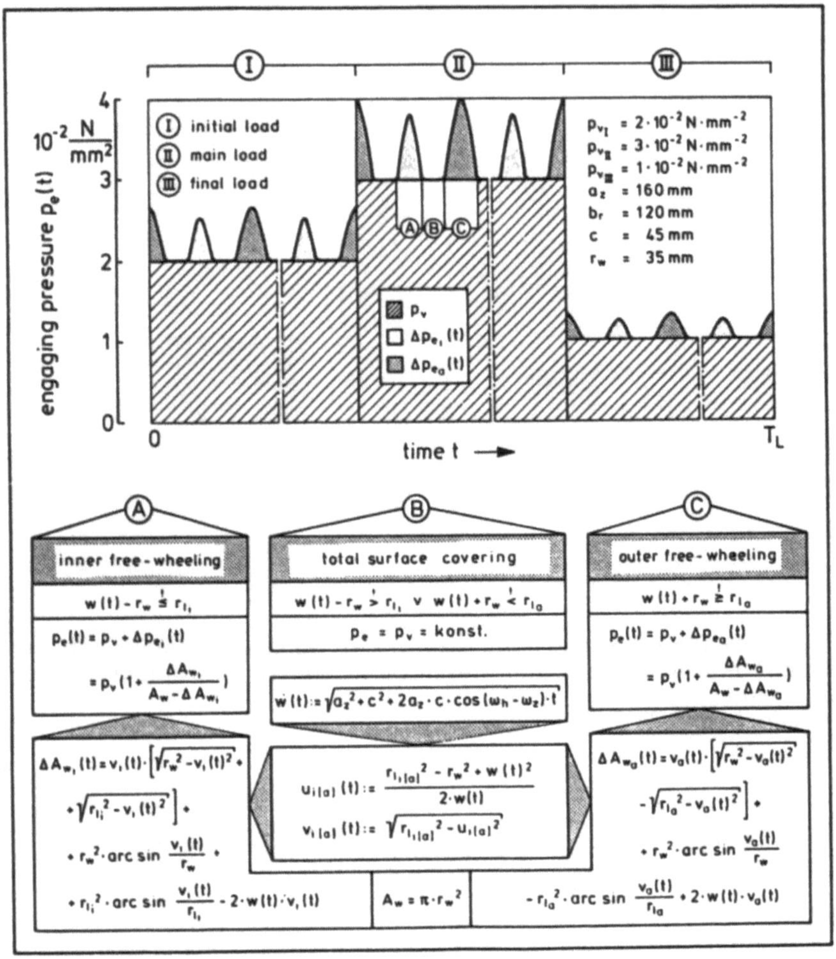

Fig. 18. Influence of the inner and outer workpiece free-wheeling on the engaging pressure
p_v lapping pressure preset at the machine (engaging pressure resulting from total surface covering), $\Delta A_{wi}(t)$ partial workpiece face which free-wheels at the inner edge of the lapping-wheel, $\Delta A_{wa}(t)$ partial workpiece face which free-wheels at the outer edge of the lapping-wheel, Aw workpiece face, r_{li} inner radius of the lapping-wheel, r_{la} outer radius of the lapping-wheel, r_w radius of the workpiece face

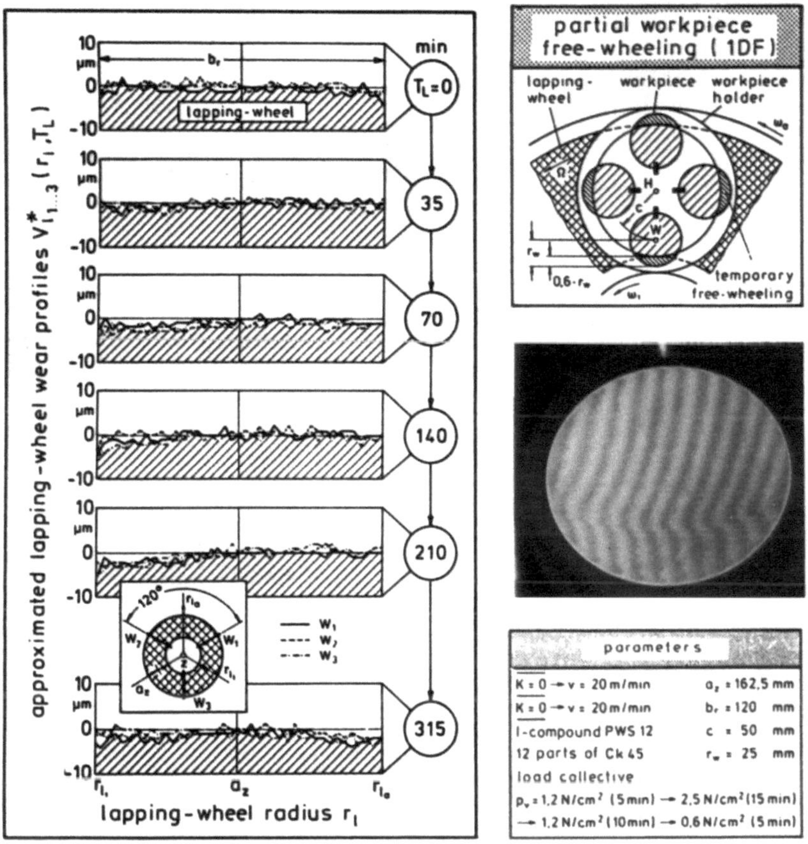

Fig. 19. Achievable workpiece form accuracy in the case of optimally chosen kinematic setting parameters and prevented workpiece self-rotation (1 DF); partial workpiece free-wheeling
[Interference image: $T_L = 210$ min]

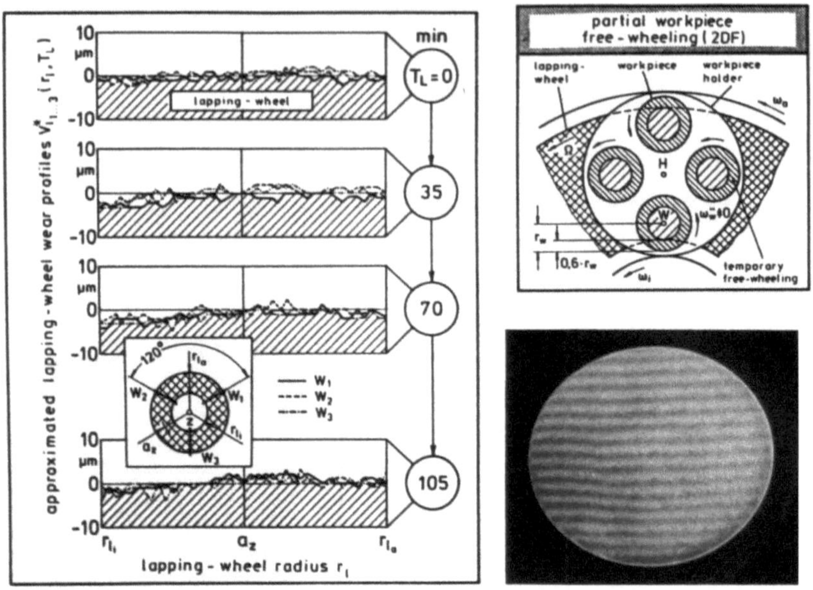

Fig. 20. Achievable workpiece form accuracy in the case of optimal kinematic setting parameters and the allowance of workpiece self-rotation (2 DF); partial workpiece free-wheeling
[Interference image: $T_L = 105$ min]

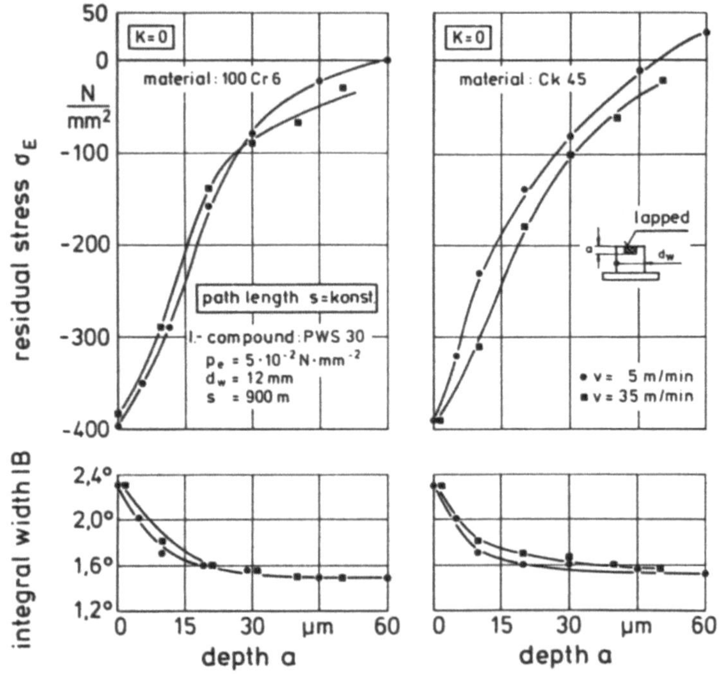

Fig. 22. Depth-courses of residual stress and integral width of the X-ray diffraction line in lapped steels Ck 45 and 100 Cr 6 as a function of the path velocity; strain-relief annealed grain structure condition

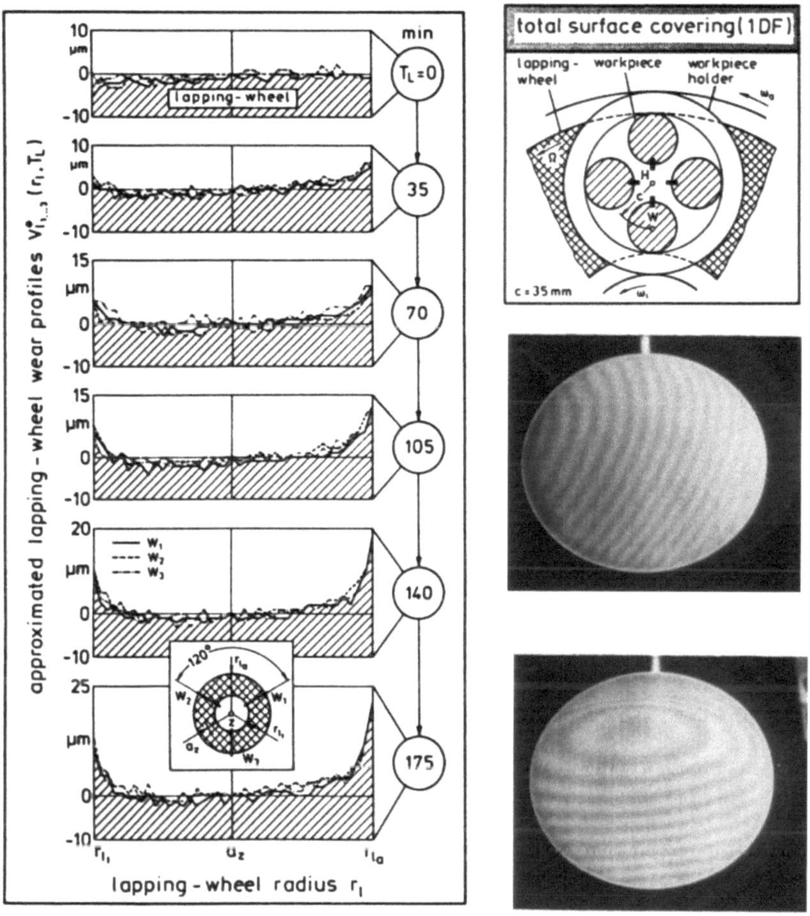

Fig. 21. Achievable workpiece form accuracy in the case of unfavourable lapping-wheel wear behaviour resulting from a total surface covering; optimal kinematic setting parameters were applied and workpiece self-rotation was prevented (1 DF) [Upper interference image: $T_L = 35$ min; lower interference image: $T_L = 140$ min]

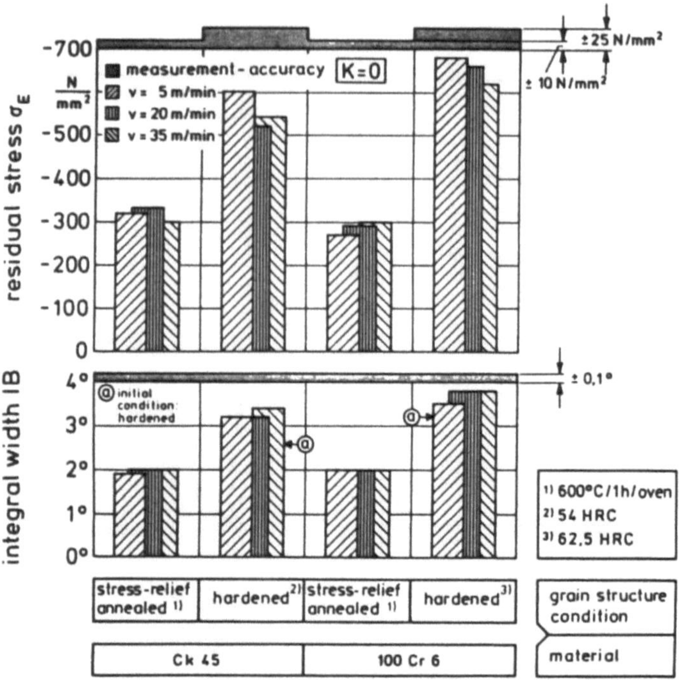

Fig. 23. Residual stresses and integral width of the X-ray diffraction line at the surface of lapped workpieces as a function of material, grain structure condition and path velocity
lapping compound PWS 12, engaging pressure 5 · 10-2 Nmm-2, constant path length 900 m

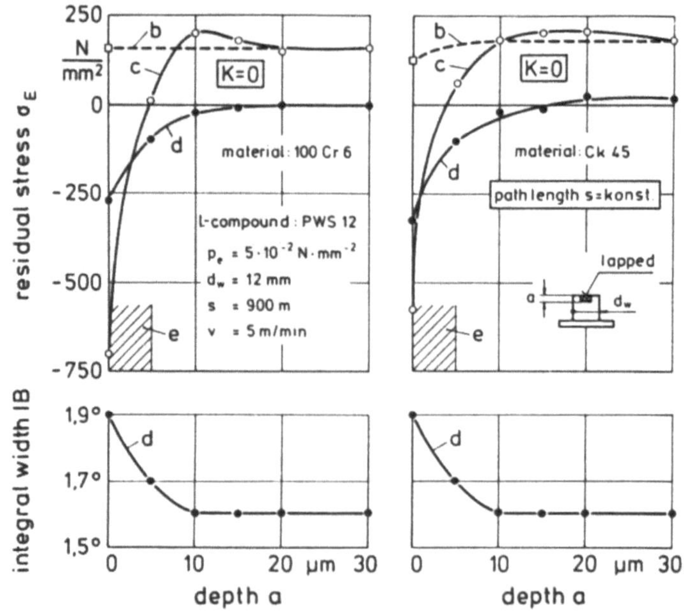

Fig. 24. Depth-courses of residual stress and integral width of the X-ray diffraction line in lapped steels Ck 45 and 100 Cr 6 as a function of the grain structure condition b initial condition: hardened (prior to lapping), c hardened grain structure condition (100 Cr 6: 62.5 HRC; Ck 45: 54 HRC), d strain-relief annealed grain structure condition, e information depth of the measurement

A)

angle of orientation $\varphi = 0°$			integral width IB [grad]	residual stress σ_E [N/mm²]
path velocity	v = 5 m/min	Ck 45	2,3	-400
		100Cr6	2,3	-400
	v = 20 m/min	Ck 45	2,3	-400
		100Cr6	2,3	-400
	v = 35 m/min	Ck 45	2,3	-390
		100Cr6	2,3	-390

B)

path velocity v = 35 m/min			residual stress σ_E [N/mm²]
angle of orientation	$\varphi = 0°$	Ck 45	-390
		100Cr6	-390
	$\varphi = 45°$	Ck 45	-380
		100Cr6	-380
	$\varphi = 90°$	Ck 45	-380
		100Cr6	-380

Table 1. Numerical values of the residual stresses at the lapped surfaces of steel materials under a strain-relief annealed grain structure condition
A) influence of the path velocity on the extent of the residual stress, B) influence of the lapping process on the orientation of the residual stress
lapping compound PWS 30, engaging pressure $5 \cdot 10^{-2}$ Nmm^{-2}, constant path length 900 m

Cutting Forces and Their Influences Upon the Surface Integrity in Single-Point Diamond Turning

T.C.G. Lo-A-Foe, J.H. Dautzenberg and A.C.H. van der Wolf

Department of Production Engineering and Automation,
Faculty of Mechanical Engineering,
Eindhoven University of Technology,
The Netherlands.

Summary

A model has been developed for determining the cutting forces.
The experiments, which have been carried out in order to evaluate the model, illustrate the tendency of a satisfying correspondence between the theory and the experimental results, especially when taking into account the forces on the clearance face. The force relationships are implemented in a numerical model for estimating the surface roughness. With the aid of this model the surface profiles of diamond-turned workpieces can be estimated within 10% and an examination of the cutting forces upon the surface roughness can be made.

1. Introduction

Single-point diamond cutting is a cutting process in which a diamond tool on a precision lathe is used for making products with a high surface quality. The tolerances of these products are 0.2 μm for the roundness, 0.1 μm per 100mm for the straightness and 0.01 to 0.1 μm for the roughness (R_z).
These accuracies are very high, but they can be achieved nowadays.
The problem is that the understanding of the cutting process is still insufficient in order to determine the optimal cutting conditions as well as to develop a simple machine tool monitoring system in mass production.
Most of the input parameters, which are the properties of the workmaterial, the geometry and material properties of the tool, the properties of the cutting fluid and the static, dynamic and thermal behaviour of the machine, play an important role in single-point diamond turning.
During the process these parameters interact with the strains, the cutting forces and the temperatures, so a lot of these will have both direct and indirect influences upon the surface quality.
The aim of this study is to quantify the influences of the input parameters upon the cutting forces and the forces upon the surface integrity.
In the past several cutting models have been introduced.

The first important model was made by Merchant [1], who was followed by several others like Lee and Shaffer [2] and Palmer and Oxley [3].
All these model have a restricted validity and several explanations have been given for the divergency. Features that have been ignored in the models are the influence of the hydrostatic pressure upon the shear flow stress [4], the forces on the clearance face [5], the nose forces [6], the side spread of the chip [7], the strain rate and the temperature upon the material properties [8], the chip curl [9] and the stress distributions on the rake face [10]. Although the incorporation of these factors results in some improvement of the models, none of them have been generally accepted yet, mainly due to the uncertainties and complexities associated with the accurate measurement of the parameters involved.

In this study a model introduced by Dautzenberg [11] is chosen because of the promising experimental correspondence and the fact that it leaves room for the incorporation of several features.
The force relationships will be implemented in a numerical model for estimating the surface roughness and this model will be evaluated for operating conditions in single-point diamond cutting.

2. The model

2.1. The forces on the rake face.

The model is based on the following assumptions (Fig 1):

- cutting is a two-dimensional plastic process.
- there are two regions of deformation.
  In the first region the chip flow is changed without any thickening of the chip and in region 2 the chip thickness increases due to the friction between the chip and the tool.
- the supplied power is completely dissipated by the two plastic processes.
- the process is continuous.
- the workmaterial behaves according to the constitutive equation:

$$\bar{\sigma} = C\bar{\varepsilon}^{n} \qquad (1)$$

where : $\bar{\sigma}$ = effective stress

$\bar{\varepsilon}$ = effective strain

n = strain hardening exponent

C = specific stress

For the global equilibrium of the tool, it holds that:

$$F_w = F_v.\sin\gamma + F_f.\cos\gamma \qquad (2)$$

$$F_n = F_v.\cos\gamma + F_f.\sin\gamma \qquad (3)$$

where: F_w = friction force
F_v = cutting force
F_f = feed force
F_n = normal force
γ = rake angle

In region 1 (Fig.1B) the chip velocity is the same as the cutting speed. Area 1 is unloaded and on the areas 2 and 3 forces and a moment are in effect.

The power balance is:

$$F_v.v - F_w.v = \int_0^{\bar{\epsilon}} \sigma.d\epsilon.A.v \qquad (4)$$

Substitution of Eq.(1) in Eq.(4) and dividing by the term C.A.v gives:

$$F_v^* = F_w^* + \int_0^{\epsilon_1} \bar{\epsilon}^n.d\bar{\epsilon} \qquad (5)$$

with: $F_v^* = \frac{F_v}{CA}$ = cutting force number (dimensionless) (6)

$F_w^* = \frac{F_w}{CA}$ = friction force number (dimensionless) (7)

$\bar{\epsilon}_1$ = effective enddeformation in region 1

Eq.(5) holds for each plastic process in region 1 i.e. bending or shear.

The integral $\int_0^{\epsilon_1} \bar{\epsilon}^n.d\bar{\epsilon}$ depends upon the number of shear planes, m, in region 1, which may vary between 1 and ∞ .

- For one shear plane (m=1) the relationship between the forces is:

$$F_v^* = F_w^* + \frac{1}{n+1}.\{ \frac{2}{\sqrt{3}}.\tan (\frac{\pi - 2\gamma}{4}) \}^{n+1} \qquad (8)$$

- For m = ∞ :

$$F_v^* = F_w^* + \frac{1}{n+1}.\{ \frac{\pi - 2\gamma}{2\sqrt{3}} \}^{n+1} \qquad (9)$$

The differences between these relationships and those of Merchant and Lee and Shaffer are shown in the Fig.2. The relationships derived from the models of Merchant and Lee and Shaffer are:

$$F_v = \tau.A.\frac{\cos(\beta-\gamma)}{\cos(\varphi+\beta-\gamma).\sin\varphi} \tag{10}$$

$$F_w = \tau.A.\frac{\sin\beta}{\cos(\varphi+\beta-\gamma).\sin\varphi} \tag{11}$$

where:

A = the area of the undeformed chip section

φ = the shear angle $= 45^o - \frac{\beta}{2} + \frac{\gamma}{2}$ for Merchant (12)

$= 45^o - \beta + \gamma$ for Lee and Shaffer (13)

β = the friction angle between the tool and the chip material

τ = the shear stress $= \frac{1}{\sqrt{3}}.C.\bar{\epsilon}^n = \frac{1}{\sqrt{3}}.C.(\frac{\cos\gamma}{\sqrt{3}.\sin\varphi.\cos(\varphi-\gamma)})^n$ (14)

The differences can be imputed to the following reasons:

- In applying the minimum energy principe Merchant assumed that, contrary to Eq.14, the shear stress on the shear plane is independent of the shear angle.
- The model of Lee and Shaffer is a slipline field solution with the assumptions that the material cut behaves as an ideal plastic which does not strain harden and that there is one shear plane which represents a direction of maximum shear stress. This gives the same relationship as the model of Dautzenberg for one shear plane (m=1) and no strain hardening (n=0), but for other values the relationships will be different.

Because single-point diamond cutting is a three-dimensional non-orthogonal process, in which the chip can be regarded as the sum of oblique chips having various thicknesses (Fig.3), the equations cannot be used without modifications.

For the sake of simplicity the following assumptions are made:

- the cutting process is a two-dimensional process with a cutting edge κ, which is the same as the chip flow angle.
- the direction of the chip flow is perpendicular to the tool edge in the middle of the width of contact between the tool edge and the workmaterial.

With these assumptions only the Eqs. (2) and (3) change:

$$F_w = F_v.\sin\gamma + (F_f.\sin\kappa + F_p.\cos\kappa)\cos\gamma \tag{15}$$

$$F_n = F_v.\cos\gamma + (F_f.\sin\kappa + F_p.\cos\kappa)\sin\gamma \tag{16}$$

When comparing the theoretical results with the experimental data [12] a moderate agreement is obtained (Fig.4). The deviations are mainly due to tool wear and will be explained in the following paragraph.

2.2. The forces on the clearance face

The previous model is based on the assumption that the total supplied energy is consumed in the primary and secondary shear zone only.
If the tool edge is infinitely sharp the forces on the clearance face and thus the energy dissipated in the burnishing process will be small.
However, tool edges in practice are not perfectly sharp and even if they are, a rapid breakdown will cause a certain dullness of the edge.
A rapid increase of the forces during cutting,especially for aluminium, is to be seen in Fig.5. This increase is mainly due to flank wear.
If the cutting time for one tool is longer than 25 minutes, as was the case in the experiments for evaluating the cutting model, the influence of the forces on the clearance face will be considerable.
In order to correct for the tool wear, the experiments have been carried out in a sequence of continually increasing contact width b_{ce}. The increase of the forces due to tool wear of one measuring cycle is approximated by the difference between the forces measured in the last and in the first section of the testpiece. The correction for each measuring cycle is the sum of extra forces of the preceding measuring cycles.
If the measured forces are corrected for the forces on the clearance face the experimental data can be fitted very well to a line which is parallel to the theoretical lines (Fig.6) and using C-values which are more than double - factor 2.3 for brass and factor 2.4 for aluminium - the values obtained from tensile tests, results in an excellent agreement between the theory and the experimental data (Fig.7).
A rough approximation of the forces on the clearance face due to flank wear can be made with a simple model which is based on the following assumptions:

1. The flank wear land is parallel to the cutting direction and
2. The normal stresses on the clearance face are uniform and equal to the the effective flow stress in region 1 of the cutting model.

With these assumptions the forces on the clearance face will be:

$$\Delta F_w = A_{VB} \cdot \sigma \tag{17}$$

$$\Delta F_v = \mu_f \cdot A_{VB} \cdot \sigma \tag{18}$$

where σ = the flow stress

A_{VB} = the area of contact between workpiece and the flank wear land

μ_f = the coefficient of friction between the tool and the workpiece at the flank face.

2.3. The force relationships

A good approximation of the forces on the rake face can be made with the simple relationship:

$$F_v^* = F_w^* + 1 \tag{19}$$

However, this is not sufficient for predicting the cutting forces, because there are two unknown variables and only one equation is available.

A second equation is:

$$F_w^* = \mu_r \cdot F_v^* \tag{20}$$

where μ_r = the mean coefficient of friction between the tool and the workpiece on the rake face.

Fig.6 shows that the mean friction coefficient is not constant, but a lot of the experimental data can be described with a constant friction coefficient, especially when brass is used as work-piece material.

Combination of Eq.(19) and Eq.(20) yields:

$$F_v^* = \frac{1}{1-\mu_r} \tag{21}$$

$$F_w^* = \frac{\mu}{1-\mu_r} \tag{22}$$

The forces on the clearance face can be estimated with the equations:

$$\Delta F_w = A_{VB} \cdot \sigma \tag{17}$$

$$\Delta F_v = \mu_f \cdot A_{VB} \cdot \sigma \tag{18}$$

These relationships are implemented in a numerical model [12] for estimating the surface profiles. This model will be discussed in the following pages.

2.4. The roughness model

An iterative process [13] is used for determining the displacements of the tool with respect to the work-piece which are caused by the operational setting of the machine, the dynamics of the machine and the cutting forces. The program starts with a calculation of the displacements of the tool with respect to the work-piece due to the operational setting of the machine and their variations caused by vibrations.
The geometry of the chip section (Fig.8), resulting from these displacements, the profile of the cutting edge and the surface profile of the work-piece before cutting, is used for calculating the cutting forces.
These cutting forces and the stiffnesses between the work-piece and the tool determine the additional displacements, which will change the position of the tool with relation to the work-piece and hence the area of the undeformed chip section and consequently the cutting forces.
So the calculation of the cutting forces and subsequent displacements is repeated until the difference between two successive positions of the tool relative to the work-piece is less than 0.1 nm.
The final position and the profile of the cutting edge are used for determining the surface profile.
The determination of the whole profile is followed by a calculation of the roughness parameters (R_a,R_q,R_t).

3. Experiments

Turning tests with aluminium and brass were carried out on a lathe with hydrostatic bearings in order to evaluate the model.
The dynamic behaviour of the lathe was determined from modal analysis using a HP 5423 A Fourier Analyzer.
The cutting forces were measured with a dynamometer having a three component Kistler-piezo-element with a resolution of a few mN.
The surface roughness was measured with both stylus and optical (modified Compact Disc sensor [14]) techniques.
The optical sensor was used for measuring the tool geometry too.
The test conditions were:

- number of revolutions : 300 - 2400 r/min
- feed velocity : 50 - 300 µm/s
- depth of cut : 5 - 20 µm
- radius of the tool : 0.2 or 1.0 mm

4. Results and discussion

A satisfactory agreement between experimental and theoretical results is illustrated in Figs.9, 10, and 11. The differences between the estimated and measured roughness values are less than 10% in most instances.

The deviations could have been due to one or more of the following reasons :

a The results of the vibration analysis were used for estimating the surface profile. Other fluctuations that could have occurred when machining the test-pieces, due to slipping of the belts were not taken into account.

b Metallurgical properties of the work-piece material were not included in the model; therefore any roughness due to voids and impurities, as well as, anisotropy of the material behaviour were omitted from the calculations.

c The accuracy of the measuring equipment was restricted to a few nm.

d Progress of the wear of the cutting edge during machining.

The influence of cutting forces upon surface roughness appeared to be quite small. Variation of these - by changing the depth of cut or using other materials (Fig.11) - does not affect the surface roughness remarkably. This is not surprising as the variation of the cutting forces is less than 50mN and the stiffness between the work-piece and the tool is about 10^7N/m, so the roughness due to variation of the forces is less than 5nm.

The differences in the surface profiles shown in Fig.11 are due mainly to differences in the phase angles of the wave forms.

On the contrary the influence of the cutting forces on the shape accuracy can be large (magnitude of order 0.1μm), especially when the tool wear or the depth of cut is large.

5. Conclusions

a. The cutting forces can be approximated with some simple relationships.

b. The surface profiles of diamond-turned work-pieces can be estimated within 10%.

c. The influence of cutting forces upon the surface roughness is relative small but their influence upon the shape can be large.

Acknowledgements

These investigations were supported by the Netherlands Foundation for Technical Research (STW), future Technical Science Branch/ Division of the Netherlands Organization for the Advancement of Pure Research (ZWO).
This support is gratefully acknowledged.
The authors are indebted to Mr. J.C. Manders , Mr. J.A.B. van Dijck and Mr. K.G. Struik for their expert assistance, and to Prof. ir. F. Doorschot and ir. T.G. Gijsbers for their support and suggestions.

References

1. Merchant, M.E.: Mechanics of the metal-cutting process. Journal of the Applied Physics,16,1945,p.267-318.

2. Lee, E.H.; Shaffer, B.W.: The theory of plasticity applied to a problem of machining. ASME Journal of Applied Mechanics,73,1951,p.405-413.

3. Palmer, W.B.; Oxley, P.L.B.: Mechanics of orthogonal machining. Proc. Inst. Mech. Engrs.,173,1959,p.623-654.

4. Santhanam, S.; Shaw, M.C.: Flow characteristic for the complex stress state in metal cutting. Annals of the CIRP,34,1985,p.109-111.

5. Spaans, C.: An exact method to determine the forces on the clearance face. Annals of the CIRP,15,1967,p.463-469.

6. Abdelmoneim, M.E.; Scrutton, R.F.: Tcol edge roundness and stable build-up formation in finish machining. ASME Journal of Engineering for Industry,96B,1974,p.1258-1267.

7. Banerjee, S.K.: The limit of continuity in some single phase materials during machining and its correlation with torsion testing. Development of Production Systems, Proc. 2^{nd}ICPR,1974,p.89-110;

8. Fenton, R.G.; Oxley, P.L.B.: Mechanics of orthogonal machining allowing for the effect of strain rate and temperature on tool-chip friction. Proc. Inst. Mech. Engrs.,183(22),1968-1969,p.417-438;

9. Dewhurst, P.: Is the machining process uniquely defined? Annals of the CIRP,27/11,1978,p.1-4.

10. De Chiffre, L.; Wanheim, T.: What can we do about chip formation mechanics? Annals of the CIRP,34,1985,p.129-132.

11. Dautzenberg, J.H.; Vosmer, J.: Cutting force relations for an arbitrary deformation model. Proc. 2nd Conf. Irish Manufacturing Committee (Ireland),1985,p.207-218.

12. Lo-A-Foe, T.C.G.; Dautzenberg, J.H.; van der Wolf, A.C.H.: A model for the micro-cutting forces of non-ferrous metals. Proc. 5th Polytechnics Symp. on Manufacturing Engineering (England), 1986,p.346-362.

13. Lo-A-Foe, T.C.G.; Dautzenberg, J.H.; Struik, K.G.; van der Wolf, A.C.H.: Cutting parameters and surface roughness in single-point diamond cutting. Proc. VII. Int. Oberflachenkolloquium Karl-Marx-Stadt,1988,p.188-195.

14. Struik, K.G.; Fan Muchang : Measurement of shape and roughness by a modified Compact Disc sensor coupled to a computer. Abstracts 4th Int. Precision Engineering Seminar (England),1987,p.81-90.

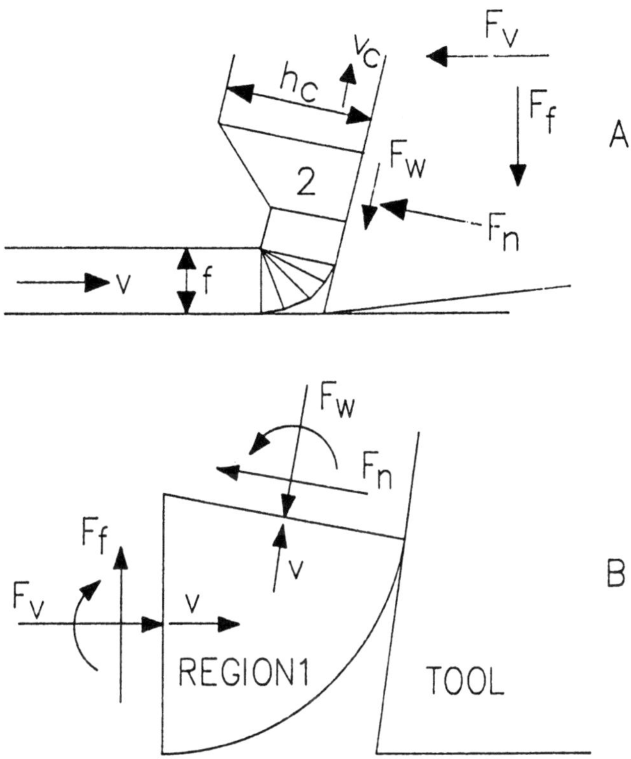

Fiq.1. Representation of the two dimensional cutting process

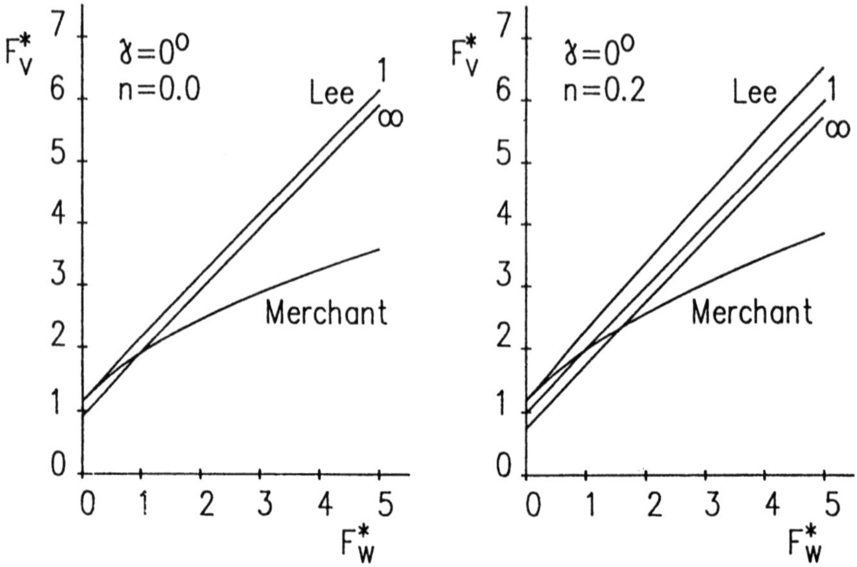

Fig.2. The cutting force number versus the friction force number for several models

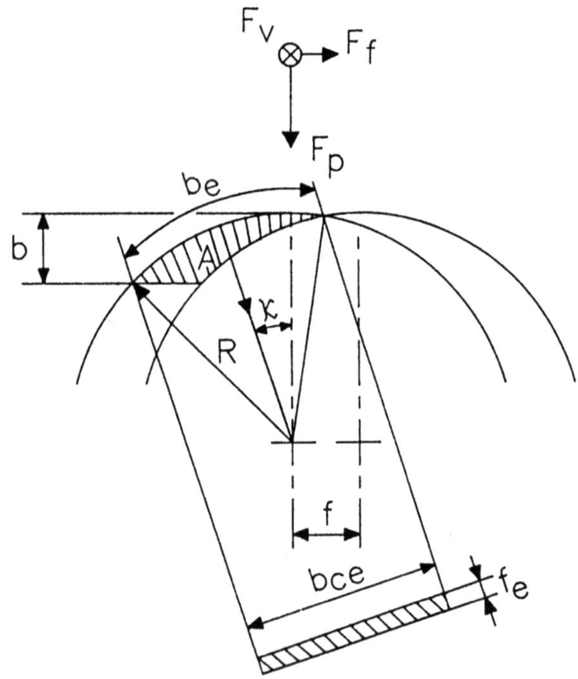

Fig.3. The geometry in single-point diamond cutting

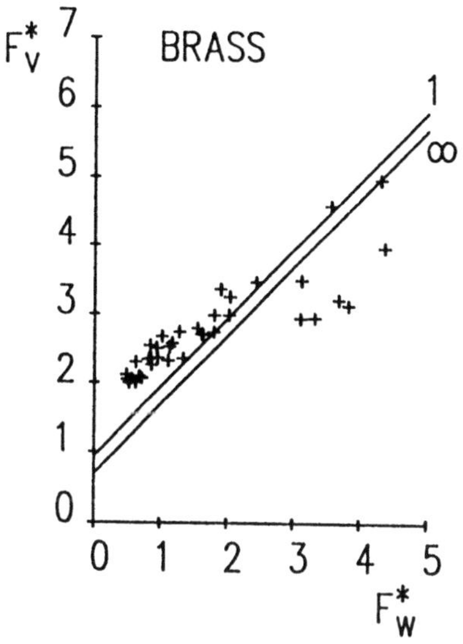

Fig.4. The cutting force number versus the friction force number without correction

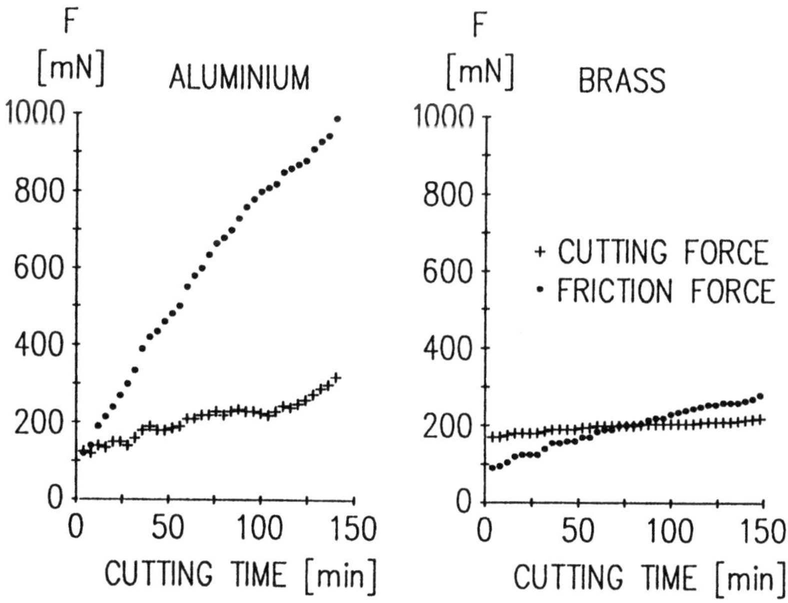

Fig.5. The cutting forces versus the cutting time

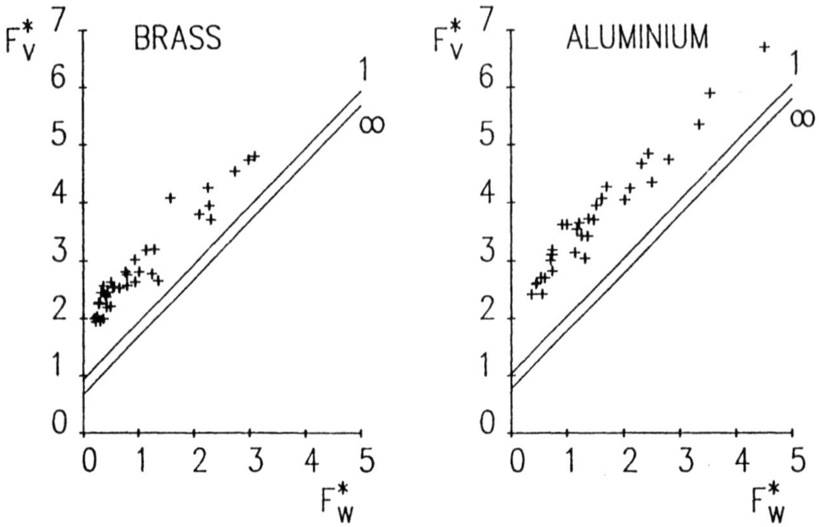

Fig.6. The cutting force number versus the friction force number corrected for tool wear

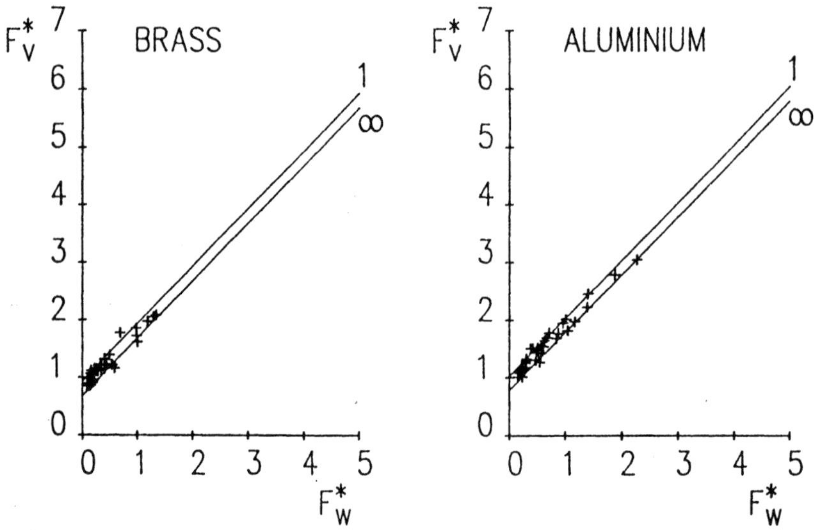

Fig.7. The cutting force number versus the friction force number corrected for tool wear and specific stresses

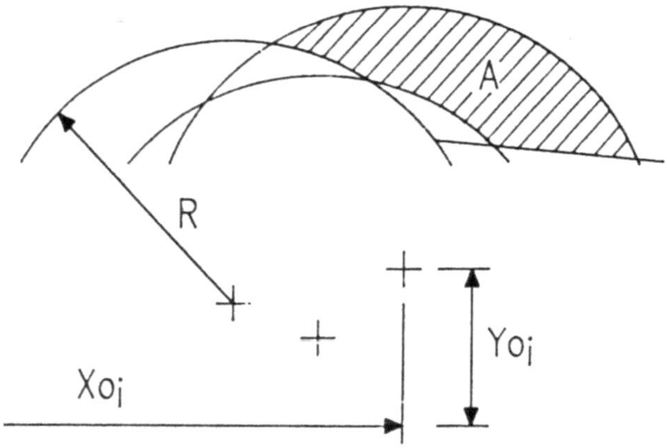

Fig.8. The position of the tool with respect to the work-piece and the undeformed chip section

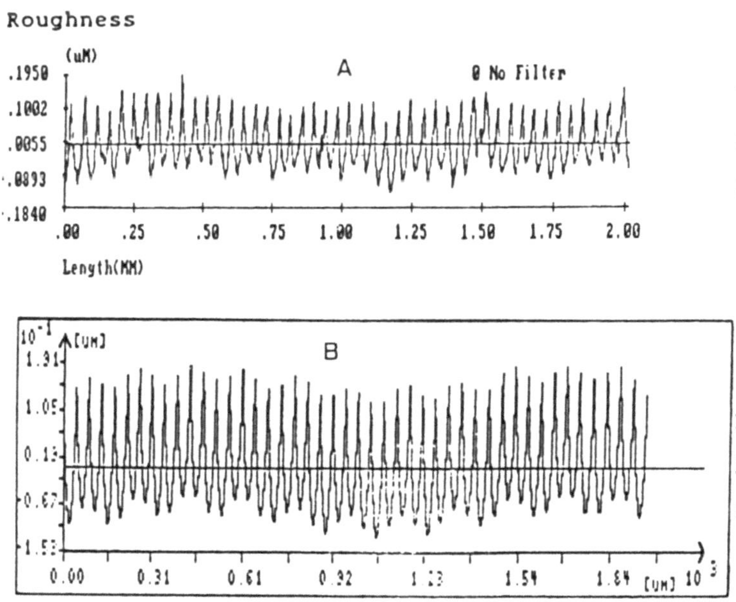

Roughness values [µm]

R_t=0.329
R_q=0.071
R_a=0.060

R_t=0.309
R_q=0.073
R_a=0.062

Fig.9. The measured (A) and estimated (B) profiles of a brass work-piece. Feed=40 µm/rev.

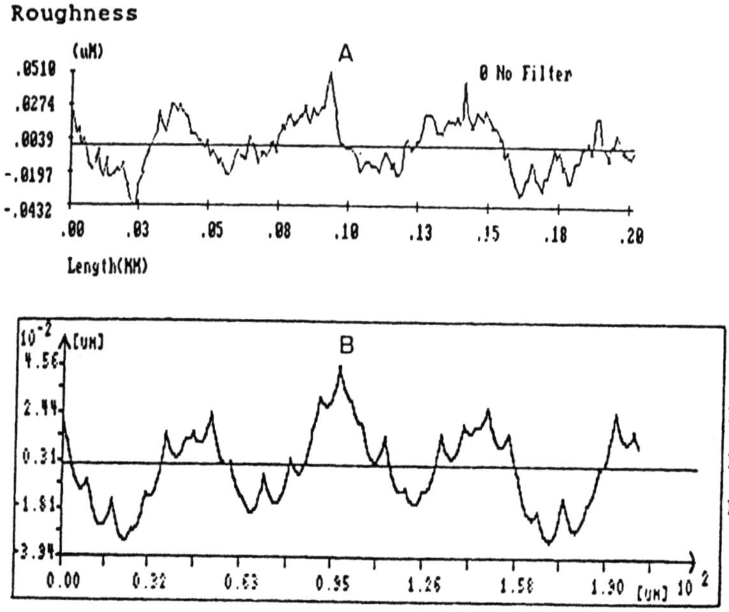

Roughness values [μm]

A: R_t=0.086, R_q=0.017, R_a=0.014

B: R_t=0.076, R_q=0.017, R_a=0.015

Fig.10. The measured (A) and estimated (B) profiles of a brass work-piece. Feed=6 μm/rev.

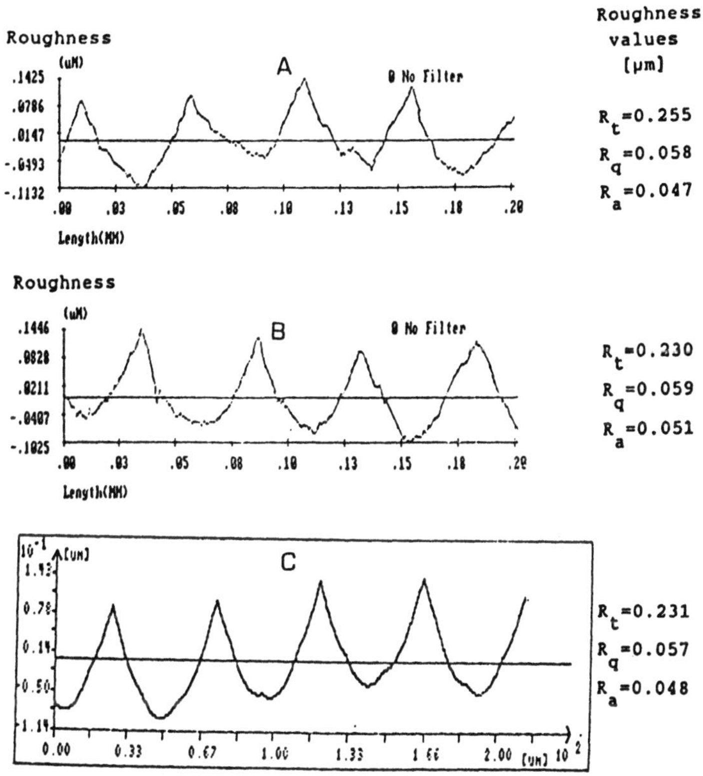

Fig.11. The measured profile of a brass (A) and an aluminium (B) work-piece and the estimated profile (C). Feed=5 µm/rev.

Recent Trends in Diamond Tool Technology

N. IKAWA, S. SHIMADA and G. OHMORI

Department of Precision Engineering
Osaka University, Suita OSAKA 565, Japan

Summary

A review is made on the recent trends in researches on single point diamond tool for ultraprecision machining. Following a brief look on some features of the properties of diamond as a tool material, the cutting edge strength of diamond tool is described in terms of the stress induced by cutting force on a cutting edge and also of the anisotropy in the resistivity to chipping. As to performance of diamond tool, the significance of the microstructure of cutting edge is emphasized and, as an evidence of unique capability of diamond tool, the extremely fine chip with a few nanometer thickness is introduced. Then the problems associated with progressive wear of diamond tool are discussed. Finally, the recent works on the scientific qualification of diamond are reviewed.

1. Introduction

With extensive industrial efforts, technology of single point diamond turning (SPDT) has reached a certain level which satisfies practical needs in production of optical and electronic components of submicron accuracy. Especially a variety of machine tools has been developed so precisely as to be capable of machining soft materials to the surface finish of a few tens nanometer.

However, technology of diamond tool, the other key of SPDT, seems to rather rely on some empirical knowledges. Although the empirical rules at present play sufficient role in ordinary industrial cases, there will be an urgent need for more scientific approach to diamond tool in an advanced technology in the near future.

Keeping this in mind, the present paper makes a review on recent scientific works which may give some basic ideas for the developments of diamond tool for an advanced use. First, a brief look is made on some features in properties of diamond as tool material. Then, the strength and unique capability of diamond cutting edge are described. Finally, along with the wear characteristics, a possibility of scientific qualification of diamond is discussed.

2. Some Features in Properties of Diamond as Tool Material

Strength of diamond will be a matter of primary concern to the reliability of cutting edge. By using the Hertzian fracture test, the features in the strength of diamond have been investigated on the fracture criteria and the size effect [1][2][3].

Fracture in diamond takes place at the condition that the tensile stress normal to (111) plane, $\sigma_{(111)}$, or the characteristic stress, which is effective to induce cleavage, reaches to a certain level (or strength) inherent to the specimen. This fact automatically leads to the presense of the "apparent" anisotropy in the contact pressure or fracture strength, P_0, at which a ring crack initiates on the specimen surface in the Hertzian test. Figure 1 shows the fracture probability $G(P_0)$ as a function of the contact pressure at the crack initiation P_0 on three different surfaces of (111), (110) and (100) on a diamond specimen in the Hertzian test. As shown in Fig.1, the different $G(P_0)$-P_0 curves are obtained on different specimen surfaces even on the same diamond specimen -- a kind of anisotropy --[2].

The strength of diamond also depends on the size of fracture -- the size effect --[2][3]. Figure 2 shows the maximum tensile stress normal to (111)plane along the contact circle at the crack initiation, $\sigma_{a(111)}$, in the Hertzian test as a function of the tip radius of the indenter, or the size of fracture consequently. The strength of diamond reaches the level of 20 GN/m^2 at the crack size of a few micrometer comparable to that in microchipping. The size effect can be explained from a probabilistic point of view [3].

The strength as measured by the Hertzian test is also affected by the surface condition of the specimen. The fine-polished surface gives higher strength as shown also in Fig.2, probably because of lesser possibility of the exsistence of the cracks induced by the polishing process.

The thermochemical effect on the strength of diamond can not be overlooked. The diamond specimen which is subject to elevated temperature higher than about 800 K deteriorates in the strength as shown in Fig.3 [4]. The deterioration is enhanced when the specimen is heated in iron powder. These results are assumed to be caused by the elongation of the surface cracks as fracture nuclei due to oxidation. The specimens heated at relatively low temperature do not show any deterioration but some improvement in the strength probably due to the healing effect of the cracks.

The other properties significant to diamond tool will be the thermochemical ones, because tool damages in general take place at elevated temperature in almost all cases in machining operations. Phaal has first reported the numerical results on etching of diamond in air atmosphere [5]. The thermal etching is accelerated by the presence of ferrous metal in contact with diamond [6]. Furthermore, the presence of the thermochemical damage or wear of diamond has been well shown by a simulated grinding test using diamond cone as a cutting edge when applied to various metals [7]. Figure 4 shows the width of wear land on the top of diamond cone as a function of the accumulated cutting distance. The significance of the thermochemical effect on the wear can be understood by the fact that the highest wear rate is seen in the case of pure iron which is softest but most reactive to diamond.

3. Cutting Edge Strength [2]

The cutting edge sharpness and its stability or durability are essential factors required for high performance of ultraprecision tool. Both may depend on the strength or the toughness of tool material. Diamond is an ideal material as a fine cutting tool because of its high strength and wear resistivity. However, cutting performance and durability of diamond tool are empirically known to show broad scattering, possibly due to the intrinsic inhomogeneity, the remarkable size effect and the anisotropy in the strength depending on the individual stone, and also due to the combined effects with the statistical fluctuation in cutting phenomena.

The cutting edge strength can be understood in principle from a mechanical point of view. As the fracture criterion is established as explained in the section 2, the distribution of the risk of fracture of the specified points under stress in a diamond tool can be expressed numerically by comparing the stress level of the effective stress $\sigma_{(111)}$ at the points with the strength of diamond.

The analysis is made for a variety of diamond tools with the different crystalline orientations. The three dimensional finite element method is employed for the analysis of the effective stress $\sigma_{(111)}$ in the cutting edge under the typical distribution of the normal and the tangential stresses on the tool rake face. The maximum normal and tangential stresses, σ_c and τ_c, take the values comparable to the microhardness and the shear strength of the work material respectively. The stress components at a specific element are computed and converted to the effective stresses $\sigma_{(111)}$ on four (111) planes which pass the element. According to the analysis, the maximum value of $\sigma_{(111)}$ appears on the rake face. Figure 5 shows the example of the distribution of the maximum value of $\sigma_{(111)}$ normalized by σ_c for each element on the rake face. The level of $\sigma_{(111)}$ is considerably lower than that of the strength of diamond and remarkably depends on the direction of the tangential stress τ_c and the crystalline orientation of diamond. According to the result of Fig. 5, the diamond tools do not tend to fracture under normal and stable cutting as experienced in practice. However, when the tangential stress directing to the cutting edge (negative shear stress) appears due to unidentified causes, the tool tends to fracture. Moreover, it can be theoretically said as a principle for the design of unti-chipping tool that the (100) plane is more suitable as the rake face than the (110) plane. This results may support the reports by Bex [8] and Hurt and Decker [9].

Although the computation gives merely a limited contribution to practical use at present, it clearly shows the significance of the selection of the crystalline orientation in design of diamond tool.

In practice, cutting edge chipping has been observed to be associated with progressive wear or attrition on a cutting edge even when cutting soft metals such as aluminum. Cutting edge chipping in these cases may be due to a concentrated load on an extreme edge by hard particles contained in soft matrix.

4. Significance of Diamond Cutting Edge

Under the ideal operation of a machine tool, the geometrical accuracy atainable by machining is determined by that of the effective thickness of cut which is the undeformed thickness of the work-material effectively removed by the cutting edge from the worksurface. The effective thickness of cut is not always equal to the nominal one which is given at the cutting edge by the infeed on a machine tool. At an extremly small depth of cut, a transition takes place in chip formation from "stable" to "unstable". The effective thickness of cut at the transition can be defined as the minimum thickness of cut. The minimum thickness of cut can be considered to determine the possible limitation of the cutting accuracy attainable [10].

The sharpness of a cutting edge is the primary factor which determines the minimum thickness of cut. Figure 6 shows an example of SEM micrograph of a diamond cutting edge carefully polished on a qualified raw diamond. Using a tool of above quality, the cutting test has been carried out to find out the extremity of micro-machining capability of diamond cutting edge [11]. Figure 7 shows SEM micrograph of electroplated copper chips produced at the nominal thickness of cut of (a) 30 nm and (b) 1 nm respectively. The morphology of the finer chip is characterized by the folding of thin chip of 20 nm thickness. The folding appears in the nominal undeformed chip thickness finer than about 30 nm.

These results may mean the extreme capability of diamond

cutting edge to machine some materials to the accuracy of nanometer level if the cutting edge conditions can be maintained.

5. Wear of Diamond Cutting Edge

Durability or stability of cutting edge is a common requirement for any types of material processing tools. The wear of diamond cutting edge appears to be classified into two types, one of which is micro chipping and the other gradual wear. Through the detailed observation of tool wear after machining Al-Si alloy by an electron microscope, Keen has reported that no other wear mechanism than mechanical process could be seen [12]. If the wear progresses only by the accumulation of small chipping, the important factors for the tool wear are the crystalline orientation of cutting edge and the microstrength of diamond, both of which are described in the section 3.

However, it is well known that diamond shows considerable weight loss by oxidation or graphitization [5] when it is exposed to high temperature and that diamond tool shows severe wear after the cutting of some kind of workmaterials especially ferrous metals. Moreover, it was reported that the atomospher of CH_4 reduced the wear of diamond tool when machining steel [13]. These results may conclude that the thermochemical process is the other important mechanism of the wear of diamond cutting edge. Although the clear explanation of this wear process has not been made, the affinity of diamond with workmaterial can play the important role. In these cases, there will be morphologically grudual wear or attrition (but not by hard particles). Figure 8 shows an example of SEM micrograph of the wear on diamond tool used for facing of Al alloy disk. In these observation, the amount of wear is so tiny that a conventional method tends to fail to make numerical evaluation. Recently, however, several works have been going on for numerical analysis of wear by using a replication techniques which make it possible to evaluate the amount of wear at submicron level [3][14]. In the replication, the profile of a cutting edge to be exammined is transferd to soft metal such as copper by grooving cut and then evaluated by a suitable stylus method. Fig. 9 shows an example of progressive change in profile of a cutting edge with cutting distance. It can be expected that the techniques such as SEM observation and the wear profile replication become the powerful tools for analysing the wear mechanism of diamond tool.

6. Qualification of Diamond

Raw diamond for ultrafine tool should be so selected or qualified that it, when shaped properly, assures highly reliable performance for cutting distance more than 1,000 km. However, the selection of raw diamond is said to rely on empirical rules which may differ depending on individual manufacturer. It will be an urgent need to develop some scientific methods of qualification of diamond. Some attempts have been made to cast a scientific light on this problem by the measurents of infrared absorption at 7.3 μm [15][16] and the laser beam scattering [17] in diamond. These have shown that defects in some forms contained in diamond affect tool life.

Natural type Ia diamond contains nitrogen impurities of less than 0.3 atomic percent in various forms of the defects. Platelet which is one of these defects is a disk like defect with diameter of less than 100 nm on (100) plane [18]. The absorption coefficient at 7.3 μm has been reported to be proportional to the total platelet area in unit volume of diamond [19][20]. A close correlation has been found between the microstrength and the infrared absorption (IRA) at the 7.3 μm band as shown in Fig.10 [3]. This correlation may give an evidence for the justifiability of experimental results on tool life by Wilks and Wilks [15] and Wong [16]. On the other hand, Fig.11 shows the correlation between the microstrength and the relative intensity of broad P2 center in the electron spin resonance spectra (ESR) [3]. The P2 center has been reported to be the defect composed of three nitrogen atoms which, as substitution for carbon, locate most closely each other in a (111) plane [21]. Although the correlation between the platelet and the P2 center has not been clarified yet, these defects can be assumed to have close relation each other in the scope of the strength measurements.

Recently, an extended cutting test has been carried out on Aluminum alloy disk by using the tools made of diamond qualified by ESR which measures the spin density of unpaired electron associated with some kinds of defects in the specimen. Figure 12 shows the volume of the progressive wear as a function of the cutting distance for four tools with different spin density[22]. The volume of the wear is calculated by the amount of the recession of the end cutting edge as profiled by the replication technique shown in Fig. 9. It can be seen that tools made of diamond with low defects density is likely to be more resistant to wear. (While the tool D with the largest spin density showed rather small wear in terms of the value calculated by the recession, it had the largest crater wear on the rake face of the side cutting edge.)

Consequently close relations between microstrength and physical parameters such as IRA and ESR in diamond and the results of the wear test may give ones the possibility of scientific qualification of diamond for ultraprecision tool.

7. Concluding Remarks

With time honoured empirical knowledges, diamond tool technology has attaind a certain level which satisfies a variety of industrial needs in production of advanced optical and electronic parts. Being stimulated by growing importance of ultraprecision machining in industries, recent scientific and systematic research works also have developed new knowledges on diamond and consequently clarified a unique potentiality of diamond tool. An example is seen in a very sharp diamond tool which is able to remove fine chip the thickness of which is at nm level, an extreme precision!. In order to utilize fully the unique capability of diamond, more extensive scientific efforts are to be made for understanding micromachining phenomena and wear mechanism of diamond. Furthermore, scientific qualification or classification of diamond will be the other important factor to establish the technology of highly reliable diamond tools for advanced use in the near future.

References

[1] N. Ikawa and S. Shimada: Microstrength Measurement of Brittle Materials, Technol. Rept. Osaka Univ, 31, 1622 (1981) 315.
[2] N. Ikawa and S. Shimada: Microfracture of Diamond as Fine Tool Material, Annals of the CIRP, 31, 1 (1982) 71.
[3] N. Ikawa and S. Shimada: Non-Destructive Strength Evaluation of Diamond for Ultra-Precision Cutting Tool, Annals of the CIRP, 34, 1 (1985) 117.
[4] N. Ikawa: Mechanical Propertis of Diamond - Strength and Wear Resistivity -, New Diamond, 4, 1 (1988) 6.
[5] C. Phaal: Structure Studies of Diamond, Industrial Diamond Review, 26, 300 (1965) 486.
[6] N. Ikawa and T. Tanaka: Thermal Aspects of Wear of Diamond Grain in Grinding, Annals of the CIRP, 19 (1971) 153.
[7] T. Tanaka and N. Ikawa: Thermal Aspects of Wear of Diamond Abrasive Grain in Grinding Mild Steel, Bull. Japan Soc. of Prec. Engg., 7, 4 (1973) 97.
[8] P. A. Bex: Diamond Turning Tool, Industrial Diamond Review, Jan. (1975) 11.
[9] H. H. Hurt and D. L. Decker: Tribological Considerations of the Diamond Single-Point Tool, Proc. SPIE, 508 (1984) 126.
[10] N. Ikawa and S. Shimada: Cutting Tool for Ultraprecision Machining, 3rd. Int. Conf. of Prod. Engg. Kyoto (1977) 357.
[11] N. Ikawa et al.: Minimum Thickness of cut in Micro-Machining, Pre-Print for JSPE 1987 Autumn Meeting (1987) 465.
[12] D. Keen: Some Observations on the Wear of Diamond Tools Used in Piston Machining Wear, 17 (1971) 195.
[13] J. M. Casstevens: Diamond Turning of Steel in Carbon-Saturated Atmospheres, Precision Engineering, 5, 1 (1983) 9.
[14] C. K. Syn, J. S. Taylor and R. R. Donaldson: Diamond Tool Wear vs. Cutting Distance on Electroless Nickel Mirrors, Proc. SPIE, 676 (1986) 26.
[15] E. M. Wilks and J. Wilks: The Mechanical Strength and Reliability of Natural Diamonds, Diamond Research 1978, (1978) 2.
[16] C. J. Wong: Fracture and Wear of Diamond Cutting Tool, Trans. ASME, Journal of Engineering Materials and Technology, 103, Oct. (1981) 341.
[17] E. M. Wilks and J. Wilks: Light Scattering in Diamonds

and Their Mechanical Reliability, Industrial Diamond Review, Jan. (1980) 8.
[18] T. Evans and C. Phaal: Imperfections in Type I and Type II Diamond, Proc. Roy. Soc., A270 (1962) 538.
[19] E. V. Sobolev, V. I. Lisoivan and S. V. Lenskaya: The Relation between the "Spike" extra Reflections in Laue Patterns of Natural Diamonds and Their Optical Properties, Soviet Physics-Doklady, 12 (1968) 665.
[20] T. Evans: Where is the Nitrogen in Type I Diamond ?, Diamond Research 1973 (1973) 2.
[21] G. Davies, C. H. Welbourn and J. H. N. Loubser: Optical and Electron Paramagnetic Effects in Yellow Type Ia Diamonds, Diamond Research 1978 (1978) 23.
[22] N. Ikawa, S. Shimada, M. Chang, H. Yoshinaga and K. Nishimura: Diamond Tool Wear as Affected by Defect Density, Pre-Print of Japan Society of Precision Engineering 1988 Spring Meeting (1988) 389.

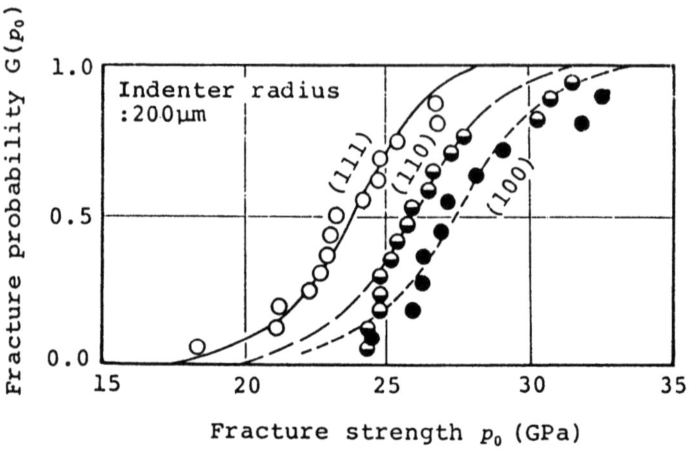

Fig.1. Fracture probalility on different surfaces of polished diamond specimens in Hertzian test

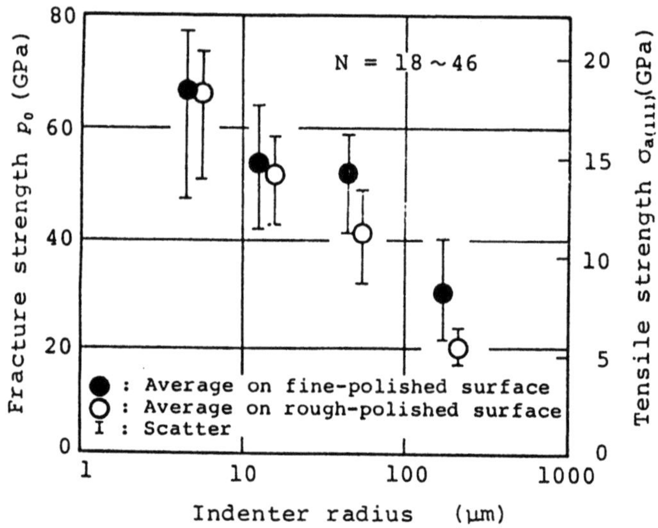

Fig.2. Hertzian strength properties of diamond on fine and rough polished specimen surfaces of (111) as a function of indenter radius.

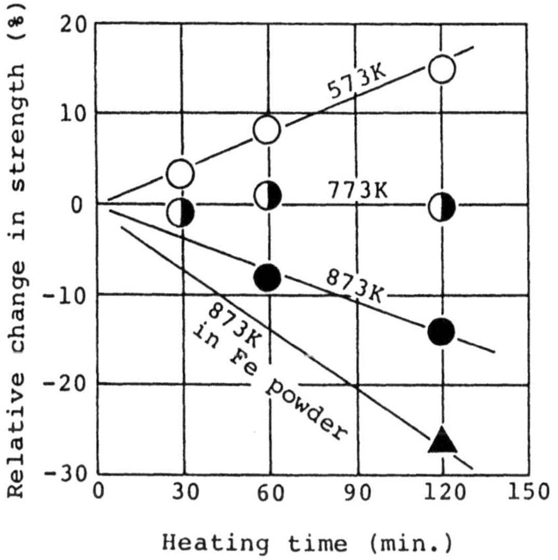

Fig.3. Change in Hertzian strength on polished diamond surfaces by heat treatment in air.

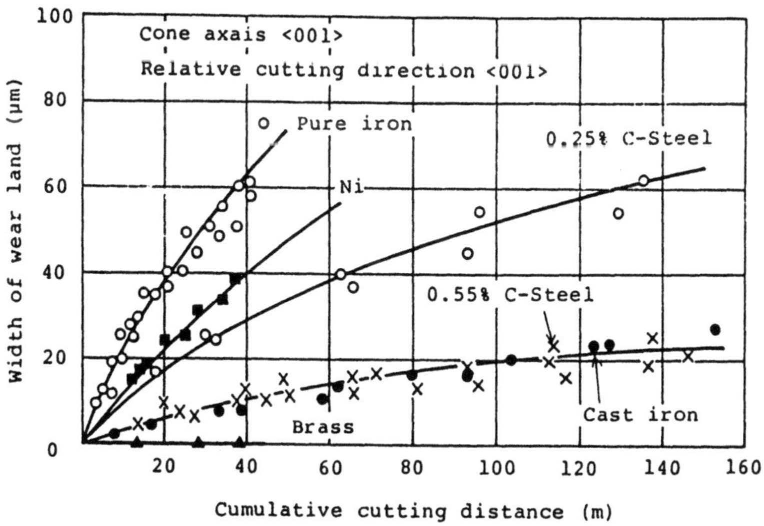

Fig.4. Wear of diamond cone as a function of cumulative cutting distance in the simulated single-point grinding of various metals.

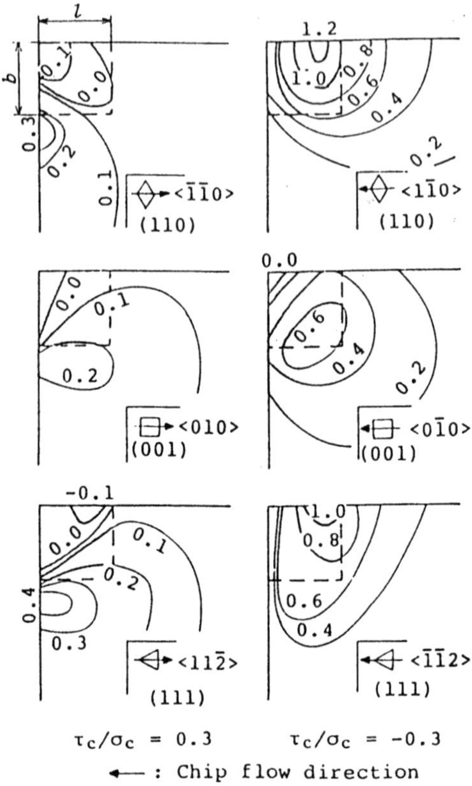

Fig.5. Effect of crystalline orientation on the distribution of maximum value of $\sigma_{(111)}$ normalized by σ_c on rake face.

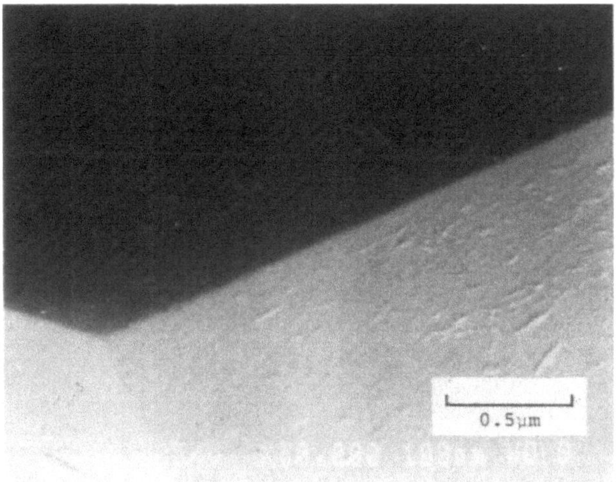

Fig.6. SEM micrograph of a fine diamond cutting edge.

(a) Nominal thickness of cut : 32.2nm

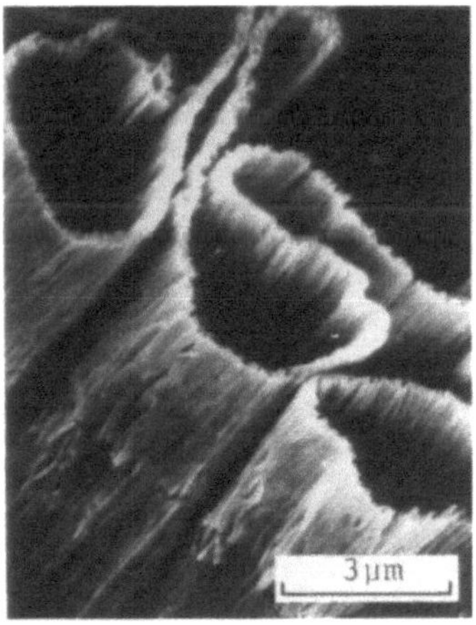

(b) Nominal thickness of cut : 1.07nm

Fig.7. SEM micrograph of chips.

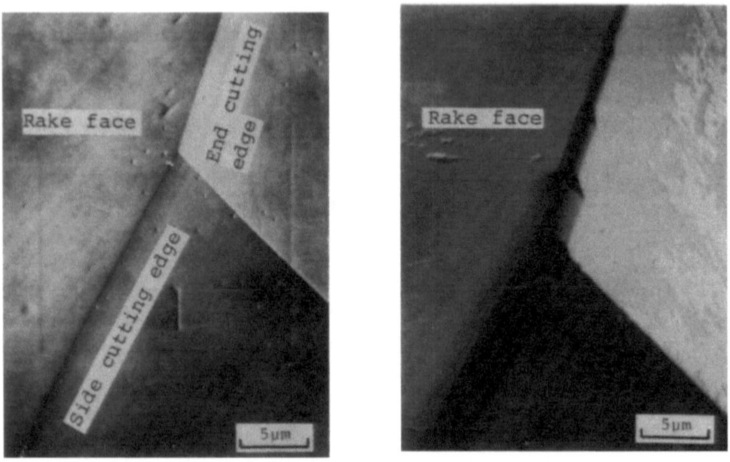

(a) Before cutting (b) After 500km cutting

Fig.8. SEM micrograph of wear pattern of diamond cutting edge.

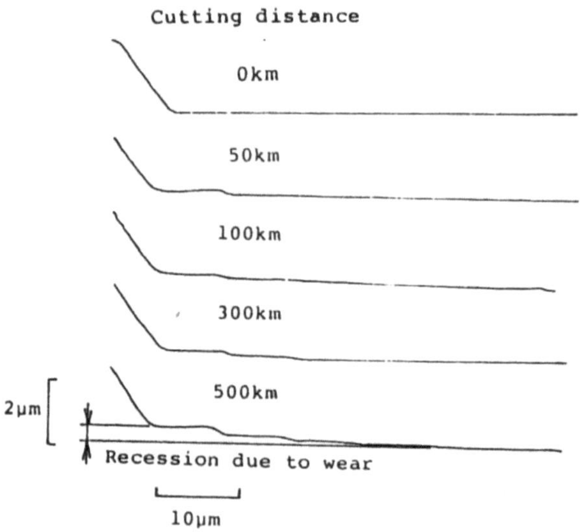

Fig.9. Progressive change in the profile of cutting edge as replicated in the grooves on OFHC-Cu specimen.

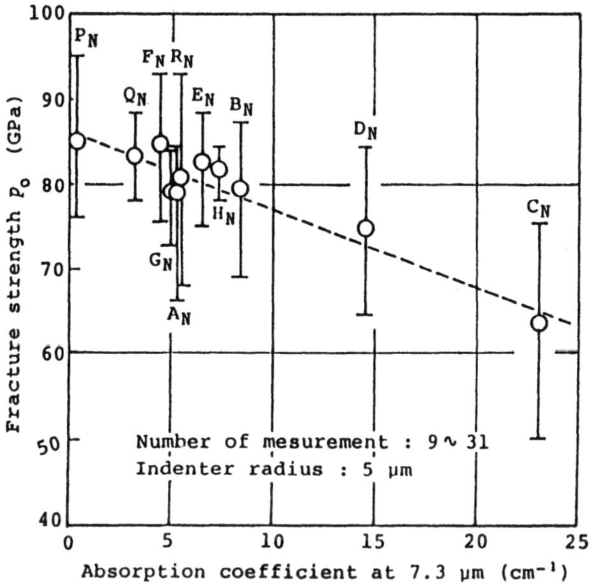

Fig.10. Absorption coefficient at 7.3 μm and microstrength of natural diamond.

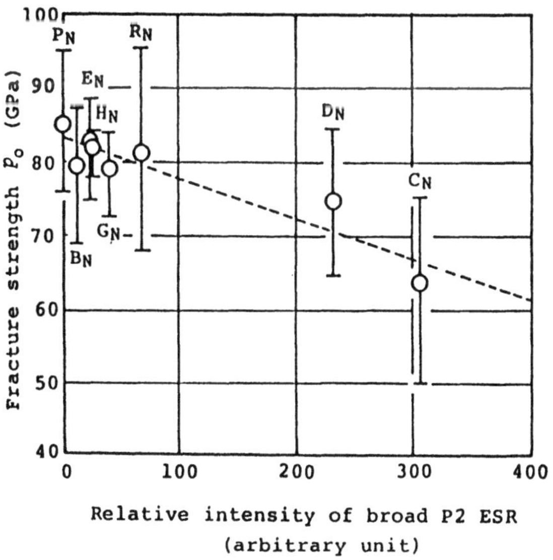

Fig.11. Relative intensity of broad P2 ESR and microstrength of natural diamond.

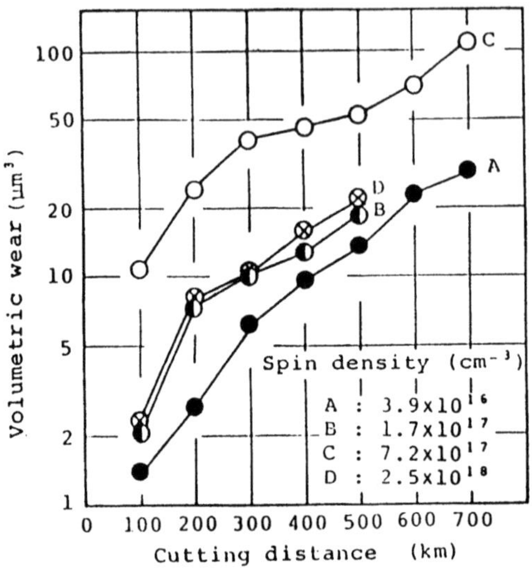

Fig.12. Volumetric wear on end cutting edge of various diamond specimens with different P2 center spin density as a function of cutting distance.

Prediction of Tolerances in Precision Cutting for Mass Production

W.J. Beekmans

Consumer Electronics Videq Lab.
Philips Eindhoven (NL.)

Abstract

Since the introduction of precision-cutting techniques in the mass production of high-quality mechanical components, the prediction of tolerances (micron and sub-micron) has become of critical importance. To obtain micron-precision dimensions in mass-production lines, the following factors in the cutting process are highly essential : the behaviour of the machine, the cutting tool quality, the stability and the material of the workpiece, the clamping method, the specific cutting technology and the environmental conditions. The influence of these factors is discussed for the production of VCR head discs. If the relationship between cutting forces and process parameters is known, analytical and F.E.M. calculations enable tolerance predictions to be made.

1 Introduction

In every videorecorder deck a scanner has been mounted to write and read video and audio signals on tape. To fulfill this function, the scanner has a tapeguiding function among other things, which is carried out by the upper and lower drums.

The upper drum, also called head disc, is a rotating part on which the video and FM audio heads have been mounted. The upper and lower drums are mass-produced precision parts.

To have a controlled production process, it is necessary that we know the factors that influence the required tolerances of the upper and lower drums.

This paper describes a method to predict the product tolerances.

The method is based on the assumption that product tolerances can be divided in :

. tolerances due to cutting forces (ΔT_f)
. tolerances due to the machine behaviour (ΔT_m)
. tolerances due to the applied clamping method (ΔT_m)

Expressed in a formula :

$$\Delta T_{tot} = \Delta T_f + \Delta T_m + \Delta T_c \qquad \text{(see figure 1).}$$

This method will be tested by predicting some of the head disc tolerances.

We shall first discuss the product tolerances and production method of the head disc.

2. Product and product tolerances.

The head disc is made of the aluminium alloy AlCu6 BiPb. The most important planes that are precision turned are (see figure 2) :

. the tape guiding surface (Ø62.003 ± 0.003 mm)
. the mounting surface for video and FM audio heads
. the hole diameter (Ø6.002 +0.004 mm).

The critical product tolerances of the head disc can be characterized as :

. dimensional tolerances : 5 μm
. form accuracy : 1 μm
. surface roughness : Ra = 0.1 μm

3. Production process.

The production process of the head discs consists of the following operations :

. impact extrusion : billets are sawn from extruded bars
. heat treatment : to improve the hardness without introducing residual stresses.
. rough-turning : to remove extruded overflows, to turn reference planes and special clamping and balancing grooves.
. drilling, tapping, milling

. cleaning
. precision-turning
. polishing head chambers and tape guiding surface
. cleaning

All precision planes are turned in one operation.
The reference and clamping planes for the last rough-turning operation are the same as those used for the precision-turning operation.

4. Tolerances due to cutting forces (ΔT_f)

Cutting forces are dependent on :

a. process parameters
- the depth of cut (a)
- the feed rate (s)

b. tool parameters
- the cutting tool radius (r_ε)
- the initial cutting edge radius ($< 0{,}01\ \mu m$)
- the cutting tool wear

c. product parameters
- kind of material

We have measured the cutting forces on a special lathe.
Some results are shown in figure 3. The main spindle of this lathe has hydrostatic bearings and can make an hydraulically controlled travel.
With respect to the above mentioned parameters the following remarks have to be made :

ad. a. The depth of cut mainly depends on the tolerances of the rough-turned workpieces.
In our mass-production process the depth of cut of each product can vary between 20 and 50 μm.

ad. b. The effect of tool wear is investigated by measuring cutting forces as a function of the cutting distance.
After a cutting distance of 50 km the cutting forces increase with a factor 10 in our case.

ad. c. Special attention must be paid to keep residual stresses as low as possible.

Depending on the product stiffness the cutting forces introduce product deflections which result in product tolerances. If the upper and lower limits of the cutting forces are known, it is possible to determine the resulting product tolerances with the help of F.E.M.-calculations. These calculations are carried out by introducing unit forces ($F = 1N$).
Some necessary experiments and calculations are running at this moment. Results will be presented at the conference.

5. Tolerances due to the machine behaviour.

Relevant properties of a precision-cutting machine are :

- . geometrical accuracy
- . static and dynamic stiffness
- . thermal stability.

To reach optimal conditions the following measures have been taken in our case :

- . hydrostatic bearings have been applied
- . the oil is circulating through the framework
- . the oil is temperature controlled ($\pm$ 0,5°C); environmental temperature fluctuations of $\pm$ 1°C are acceptable.
- . the cutting machine runs for about half an hour before precision-turning operations are started.
- . the cutting machine has been installed on rubber blocks to prevent that environmental vibrations influence the cutting process.

In our approach however, we consider the cutting machine more or less as a "black box".
We determine the product tolerances due to the machine (ΔT_m) by turning test workpieces with a static stiffness $\geqslant$ 20 N/μm (see figure 4).
The tolerances of these workpieces are caused by the machine behaviour (see figure 5).
The different sections of the test workpieces are turned

with 30 minutes time intervals to test the thermal stability of the machine.

6. Tolerances due to the clamping method.

In mass-production the workpieces are clamped with specific clamping devices.
Product deflections may be introduced due to the clamping method. After relieving the clamping device these deflections cause extra product tolerances (ΔT_c).

The clamping method can be optimized with the help of F.E.M.-calculations. The clamping method is optimal when the product deflections are minimal.
The head disc is clamped axially in a special clamping groove.
By comparing tolerances of products clamped with the clamping device with tolerances of products clamping with wax, it is possible to check the F.E.M.-calculations.
Results of our experiments will be presented at the conference.

Concluding remarks.

The demand for shortening developing throughput times, makes it necessary to predict the feasibility of precision parts.
We are convinced that the presented method is a useful tool to achieve this goal.
A very close cooperation between product and process-developers is necessary for the application of this method.
Some experiments still must be done. Once again, more figures about the prediction of tolerances of precision parts will be presented at the conference.

Acknowledgements.

The author wishes to thank Ir. C.T.J. Stilma (Centre for Technology, Philips Eindhoven) for his helpful discussions during the preparation of this manuscript.

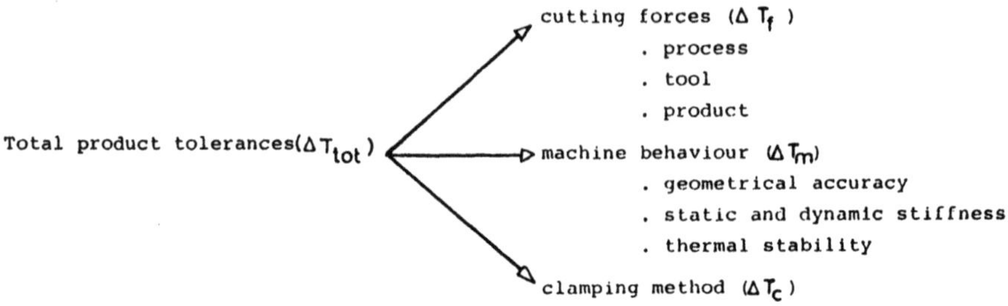

$$\Delta T_{tot} = \Delta T_f + \Delta T_m + \Delta T_c$$

Fig.1. Factors influencing product tolerances

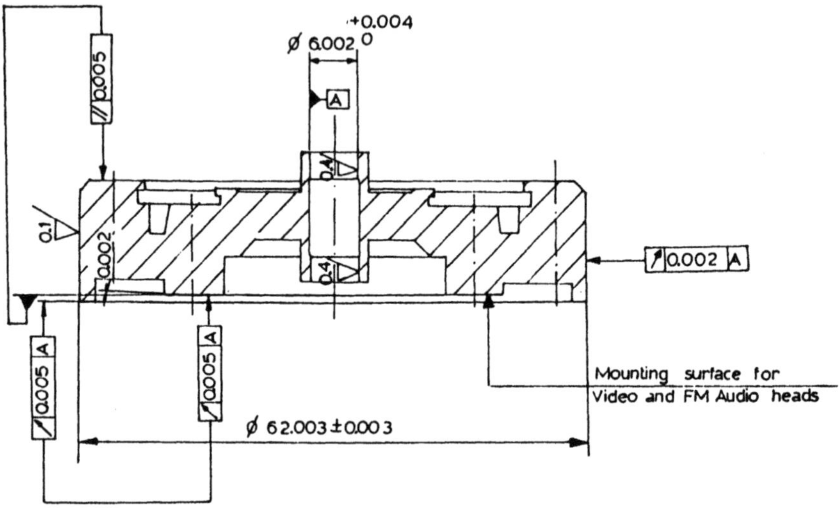

Fig.2. Tolerances of head disc VHS-scanner

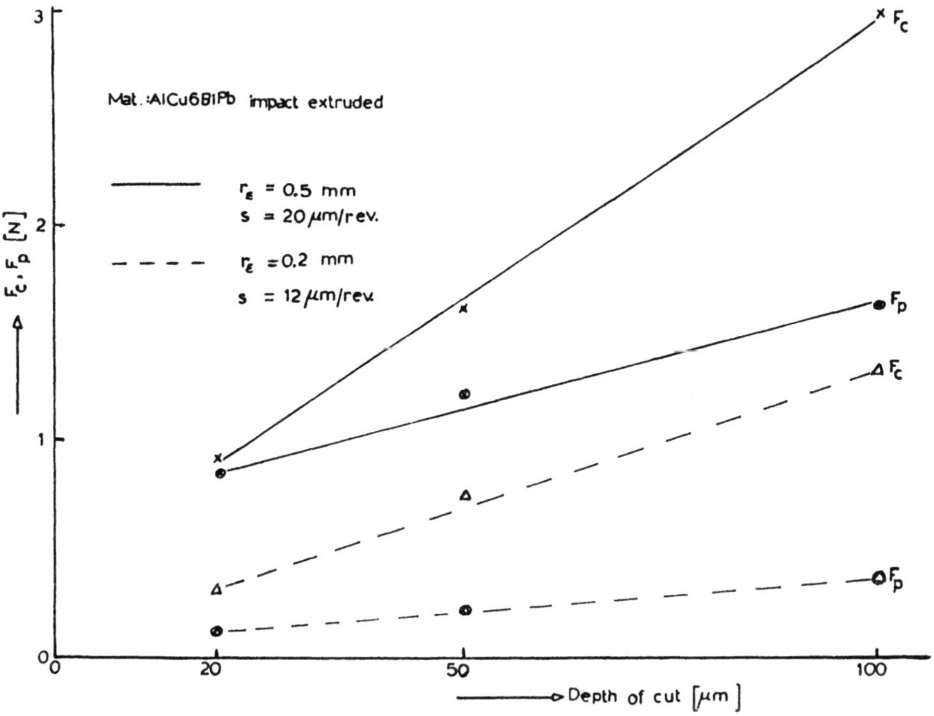

Fig.3. Cutting forces as a function of technological parameters

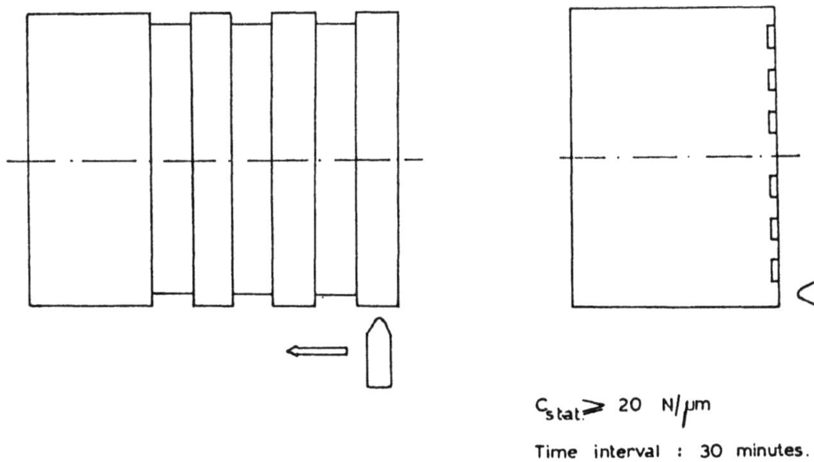

Fig.4. Test workpieces to determine tolerances due to the machine behaviour

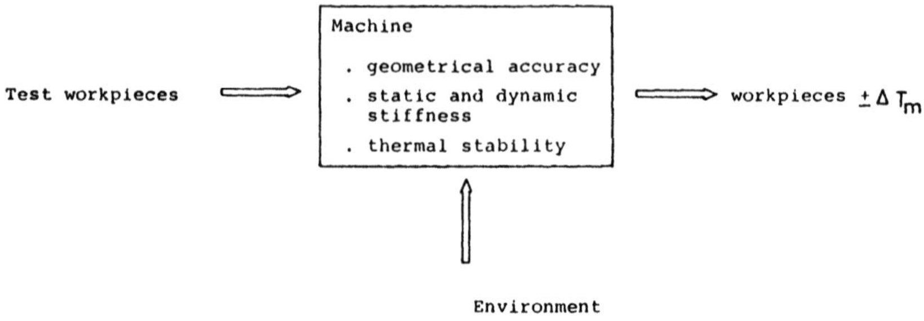

Fig.5. Tolerances due to the machine behaviour

Session 2

Machine Development and Improvement

Limits of Workpieces Accuracy Caused by the Geometrical and Dynamical Behaviour of Ultraprecision Diamond Turning Machines

M. Weck
R. Hartel
T. Bispink

Fraunhofer-Institut für
Produktionstechnologie (IPT),
Aachen, W-Germany

Summary

The influences of straightness errors and machine resonances on the surface of faced mirrors are discussed. Measurement results of the machine behaviour are compared with the surface-shapes and micro-structure of produced workpieces. A correlation between the measurement results of the machine and the analysed workpiece surfaces is presented in this paper.

1. Introduction

Numerically controlled ultraprecision lathes are used to machine components like aspherical laser optics, video drums and computer discs, figure 1.

Basically copper, aluminium and brass materials are diamond turned to achieve a surface roughness of a few nanometers and a contour accuracy in the submicrometer range.

Extremely sharp natural diamond tools remove the material with a chip thickness of some micrometers from the workpiece surface.

With stylus measurement devices or non contacting laserinterferometer surface measurement systems it is possible to check the quality of the produced profile and the micro-roughness of the surface. Accuracies of the figure smaller than 0.1 µm/100 mm and roughnesses of rms = 2 to 5 nm are state of the art /1,2,3/.

Figure 2 shows an ultraprecision cross-slide lathe. It has a granite machine bed. The x-carriage has air bearing slides and is moved by a capstan drive system with a torque-motor. The z-carriage has precision needle roller-slides and is driven by a rollerscrew. The airbearing spindle has a double-sphere-shape. The machine is controlled by an NC-controller. The feedback is realized by a laserinterferometer position transducer with a specially designed interface which is connected with the controller. The influences of the air on the wavelength of the laserlight is compensated by the interface. The resolution of the position transducer and the controller is 0.01 µm.

2. Influences on the workpiece accuracy

Figure errors and surface roughness of diamond turned workpieces are caused by influences of the cutting process, environmental conditions and the machine tool characteristics, figure 3.

Considering only the influences of a diamond turning machine with temperature controlled conditions, the figure accuracy is mainly affected by the geometric and kinematic behaviour. The surface roughness is influenced by the dynamic behaviour of the machine. According to this fact both machine tool characteristics and their influences on the machined surfaces of metal optics are analysed and described in this paper.

To examine the behaviour of ultraprecision machines different test procedures using high precision measurement devices must be applied.

3. Analysing the influences of the geometric machine behaviour on the workpiece accuracy

The geometric behaviour describes the form and position errors of moved machine parts like carriages or main spindle systems. The various geometric errors are specified in figure 3.

Not all errors have a perceptible influence on the workpiece accuracy, therfore an error budget is a helpful method in predicting their effects /4/.

Spindles and machine carriages can be aligned using the results of cutting tests, so that parallelity and orthogonality errors can be minimized to the limits of the measurement device's capability.

The geometric errors of spindle systems like axial, radial, angular runout and imbalances can be detected on the machined workpiece surfaces. Every spindle creates a special figure on the surface, therefore the machine is recognizable through an analysis of a workpiece it produced /5/.

Figure 4 presents the two most important kinematic arrangements of 2-axis-diamond turning machines namely the cross-slide-lathe and the T-base-lathe.

For both machines an error budget is made, figure 5.

The position error of ultraprecision lathes is very small, because the used laserinterferometer position transducers are highly accurate and the measurement planes are very close to the tool-tip plane, so that the Abbe offset errors are minimized. At the Fraunhofer-Institut for Production Technology in Aachen several table motion systems of different ultraprecision machines have been examined. The measurement results demonstrated that the systematic position errors of ultraprecision machines are in the range of the measurement resolution.

The horizontal straightness errors and roll motions of the x-slide create figure errors on a faced workpiece surface. To examine the influences of these errors on the workpiece shape the geometrical errors of the x-slide system in the cutting area will be compared with the profile accuracy of a machined plane mirror.

By measuring the horizontal straightness errors in the tool-tip plane the effects of the roll motion on the z-errors are included in the straightness measurement results. Therefore only the straightness measurement in the tool-tip plane must be performed.

For these measurements of an ultraprecision lathe x-slide-system, a Hewlett Packard laser with a Wollaston prism interferometer was used.

Figure 6 shows the operating principle and the test-set-up of the straightness measurement system /6/. Angular motions of the straightness reflector will cause measurement errors, so that the straightness reflector has to be fixed on the machine bed and the Wollaston prism must be moved with the carriage.

By averaging several measurements using the mirror reversing method, careful alignment of the laser beam and refering the measurement results to a regression reference the predicted measurement system errors were less than 7 nm and smaller than the device resolution of 10 nm /7/.

The straightness errors have been measured quasi continuous during the carriage motion at 800 points on a travel of 100 mm. Six measurements have been made and averaged afterwards by a computer system.

The results of the straightness measurements are shown in figure 7. To establish the correlation between straightness errors and workpiece errors, the cutting area for a testpiece of 160 mm diameter, which was diamond turned on the examined lathe, is marked in the figure.

The straightness errors which are caused by the form errors of the guideway are represented by the smoothed mean line .

Their magnitude in the cutting area is 0.044 µm and can be regarded as minimal.

Lateral forces of the ball-screw on the carriage lead to periodical straightness errors with amplitudes of about 10 to 20 nanometers. They superimpose the guideway straightness errors.

To examine the influences of the straightness errors on the workpiece accuracy a plane mirror was diamond-turned on the analysed machine. The technology data were optimized to minimize the influences of the cutting process.

The contour accuracy of the testpiece with 160 mm diameter was analysed with the Wyko 6000 phase laserinterferometer system. The measurement system has a resolution of about 0.6 nm and an accuracy of about 6 nm /8/. The data were evaluated by a computer and can be drawn in a 3-dimensional or a profile print. Figure 8 shows the measurement set-up, the operating principle and an example of a 3-dimensional surface plot.

To analyse the influence of the tool path on the workpiece a radial surface profile from center to the boarder was examined. The spindle runout errors were substracted from the measurement results by digital filters.

Figure 9 shows the measurement results of the radial profile. The profile has a peak to valley of 68 nanometers. The mean contour error is 48 nanometers. Shorter period profile errors with amplitudes of about 5 nanometers superpose the mean error.

In figure 10 the results of the straightness measurements and the figure errors of the radial testpiece profile are compared.

The mean shapes of the profiles are similar and their sizes are in the same order of magnitude.

Also influences of the short period straightness errors can be detected in the surface profile. However, the amplitudes of the surface errors are much smaller than those of the straightness errors and the periodicity is not always constant. Reasons for inconsistancy are other influences on the figure accuracy like material effects or spindle runout errors, which superimpose the straightness errors.

4. One approach to present the effects of the dynamic relations during microturning

Frequency-dependent displacements between workpiece and tool caused by dynamic weakpoints of the ultraprecision lathe are responsible for a micro-waviness of the surface texture.

To test the compliance behaviour of an ultraprecision lathe a piezo-electric driven shaker system is fixed between the toolpost and the workpiece, as shown in figure 11.

Exciting forces and relative displacements are measured in the frequency range up to 1000 Hz and in a second test up to 12000 Hz.

The resulting compliance frequency response in infeed direction (z-direction) is shown in figure 12. Measured mode shapes of the ultraprecision lathe indicate, for example, one vibration mode at 380 Hz of the aerostatic spindle unit.

In the high frequency range a weak point is recognized at 7800 Hz. Analysed with the mode shape measurement, tool vibrations in infeed direction can be detected.

The machined metal optic surface textures are examined by an advanced microscope phase interferometer measuring device.

The wavelength of the surface texture corresponds to the local cutting speed and the vibration frequency at the cutting area. The analysed surface texture of one machined workpiece is shown in figure 13.

Considering the relationship between wavelength and cutting speed at the measured positions, a good correlation between wavelength and resonance frequencies can be seen.

The measured area of the surface interferometer system has at least a size of 1 mm by 1 mm. To display one lower frequency wavelength in the window of the microscope interferometer it is necessary to measure near the turning center. High frequencies are easy to detect in the outer area.

The amplitude of the surface texture is less than Ra = 10 nm.

The fingerprint of the ultraprecision lathe on a metal optic near the turning center ø $\leq$ 4 mm is pointed out in figure 14. The small pages of photographs taken by Normaski technic from the surface are arranged to express the influence of the microturning process.

The surface texture P-V-heighths are less than 200 nanometers.

There are clearly marked areas where the tool removed the material from the surface and areas where the tool did not cut. Calculating wavelength and cutting speeds we see that low frequencies at 23 Hz are dominant. The reasons for this surface texture will be examined in contouring research work. Especially faulty machined surface contours caused by the spindle runout is one of the topics in future work.

5. Conclusion

At the examples of plane mirrors the influences of the geometric and the dynamic behaviour of ultraprecision diamond turning machines on the workpiece accuracy have been examined.

The effects of the straightness errors on the workpiece profile are demonstrated. The relationship between the resonance frequencies of the machine and the waviness of a machined metal optic surface has been described.

The presented results are a first approach to point out these correlations. To confirm their validity further measurements of different machines have to be made.

6. References

1. Taniguchi, N.: Current Status in, and Future Trends of, Ultraprecision Machining an Ultrafine Materials Processing. Annals of the CIRP Vol. 32/2/1983

2. Becker, K.; Heynacher, E.: Bearbeitung optischer Flächen mit Einzeldiamantwerkzeugen. Feinwerktechnik & Meßtechnik 95 (1987) 4

3. Gijsbers, T.G.: COLATH, eine numerisch gesteuerte Drehmaschine sehr hoher Präzision. Philips Technische Rundschau, 39. Jg. 1980/81, Nr. 8

4. Donaldson, R.R.: Error Budgets. Technology of Machine Tools, Machine Tool Task Force, Vol. 5, Section 9.14, UCRL-52960-5, Lawrence Livermore National Laboratory, October 1980

5. Raja, J., Whitehouse, D.J.: An investigation into the possibility of using surface profiles for machine tool surveillance Int. J. Prod. Res.; Vol 22., No. 3, 1984

6. Weck, M: Geometric and Kinematic Errors. Technology of Machine Tools, Machine Tool Task Force, Vol. 5, Section 9.12, UCRL-52960-5, Lawrence Livermore National Laboratory, October 1980

7. 5528 A Laser Measurement System. Manual from Hewlett Packard, Santa Clara, CA (USA)

8. Wyko 6000 Laser Interferometer System. Company brochure from Wyko Corporation, Tucson, Arizona (USA)

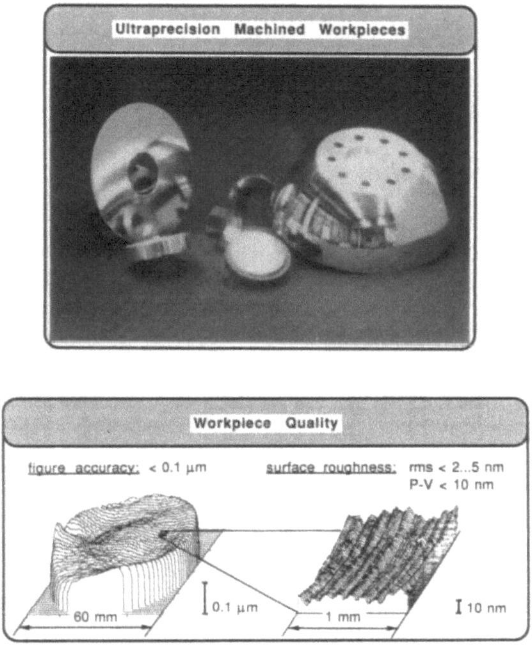

Fig.1. Ultraprecision machined workpieces

Fig.2. Ultraprecision diamond turning machine

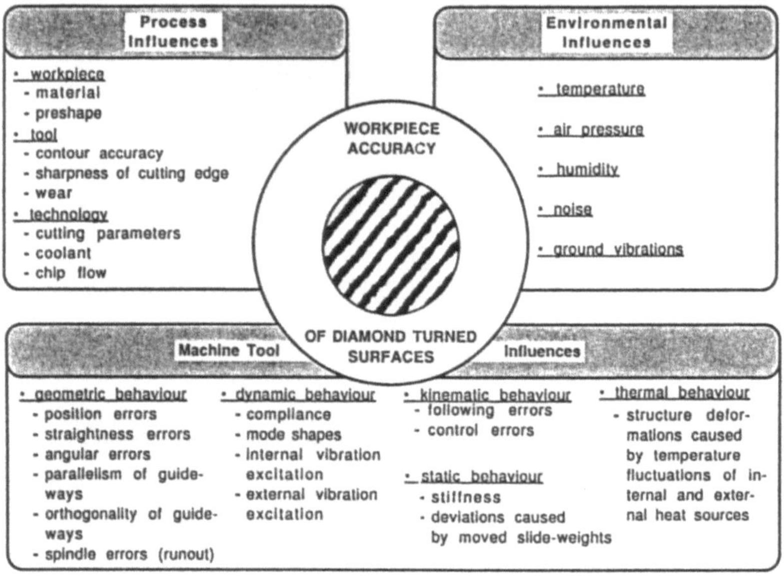

Fig.3. Influences on workpiece accuracy

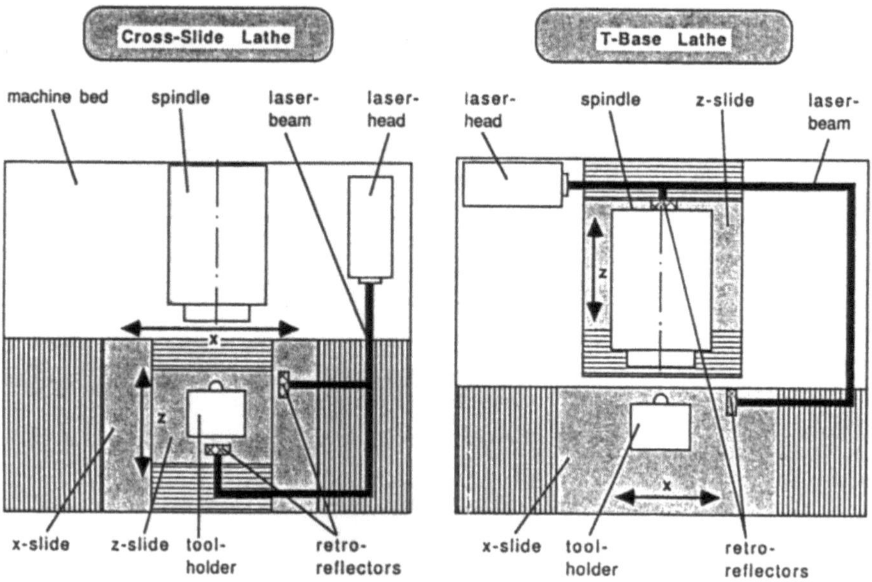

Fig.4. Arrangements of laserinterferometer controlled cross-slide an T-base-diamond turning machines

Slide	Geometric Error	Workpiece-Error Cross-Slide Lathe	Workpiece-Error T-Base Lathe
x-slide	position error	x-error	x-error
	horizontal straightness error	z-error	z-error
	vertical straightness error	-	-
	pitch	-	-
	roll	z-error	z-error
	yaw	x-error *	-
z-slide	position error	z-error	z-error
	horizontal straightness error	x-error	x-error
	vertical straightness error	-	-
	pitch	-	-
	roll	x-error	x-error
	yaw	x-error	x-error z-error **

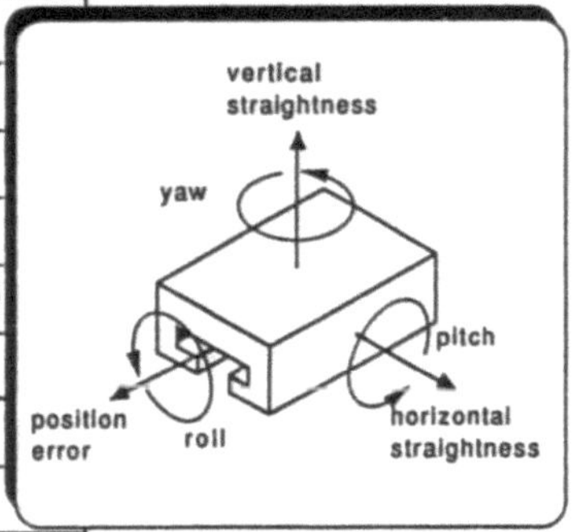

* dependent on z-table position

** dependent on x-table position

(shaded) critical for facing-operations

Fig.5. Error budget of cross-slide and T-base-diamond turning machines

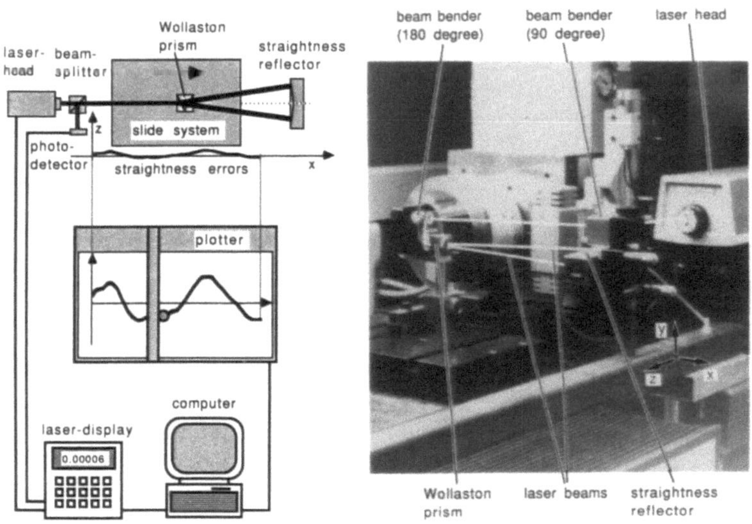

Fig.6. Test-set-up for the laserinterferometrical measurement of straigtness errors

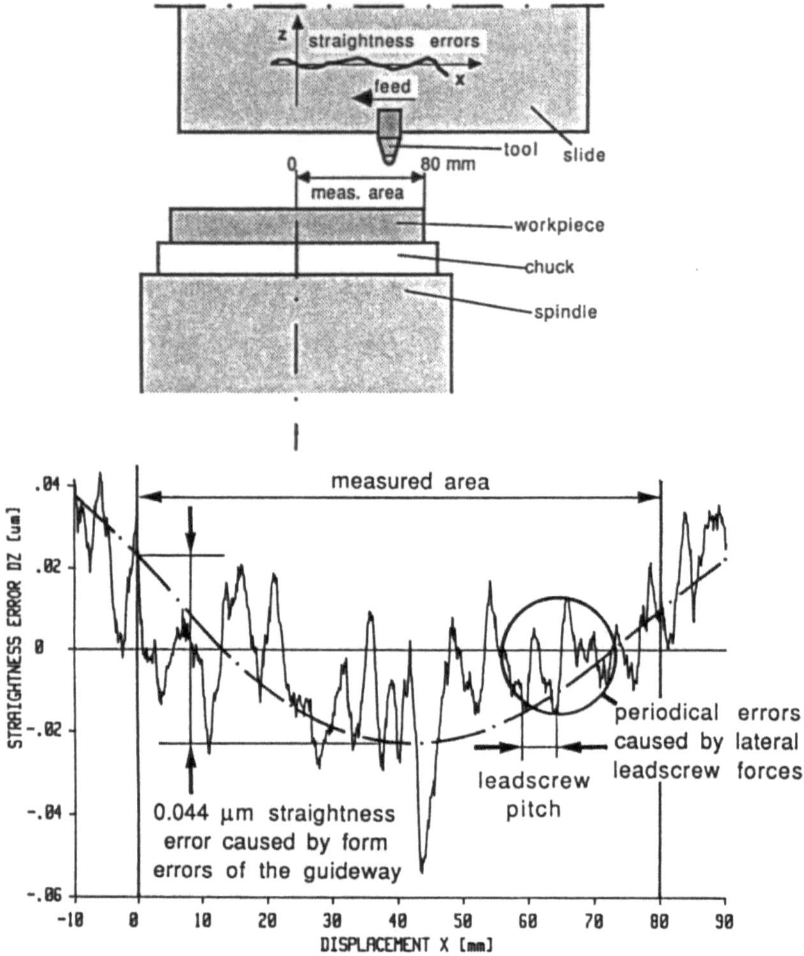

Fig.7. Measured horizontal straightness errors of the x-table of an ultraprecision lathe

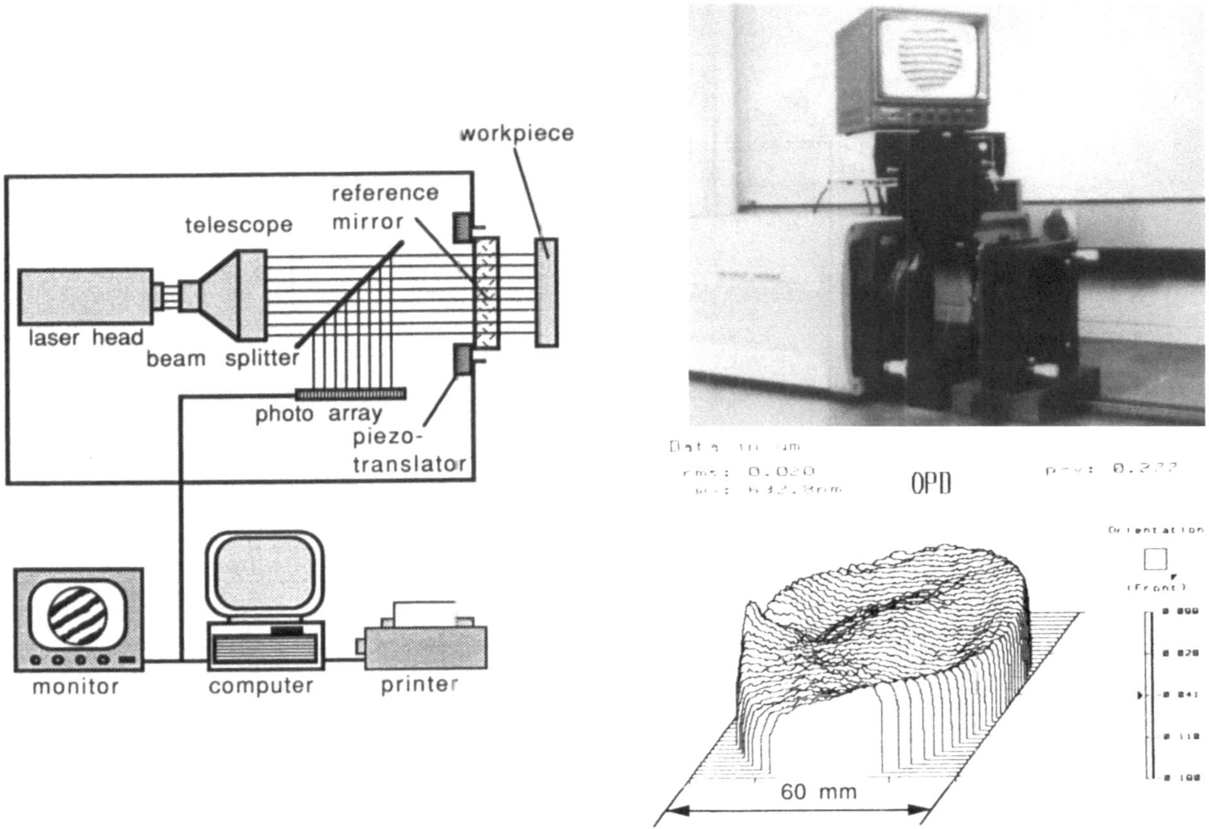

Fig.8. Laser phase surface interferometer

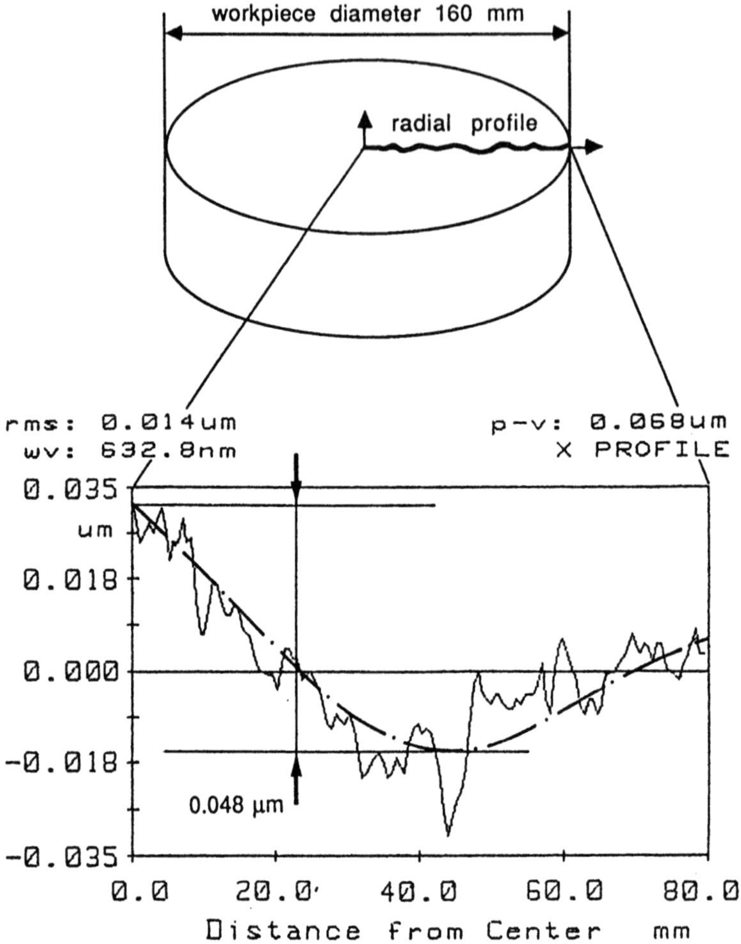

Fig.9. Measured profile of an ultraprecision machined aluminium workpiece surface

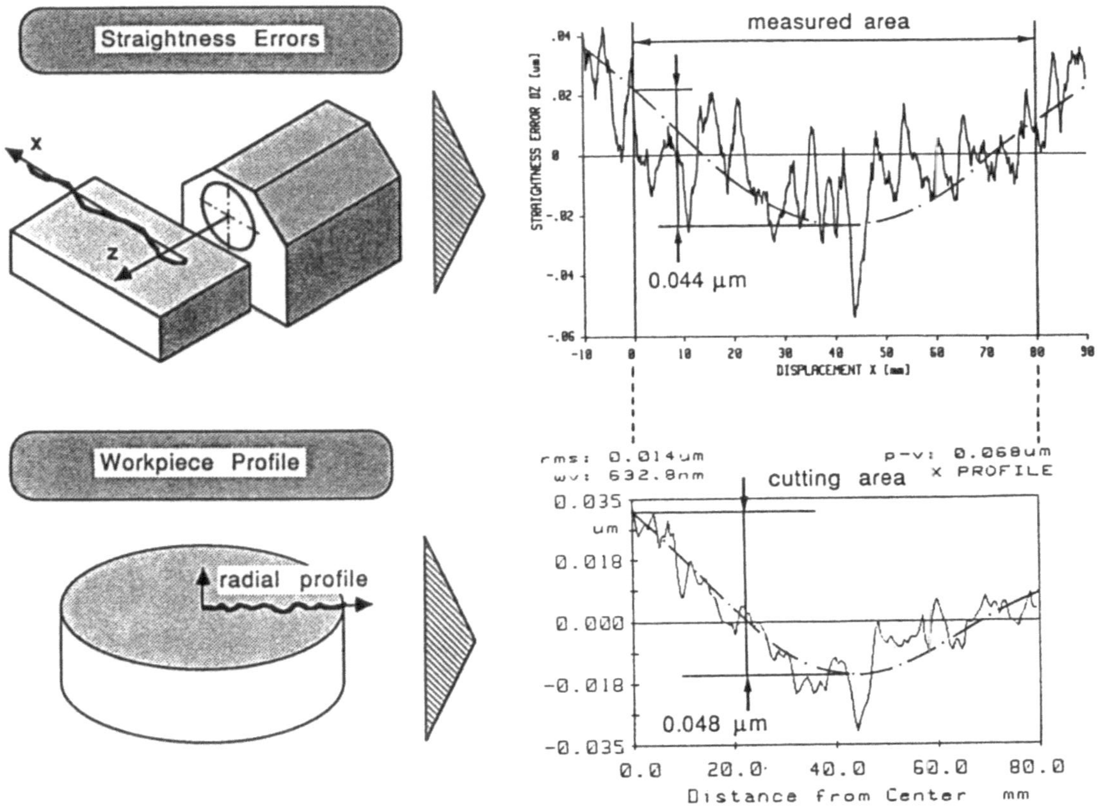

Fig.10. Correlation between straightness errors and the workpiece profile

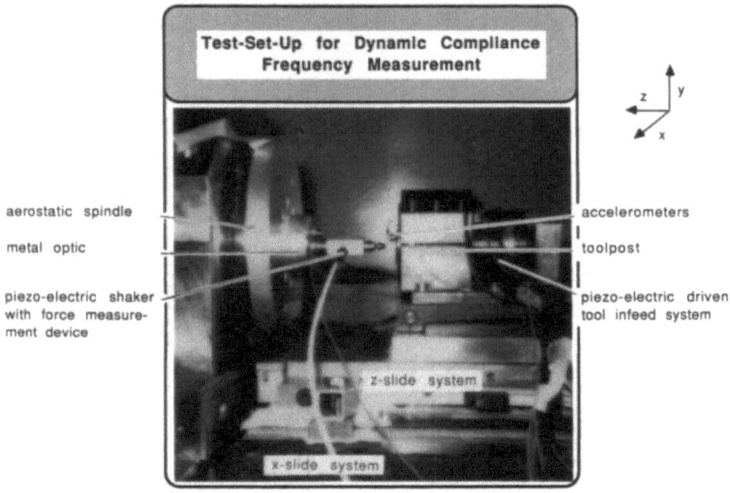

Fig.11. Dynamic testing device to measure compliance behaviour

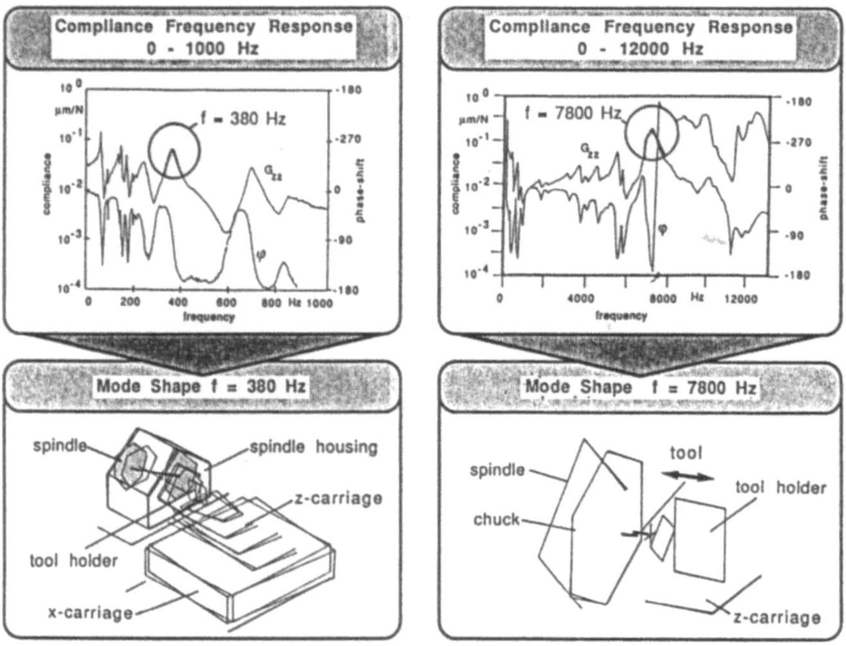

Fig.12. Critical compliances and mode shapes of the analysed ultraprecision lathe

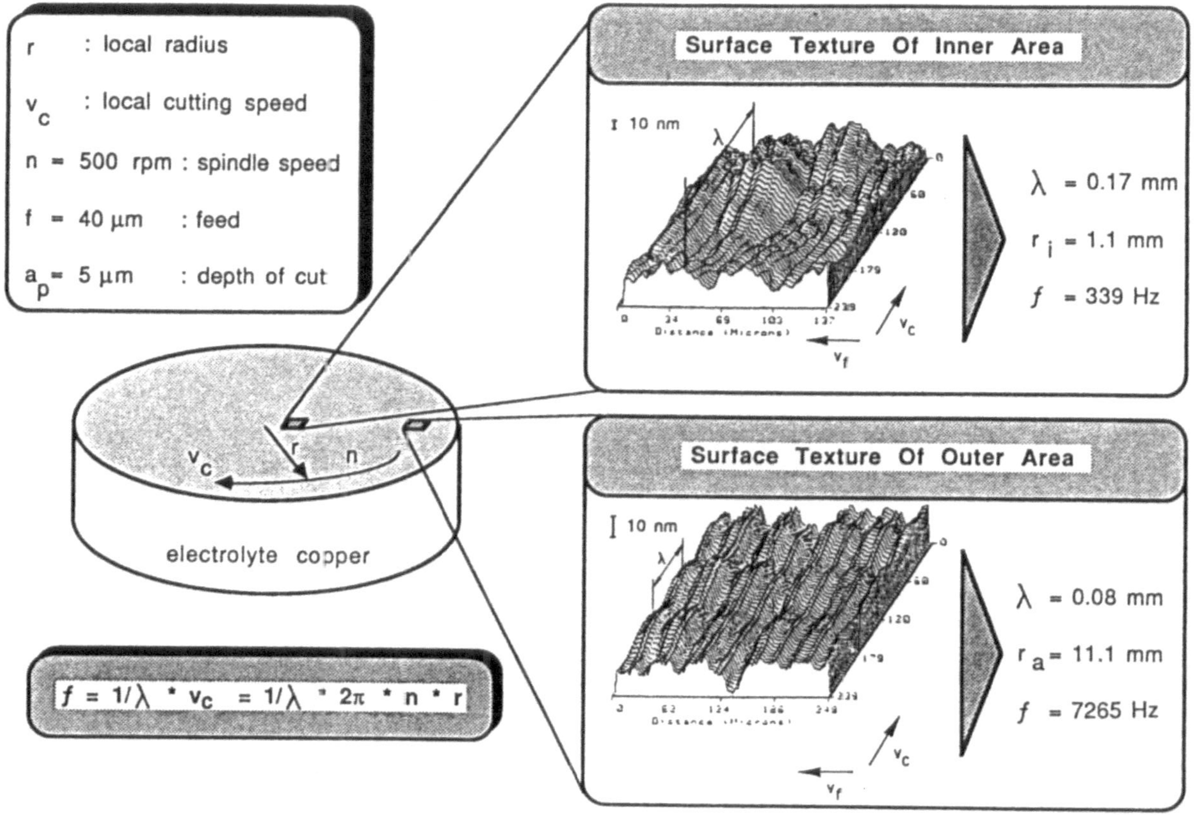

Fig.13. Influences of machine vibrations on the surface texture of metal optics

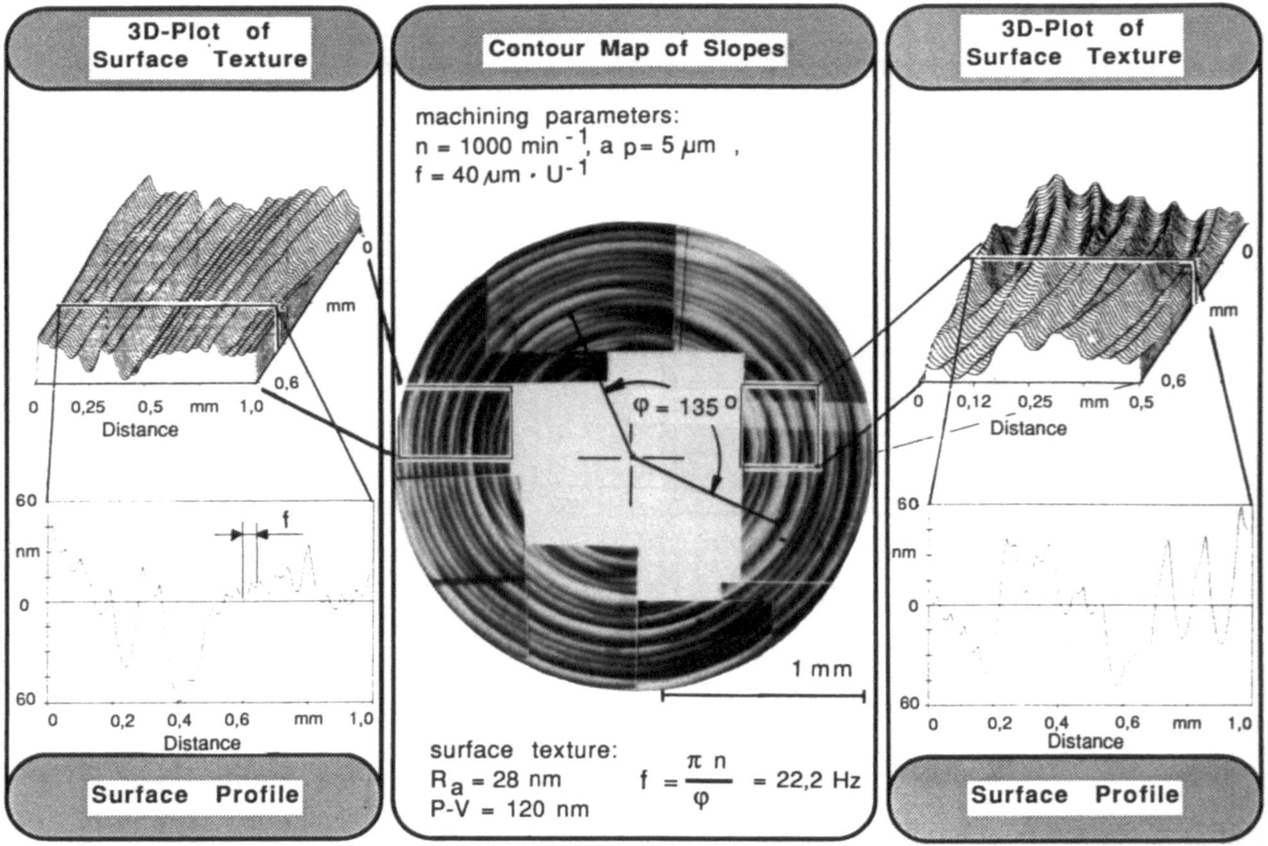

Fig.14. Surface texture near the turning center influenced by low frequency vibration

New Material Application to Ultraprecision Lathe

M. TSUTSUMI
Tokyo University of Agriculture and Technology
Koganei, Tokyo, Japan.
K. UNNO
Toyoda Machine Works Ltd., Asahimachi, Kariya, Aichi
M. YOSHINO
TOTO Co. Ltd., Honson, Chigasaki, Kanagawa
F. YAMAUCHI
Nippon Electric Co. Ltd., Miyazaki, Miyamae, Kawasaki

Summary
We have developed an ultraprecision lathe using alumina-ceramics and ferrite-resin concrete as the main structural material. These new materials have the characteristics of low thermal conductivity, and the ceramics is light weight and the ferrite-resin concrete is high damping. Some experiments have been carried out to evaluate the motion accuracy of the slide table and the thermal deformation. From the results, it was found that the straightness error is less than 30 nm, and the thermal deformation of the bed and the thermal expansion of the spindle are negligibly small.

Introduction

The higher machining accuracy is needed for providing the computer memory disk and the optical mirrors or lenses, and many kinds of ultraprecision machine tools are developed in world wide. At present, the demand of two axis controlled machine is increasing to produce non-spherical mirrors for high technology.

The significant factors to be considered in the structural design are to minimize the production errors and the structural deformation due to the temperature change and the various forces. However, these factors could not be improved if the traditional materials such as cast iron or carbon steel have been applied to the machine tool structures. On the other hand, new material such as advanced ceramics has the possibility to enhance the performance of the machine tools.

This paper describes the outline of an ultraprecision lathe using advanced ceramics and ferrite-resin concrete, which are new structural material. And some experiments have been carried out to evaluate the motion accuracy of the slide table and the thermal deformation. From the results, it was found that the straightness error is less than 30 nm, and the thermal deformation of the bed and the thermal expansion of the spindle are negligibly small.

Design of Ultraprecision Lathe

The schematic diagram and the specifications of the developed machine are shown in Fig.1 and table 1, respectively. The machine is mainly consisting of three units; these are a ferrite resin concrete bed, a work table unit and a spindle unit.

The ferrite resin concrete is one of the resin composite material consisting of polyester resin and ferrite powder, which are developed by NEC Co. laboratory. Ferrite of which principal ingredient is Fe_3O_4 is produced in large quantity through the disposal process of waste-water in the heavy industries. The mean size of particles of supplied ferrite powder is about 0.8 μm.

The mechanical properties are shown in table 2. The damping capacity is very high as compared with cast iron, and the thermal conductivity is also small, as shown in the table.

The ferrite-resin concrete bed is supported by four pneumatic isolators with subsidiary chambers to reduce the resonant frequency and its level is automatically controlled by the servo-mechanism. Measured resonant frequency is under 1 Hz.

Two steel plates are bonded onto top and bottom surfaces in order to connect main three units, pneumatic isolators and other mechanical elements by means of bolts. In addition, as the ferrite resin concrete is visco-elastic material, the creep deformation could not be disregarded in the ultraprecision machining. Also these two steel plates play a role of reinforced members.

The guideways and slide tables, which are produced by TOTO Co. Ltd., are made of alumina-ceramics of 87% purity. The mechanical properties are shown in table 2. The alumina-ceramics has the characteristics of low thermal conductivity, low coefficient of thermal expansion, high Young's modulus, and light density as shown in the table.

Fig.2 shows the slide table. The table is supported by hydrostatic step bearings and driven by the high quality ball screw-thread through a hydrostatic coupling, which is a non-contact type coupling to eliminate the eccentricity of the ball screw shaft in rotating. The hydraulic oil returns to the tank through the inside of the cast iron base.

Twelve oil pockets, which are machined by a diamond cutter using a machining center, are arranged on one side of the table and their depth is 0.3 mm. The bearing surfaces are finished by lapping after grinding. The finished surface roughness is about 0.1 µm in maximum height and the straightness of the bearing surfaces is less than 0.1 um in overall length of the slide table as well as guideways.

Weight of the ceramics table is 45 % of the cast iron table of same configuration as the density is small as shown in table 2. Therefore, the dynamic response of the ceramics table is superior to that of cast iron.

Fig.3 shows a cross section of guideways. The guideways are consisting of three rectangular beams, which are fixed tightly with each other by stainless steel bolts of specifications, M 12, after finishing the surfaces. Female screw threads can not make directly into the ceramics beam, because it is brittle material. Therefore, threaded metal blocks have to be inserted to assemble the guideway by means of bolts. And also adhesive bond is used to fit tightly the blocks to the ceramic body as well as the throttle nozzles for the hydraulic bearings. The assembled guideway is fixed on the cast iron base which are precisely finished by hand scraping.

The spindle consists of two materials, as shown in Fig.4. The appearance of the spindle is alumina-ceramics of 97% purity of which characteristics are shown in table 2. A vacuum chuck is mounted on the front end of the spindle and the rotor of AC inverter motor is directly mounted on the other end of it. These elements are fixed on the part made of carbon steel. In addition, to prevent the brittle fracture of the ceramics spindle, the preload in axial direction is applied by using a steel rod with a through hole, and then the adhesive bond is filled up between the steel rod and the ceramics spindle.

The spindle is supported by the hydrostatic bearings; 105 mm diameter thrust bearings and 60 mm diameter radial journal bearings. The hydraulic oil used for the bearings and the cooling oil for AC motor return to the oil tank through a same tube.

Measured performance

Fig.5 shows the straightness over 130 mm of travel of X axis slideway, measured in the horizontal direction at the height of 20 mm above the worktable top, using an electronic comparator and a straightedge. The measured data by this method include both straightness forms of travel of slide table and of the straightedge. Therefore, the surface form of the straightedge is corrected mathematically. Over 130 mm of travel, the deviation is 30 nm in the form of a smooth curve as shown in the figure. By using the machine control computer in the feed back loop, it is possible to correct for repeatable straightness error.

Fig.6 shows the motion of the table in horizontal plane due to the eccentricity of the ball screw shaft. The amplitude of peak to valley is less than 20 nm. It can be said from the result that the eccentricity of the ball screw shaft can not be eliminated even if the hydrostatic coupling is used.

It is, in general, required that the resonant frequency of the drive system for an ultraprecision machine tool is larger than

100 Hz [1]. The inertance transfer function of the slide table has been measured. The results are shown in Fig.7. It can be seen from the figure that the resonant frequencies of X axis and Z axis tables are less than 100 Hz in each feed directions. This is due to the fact that the hydrostatic coupling has low stiffness. However, another resonant frequency of the drive system was not found in the range 0 Hz to 5 kHz.

From above results, it may be concluded that the hydrostatic coupling is not suitable for a drive system of ultraprecision machines. Therefore, it is better to use another drive system such as the friction drive [2].

Thermal behavior and stability

Precise temperature control is essential to the accuracy of precision machine tool. And heat sources have been removed from the machine, and directly cooled [1].

The alumina-ceramics has good characteristics for temperature change; low thermal conductivity and low thermal expansion. However, some mechanical elements of cast iron or carbon steel used in the developed machine are sensitive to the temperature change. Therefore, the temperature of the machine and its surroundings have to be controlled as well as other precision machine.

Main heat source in this machine is in the hydrostatic bearings. The amount of temperature rise of the oil supplied to the bearings is hold about 1.0°C. The pressurised oil temperature is precisely controlled in the oil tank (0.01°C).

Fig.8 shows the temperature change of the room and the outlet oil. The developed machine is installed in the heat isolator room which are controlled by air conditioner. The room temperature change is considerably large, because the ability of the air conditioner is not so good. However, the temperature of outlet oil rises smoothly as time proceeds and then converges after 3 hours. The amount of temperature rise is about 1.0°C. In addition, as shown in Fig.9, the temperature rise of the surface of the ceramics table is equal to that of the outlet oil, and also the temperature change can be ignored. Therefore, it is possible to produce the insensible structures to the temperature change by using advanced ceramics.

Fig.10 shows the temperature distribution of the top surface of the bed. In this case, the room temperature is kept at 20°C, and each temperature of supplied oil and outlet oil are kept at 19°C and 20°C, respectively. As shown in the figure, the maximum difference is 0.3°C in stable conditions.

Then, the relative displacement between the spindle unit (Zaxis) and the work table unit (X axis) has been measured, changing the supplied oil temperature. The measured results are shown in Fig.11. It is found form the figure that the relative displacement is considerably small in case of 19°C. Fig.12 shows the thermal expansion of the spindle in Z direction. It may be concluded form these results that the outlet oil temperature have to be controlled to reduce the thermal deformation if the hydrostatic bearing are used.

Conclusion

An ultraprecision lathe using new structural material has been developed, and some performance test has been carried out. This machine is not a complete one. As a next stage, we are now developing CNC controller using a laser feed back system and then we will try to carry out the cutting test of optical mirrors in a few year.

References

1. Atkinson, D. P.; McClure, E. R.; Saito, T. T.: Interactions between Science and Precision Engineering. Proc. 6th Int'l Conf. on Prod. Eng. Osaka, (1987) 43-51.

2. Donaldson, R. R.; Maddux, A. S.: Design of a High-Performance Slide and Drive System for a Small Precision Machining Research Lathe. Annals of the CIRP. 33/1 (1984) 243-248.

Table 1. Specifications of the developed machine

Bed		
Length	mm	1,800
Width	mm	1,500
Thickness	mm	400
Height	mm	700
Turning length	mm	150
Swing over table	mm	200
Swing over bed	mm	500
Facing diameter	mm	300
Z axis travel	mm	200
X axis travel	mm	200
Spindle diameter	mm	60
Center height on bed	mm	520
Range of spindle rotational speed	min^{-1}	0~6,000
Mass of machine	kg	6,400

Table 2. Mechanical properties of alumina-ceramics and ferrite-resin concrete

Material		Carbon steel	Cast iron	Alumina-ceramics 97 %	Alumina-ceramics 87 %	Ferrite-resin concrete	Cement concrete
Young's modulus	MPa	205	100	370	280	30	24
Density	g	7.8	7.0	3.9	3.4	3.4	2.3
Tensile strength	MPa	470	180	200	-	16	3
Compressive strength	MPa	-	600	2000	-	140	30
Coefficient of thermal expansion	$X10^{-6}/K$	12	11.5	4.5	4.2	13	11
Thermal conductivity	W/(m·K)	53.5	52.5	25.6	16.3	2.8	1.0
Thermal diffusivity	m/s	147	163	75	54	-	5.4
Logarithmic decrement	$X10^{-3}$	0.5	2	-	0.6	50	40

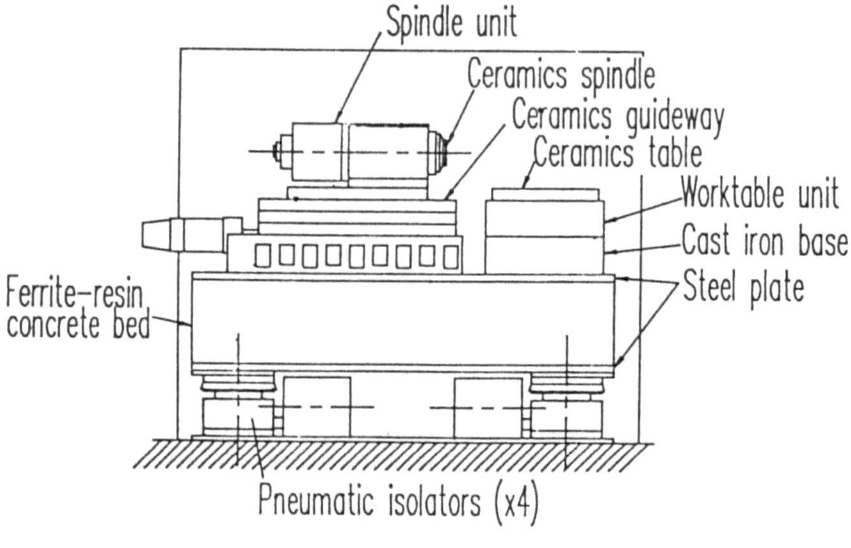

Fig.1. Schematic diagram of developed ultraprecision lathe

Fig.2. Slide table made of alumina-ceramics (87%)

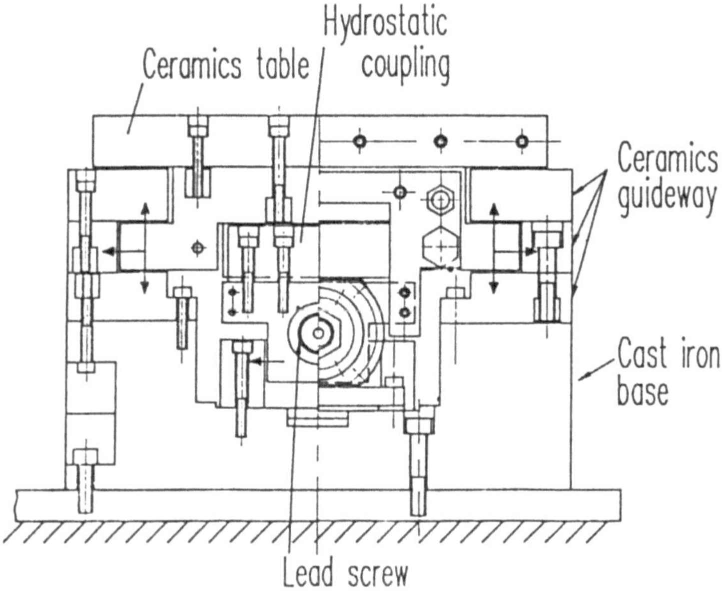

Fig.3. Cross section of guideway

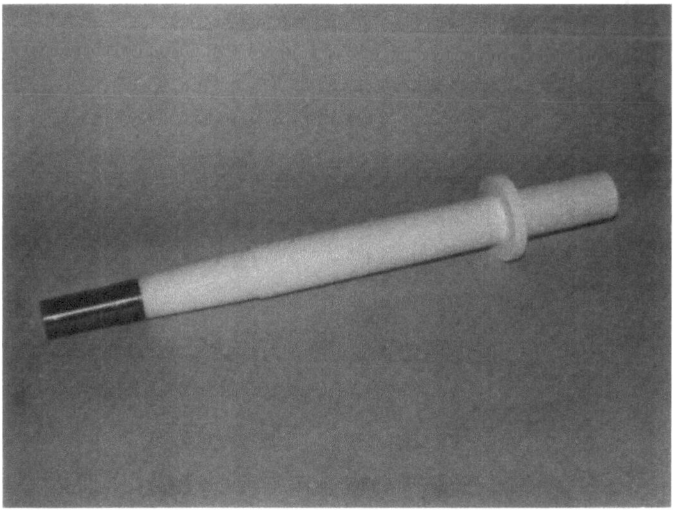

Fig.4. Spindle made of alumina-ceramics (97%)

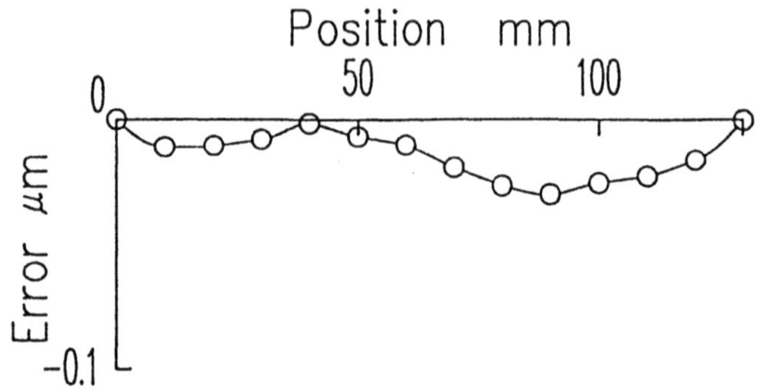

Fig.5. Straightness of travel of guide way
(X axis, worktable unit)

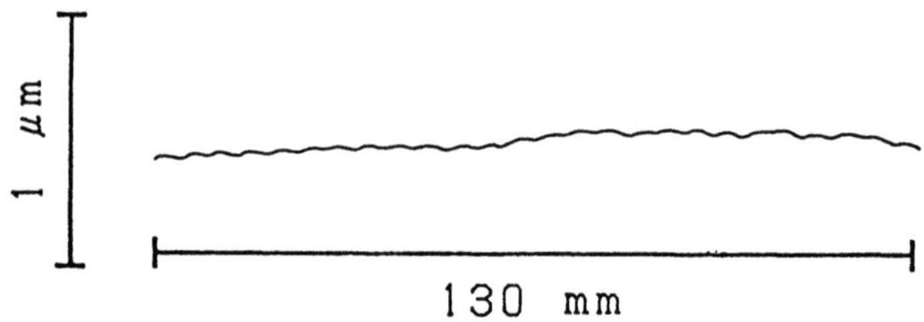

Fig.6. Motion error due to the eccentricity of
leadsrew shaft

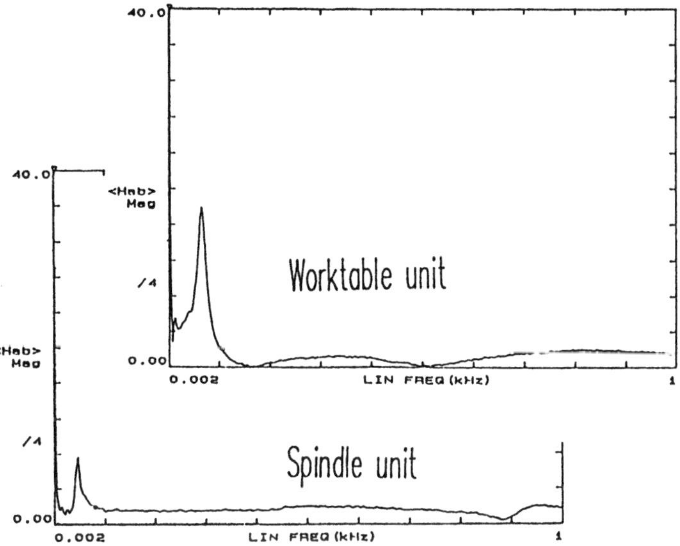

Fig.7. Inertance transfer function of the slide tables

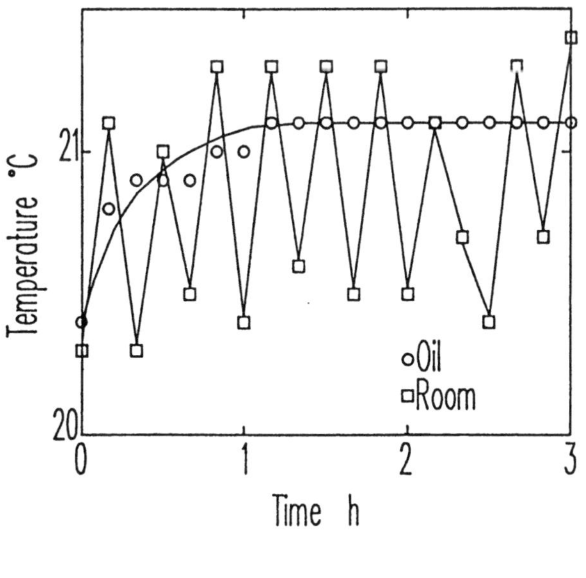

Fig.8. Temperature change of the room andthe autlet oil

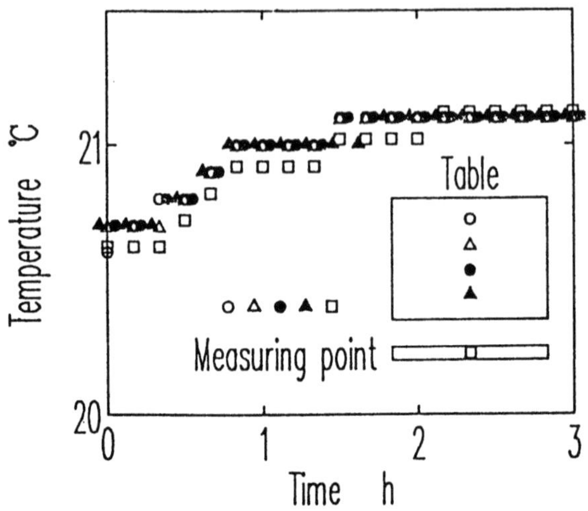

Fig.9. Temperature change of the worktable

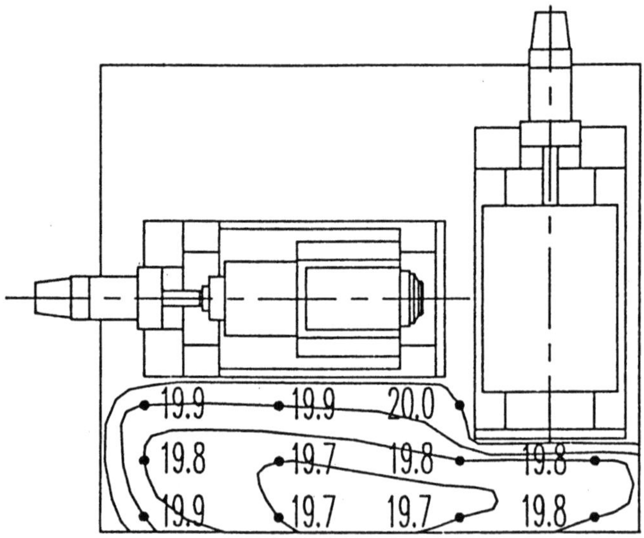

Fig.10. Temperature distribution of the top surface of bed (room temp. 20 C, supplied oil temp. 19 C, outlet oil temp. 20 C)

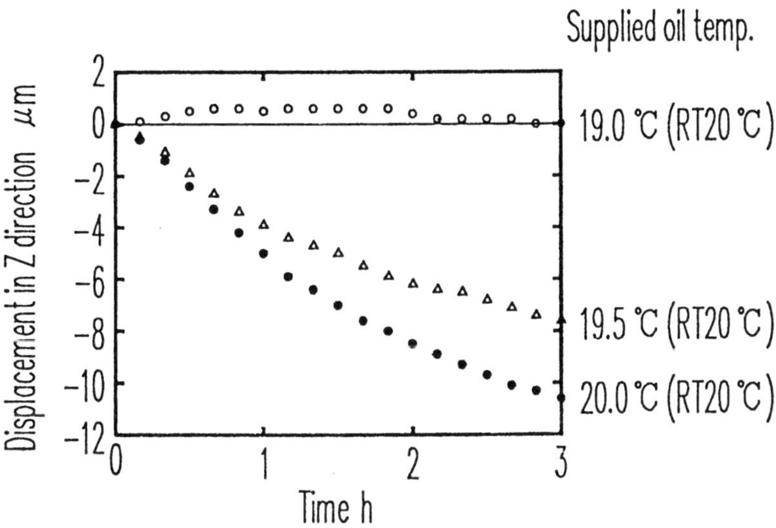

Fig.11. Relative displacement between spindle unit and worktable unit (RT ; room temperature)

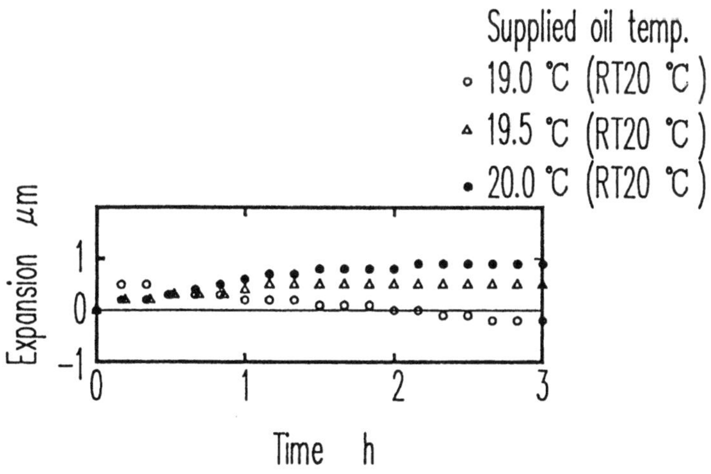

Fig.12. Thermal expansion of spindle (RT ; room temperature)

Machining of Aspherical Surfaces

H. Hashimoto

Mechanical Engineering Dpt.
Kanagawa Institute of Technology

Summary

Research on fabricating aspherical surfaces, conducted by Nikon, is reviewed. Based on ductile-regime grinding, the material removal process of optical glass is first discussed. Then, cam-guided machines and CNC machine developed are presented, leading finally to developments of aspherical surfaces of Wolter type X-ray optics.

1) Introduction

It is well known that properly designed and fabricated aspherics may greatly improve the performance of certain optical systems, but they are not widely employed because of difficulties in the large-scale production of aspherics of low cost and high quality. Until recently, a number of methods have been proposed, several of them based upon a principle of fabricating aspherics by a machine equipped with a diamond wheel. However, difficulties in these methods are as follows: (1) High precision is required of the machine. (2) It is extremely difficult to fine-grind and polish the surface without destroying the basic shape generated by the machine, and a laborious process of combined skilled hand correction and measurement is necessary, as grinding causes brittle fractures with deep cracks. Hence, ductile-regime grinding was investigated and various grinding machines were developed.

2) Ductile-regime grinding of optical glass

The depth of cracks on the inside normal surface which has brittle fractures are generally more than 5 times the surface roughness. Ground surfaces with #400 diamond wheel have surface roughness Rmax 10-20 μm.

In order to realize so called ductile-regime grinding, various bonds and grains are investigated. As a result, it was found that resin bonded Sic wheel can generates ductile-regime surface with surface roughness about Rmax 0.1 μm. However, the Sic wheel wears fastened tipping and fracture characteristics, caused by the deterioration of the wheel, produces the pits sometimes found on the ground surface. Further investigations with a #3000 metal bond diamond wheel found that a surface roughness Rmax 0.05 μm and a profile accuracy below 0.5 μm could be attained.

3) Machines developed

a) Cam-guided machines

On the basis concept that the precision surface contour is accurately generated in grinding with a surface finish to retain its shape in the subsequent process, the machines were designed and constructed for fabricating axially symmetric aspherics with vertex radius of up to ± 150 mm. These machines are also used as a measuring device for the aspherics contour. The principle is to use resinoid bonded wheel for grinding after rough grinding with a metal bonded diamond wheel. The resinoid bonded wheel is dressed on a tool spindle by a diamond dresser. In operation, the location of the diamond dresser is used to provide feedback corrections for the grinding wheel. Through experiments, it was found that precision aspherics within a few micro-meters can be obtained with a surface finish which permits polishing in a short time without significant deterioration.

b) CNC machine

The ultra-precision machining/measuring machine developed uses aerostatic bearings for the slides and spindles. It is a CNC machine; turning, grinding and measuring of the optical and related elements is achieved by machine selection of the appropriate tool which is then attached to the rotary table of the machine.

The machine has 3-axis control: the linear X and Z axes, which have a closed-loop laser feed-back facility, and the rotational C axis. It is capable of turning spherical and aspherical surfaces using only a single point on the cutting tool tip, even if the tip is not of true radius. Disc-and cylindrical-type grinding wheels are used because they can be set true with comparative ease and precision. Construction is of the T-bar type, making the vertical distance between the tip of the working tool and the machine base small. With this feature, the X and Z slides can function and maintain their performance independent of each other, and position detection satisfying Abbe's principle becomes feasible.

The slides are sufficiently rigid to withstand the normal components of the grinding and turning forces, as well as the bending moments involved. Friction drive units are employed for the slides with the aim of smooth operation and of minimizing any backlash as well as realizing a fast servo any backlash as well as realizing a fast servo response. To minimize vibration, the machine base is made of granite and a suspension type platform is employed. The motor driving the main spindle is separated from the machine base and the spindle is balanced dynamically. A slide-type magnetic coupling is used to locate the main spindle on the slides. To enhance grinding precision, the aero-static grinding tool spindle is powered with a motor and is balanced dynamically. The turning and dressing unit also employs aerostatic bearings.

4) Fabrication of X-ray optics

Wolter type 1 X-ray optics were developed by Ultra-precision machining and precision replica technique. Fabrication process is as follows;

(1) Tungusten carbide was selected as mandrel material.
(2) Mandrel was ground by Ultra-precision CNC machine, and then annealed.
(3) Computer controlled polishing machine was used to obtain a super smooth surface.
(4) Pyrex glass replica was made at 750 ^{o}C in vacuum. And gold was coated on inner surface of the replica.

Measurements of resolution were made by X-ray test apparatus.
As a result, resolution below a few micrometer for 8.3A X-rays.

A Diamond Turning Apparatus for Fabrication on Non-Rotationally Symmetric Surfaces

Peter J. Falter

Thomas A. Dow

Precision Engineering Center

North Carolina State University

Raleigh, NC 27695-7918

This paper will describe an attractive method of producing non-rotationally symmetric surfaces on a diamond turning machine. The technique utilizes a piezoelectric-driven, high-bandwidth tool servo, workpiece based metrology, and high speed digital control to successfully cut features on specimens turning at several hundred rpm. The use of single point turning to cut such asymmetries could drastically affect the cost of producing some optical and mechanical components. Methods of incorporating tool servo technology into existing diamond turning machines is also being investigated.

INTRODUCTION

Current diamond turning and precision grinding provide the capability for fabricating technology workpieces to very high accuracies. While a variety of geometries may be produced by such fabrication methods forms are restricted to surfaces of revolution [12,13,2]. There is a need to produce mechanical and optical surfaces which are not members of this category. Examples of these are cams, pistons, pump parts, wavefront correction elements, and specialized optical encoders.

Non-rotationally symmetric mechanical parts are typically produced on a cam-driven lathe or grinding machine. Such machines are limited in their accuracy and repeatability by the positioning resolution and frequency response of the actuation system and require considerable set-up time for a new geometry. Tomizuka [17] has experimented with the control of a noncircular turning operation that might be applied to such components. In the optics field, asymmetries are generally produced by spot polishing or etching operations. Single point turning could greatly speed the production of asymmetric metal optics. One example of an optical system component which could benefit greatly from such an operation is the wavefront corrector plate. The turning of such plates has been attempted at Lawrence Livermore on their Large Optics Diamond Turning Machine (LODTM) [3,11].

These two examples represent extremes in feature size, tolerance requirements, and application, but they share the following needs:

- Tool actuation
- Positional metrology
- Dynamics characterization
- High speed control

In mid-1986 an attempt was made to machine rotationally asymmetric specimens on the original PAUL[1] machine shown in Figure 1(a). This apparatus consists of a pair of air-bearing spindles with their axis perpendicular to the table and parallel to each other. Spindle 1 holds the workpiece and is driven by an integral dc motor. Spindle 2 supports the cutting arm which holds the tool and is moved by a dc motor driven lead screw across the workpiece. As reported earlier [4], sine waves whose height was proportional to the radius and rotationally positioned steps were machined on electroless nickel and aluminum substrates.The tool position was varied during such cuts using the piezoelectric stack located at the rear of the tool arm and rotational information from an incremental encoder on the workpiece spindle. The cuts were made with open loop control at spindle speeds up to $200 rpm$.

In evaluating the results of these experiments several problems with the apparatus became apparent. First, the piezoelectric stack was forced to move a $5kg$ tool arm in order to change tool position. This large moving mass caused considerable forces to be generated in the structure and limited the useful bandwidth of the machine to

[1]Parallel Axis Ultraprecision Lathe

about $200Hz$. Secondly, the current required to drive the piezoelectric stack at higher frequencies was beyond the capability of the high voltage amplifier used. Thirdly, feedback of tool position relative to the workpiece was needed. Finally, a system bandwidth of as much as $1KHz$ would be necessary to produce sharp features on parts rotating at several hundred rpm. This high bandwidth would require a controller with a cycle time less than $100\mu s$ ($10kHz$).

These conclusions and the other items listed above were considered in the selection of a scheme to machine nonrotationally symmetric parts at high spindle speeds. Tool positioning is performed by a piezoelectric tool actuator and the tool motion is measured relative to the spinning workpiece using a pair of capacitance gages. Control of the servo is performed using a high speed digital control subsystem developed at the Precision Engineering Center (PEC). These apparatus are discussed below along with the performance of the overall system.

APPARATUS

The original machine geometry shown in Figure 1(a) was modified to correct the problems discussed in the last section. The modifications consisted of a new cutting arm which supports the fast-tool servo described in the next section. With this design, the moving mass was greatly reduced and the frequency response improved. A reference surface was also added to the workpiece spindle as a means of providing tool/workpiece metrology information.

Tool Servo Design

The FLATS (Fast Low Amplitude Tool Servo) is shown in cross-section in Figure 2. This device is similar to other fast tool servos intended for use on precision machine tools [7,10], but it has both significantly higher travel ($0 - 15\mu m$) and bandwidth ($> 1000Hz$). The servo is driven by a high-voltage high-power amplifier built in the PEC especially for this purpose. The high bandwidth is necessary to produce sharp surface features and it also shortens the time required to produce any given geometry. This improves throughput and reduces the effects of thermal drift on the accuracy of the finished part.

Metrology System

To control the cutting process, both position of the tool and position of the workpiece must be known. For this reason, the metrology system plays a critical role in the control strategy. Position of each of these components are measured relative to the tool arm which is a massive (35kg) steel assembly. Measurement of tool position is achieved using two capacitance gages. These sensors were chosen because of their high bandwidth, high sensitivity, and ease of implementation. The first sensor is built into the FLATS and senses the axial motion of the tool relative to the servo body. A direct measurement of position is desirable since piezoelectric actuators exhibit nonlinear gain and hysteresis in their displacement versus voltage characteristics. The second capacitance gage, shown in Figure 1(b) is mounted in the arm which holds the tool servo. This gage measures the motion of the arm relative to a reference surface machined on the faceplate of the workpiece spindle.

Rotational information is taken from an incremental encoder mounted on the workpiece spindle. This encoder has 1024 lines with quadrature outputs and a once per revolution marker pulse.

Control System

The machine controller is designed to position the tool relative to the workpiece in a prescribed fashion while avoiding machine resonances. To perform this task, a control algorithm is being developed based on the machine's dynamics and the cutting process model described in the next two sections. Control implementation will be performed using a digital control system developed at the PEC. This apparatus consists of several specially designed CPU and interface boards mounted in an Intel 310 (PC) chassis. The high speed calculations are performed on a TMS32010 digital signal processing chip and the more standard computer activities are handled by the 8086 processor in the 310 system. This arrangement allows program development to be performed in a familiar MS-DOS environment and the control code to be compiled and downloaded to the TMS32010. The system's interface board will accept several channels of analog and digital input/output. In addition an external counter and an interrupt line were added to allow the controller to correctly utilize information from the rotary encoder. Figure 3 shows the flow of information in the present servo control implementation.

DYNAMIC CHARACTERIZATION

Dynamic response of the supporting structure is an important consideration for this machine. Because both the workpiece and the FLATS are supported by this structure, motions due its dynamic response are significant when submicrometer cutting accuracies are required. These dynamics must be well characterized in order to design a controller that compensates for the unwanted motions.

Structural Response

Figure 4 shows the response of the structure to a normal tool force (F_{z1}) as displacement measured by capacitance gage 2 (z_2). The peaks present are: 94 Hz, pitching of arm about the center of the rear spindle; and 144 Hz, vibration of the table on which the apparatus rests. Also present is valley at 106 Hz. This mode corresponds to the rolling of the arm about the center of the rear spindle. Such a vibration primarily causes a lateral motion of the second capacitance gage and since the transducer measures axial motion, little or no output is generated. The sharpness of the peaks in the frequency response indicates that the system is only lightly damped. Higher damping would aid in the control of the machine by reducing the effect of cutting process damping on the response of the structure and making the model coefficients less sensitive to small changes in the dynamics. Increased damping may also improve the quality of the finished part [8].

Cutting Model

Models of the cutting process have been published for over forty years and yet no adequate general solution may be recommended for even the simple case of orthogonal cutting. Linear and nonlinear models have been proposed for cutting dynamics, primarily in the area of chatter [14,9,16,6]. The forces in turning are known to be nonlinear, to demonstrate some hysteresis, and to lag changes in the depth of cut [16]. Including all of these effects would require a model much too complicated for use in a high speed controller.

Recently, efforts have been made to model the orthogonal cutting process using a variety of data fitting algorithms and models. Among these results are those of Carroll [1] which indicate that the plowing force is proportional to the cut area in low speed diamond turning tests. Srinivasan [15] reports that the plowing force may be modeled as linear with depth of cut and the thrust force varies linearly with the depth of cut and the penetration rate of the tool. Tomizuka [19] also uses a first order model for the thrust force. Most recently, Franse [5] claims that cutting behavior of diamond turned copper indicates that different values of the penetration rate constant must be used depending upon whether the tool is moving into or out of the workpiece.

Based on the above arguments the model chosen is:

$$F_c = F_{cu} u \quad (1)$$

$$F_t = F_{tu} u + F_{t\dot{u}} \dot{u} \quad (2)$$

$$u = u_o - z_{1a} - z_1 \quad (3)$$

where

F_c = *plowing force*

F_t = *thrust force*

u = *depth of cut*

$\dot{u}$ = *penetration rate*

u_0 = *nominal depth of cut*

z_1 = *motion of tool relative to servo body*

z_{1a} = *vertical motion of tool servo body relative to workpiece*

The coefficients F_{cu}, F_{tu} and $F_{t\dot{u}}$ are functions of the nominal depth of cut, feed rate, tool nose radius, and material properties. An additional force arising during cutting is generated by the inertia of the moving elements of the FLATS. This force is:

$$F_m = -m_m \ddot{z}_1 \quad (4)$$

The value of z_{1a} in Equation (3) is estimated from the displacement z_2 measured by the second capacitance gage as:

$$\hat{z}_{1a} = \alpha_{12} z_2 \qquad (5)$$

where α_{12} is the ratio of distance from the rear spindle to the second capacitance gage to the distance from the rear spindle to the tool. This estimate assumes that the motion is entirely due to the rocking of the arm about the center of the rear spindle.

Structural Model

Control of the cutting apparatus requires a model of the structural dynamics and the effects of the forces given by Equations (1) to (4). The response of a discrete fourth-order model of these dynamics for a sampling rate of 100 μs is plotted in Figure 5. Such a fourth-order model can match the two lowest modes of vibration measured in Figure 4 and a comparison of the measured and modeled response shows excellent agreement. The discrete model will serve as the basis for determining the coefficients of the control algorithm which will be designed to minimize the unwanted effects of the resonances in the structure by controlling the frequency content which is input to the servo.

The control computer described earlier is capable of executing a fourth-order compensator at sampling rates as small as $20\mu s$. However, the algorithm coefficients become smaller as the sampling time is reduced and quantization errors are magnified. Therefore, a sampling rate of $100\mu s$ was selected for the control algorithm.

INITIAL EXPERIMENTATION

Cutting tests show that the modifications to the PAUL have improved the dynamic performance [4]. The low moving mass of the tool servo and the large mass of the arm have enhanced the performance at high frequencies where inertial forces dominate the system response. The increased bandwidth of the servo aids in the cutting of high spatial frequency features.

As a test of the system's ability to produce sharp repeatable features, open-loop code was written to generate words on the surface of a machined specimen. Figure 6 is a photomicrograph of a section of the letter *N* in the word it PRECISION. This cut was made in brass at $94rpm$ and the depth of the letters was $0.33\mu m$. The repeatability of the triggering is evident in the clean transitions into and out of the workpiece. Some ringing can be seen on the trailing edges of the steps. This is caused by the servo vibrating at its natural frequency of $13kHz$. This ringing could have been avoided by filtering the input to the servo amplifier, but no filtering was used in this case.

DISCUSSION

A fast tool servo system has been designed and developed for precision fabrication of non-rotationally symmetric optical surfaces. This apparatus uses closed-loop feedback control to position the tool relative to the workpiece. It has a range of $15\mu m$ and a bandwidth above 1 KHz. A metrology frame has been included to provide feedback information to the controller. This frame utilizes a pair of capacitance gages to simultaneously measure the position of the tool and the workpiece; the relative position of which determine the depth of cut.

A control system has been implemented using a high-speed microprocessor and a digital control algorithm. This algorithm is based on the structural dynamics of the diamond turning machine and the model of the cutting process. The open loop response of fast-tool servo exhibits high bandwidth with little overshoot and/or oscillation. The control algorithm is currently being evaluated and the coefficients which define the material response are being measured. The final apparatus should dynamically control the fabrication process and produce the desired non-rotationally symmetric shape. It can also be used to correct for non-rotationally symmetric errors in the spindle or high frequency external disturbances.

Bibliography

[1] J. T. Carroll, T. A. Dow, and J.S. Strenkowski. Tool Force Measurement and Prediction in Diamond Turning. August 17-22 1986. SPIE Proceedings, 30th Annual International Technological Symposium on Optical and Optoelectronic Science and Engineering.

[2] Donald L. Decker. The Future in Diamond-machined Optics. *Proc. SPIE*, 676:2–7, 1986.

[3] R. R. Donaldson. *Large Optics Diamond Turning Machine Technical Report.* Technical Report, Lawrence Livermore National Laboratory, April 1981.

[4] P. J.Falter and T. A. Dow. Design and Testing of a Small–scale Diamond Turning Machine. *Precision Engineering*, 9(4):185–190, 1987.

[5] Jelm Franse and R. J. P. Schrama. The Precision Cutting Process as a Non Linear Closed Loop System. 1987. Philips Research Laboratories, Eindhoven, Netherlands.

[6] T. Kaneko, H. Sato, Y. Tani, and M. O-hori. Self-excited Chatter and its Marks in Turning. *ASME Journal of Engineering for Industry*, 106:222–228, August 1984.

[7] E. Kouno. A Fast Response Piezoelectric Actuator for Servo Correction of Systematic Errors in Precision Machining. *Annals of the CIRP*, 33(1):369–372, 1984.

[8] Karl Modemann and M. Weck. Checking Surface Modulation on Metal Optics by Analyzing Ultraprecision Machines. 1987. Fraunhofer–Institut für Produktionstechnologie – IPT.

[9] H. Ōta and K. Kōno. On Chatter Vibrations of Machine Tool or Work Due to Regenerative Effect and Time Lag. *ASME Journal of Engineering for Industry*, 95:1337–1346, November 1974.

[10] S. R. Patterson and E. B. Magrab. Design and Testing of a Fast Tool Servo for Diamond Turning. *Precision Engineering*, 1985.

[11] S. R. Patterson and T. T. Saito. Diamond Turning Wavefront Correctors: opening new optical design flexibilities. *Proc. SPIE*, 676:10–16, 1986.

[12] Theodore T. Saito. Precision Machining Application and Technology: an overview and perspective. *Proc. SPIE*, 433:48–53, 1983.

[13] Gregory M. Sanger. Optical Fabrication Technology, the present and future. *Proc. SPIE*, 433:2–18, 1983.

[14] J. F. Sarnicola and G. Boothroyd. *Machine Tool Chatter: Effect of Surface Slope on Machine Forces During Wave Removing.* Technical Report 73–WA/PROD–5, ASME, 1973.

[15] K. Srinivasan and C. L. Nachtigal. Investigation of the Cutting Process Dynamics in Turning Operations. *ASME Journal of Engineering for Industry*, 100:323–331, August 1978.

[16] Richard J. Szakovits and A. F. D'Souza. *Metal Cutting Dynamics with Reference to Primary Chatter.* Technical Report 75–WA/PROD–31, ASME, 1975.

[17] M. Tomizuka, M. S. Chen, S. Renn, and T. C. Tsao. Tool Positioning for Noncircular Cutting with Lathe. 1985. Mechanical Engineering Department, UC Berkeley.

[18] Masayoshi Tomizuka, D. Dornfield, M. S. Chen, and T. C. Tsao. Noncircular Cutting with Lathe. In *1986 NAMRC*, pages 201–206, 1986.

[19] Masayoshi Tomizuka, S-J. Zhang, J-H. Oh, and M-S. Chen. Modeling of Metal Cutting Process for Digital Control. In *1985 NAMRC*, pages 575–580, 1985.

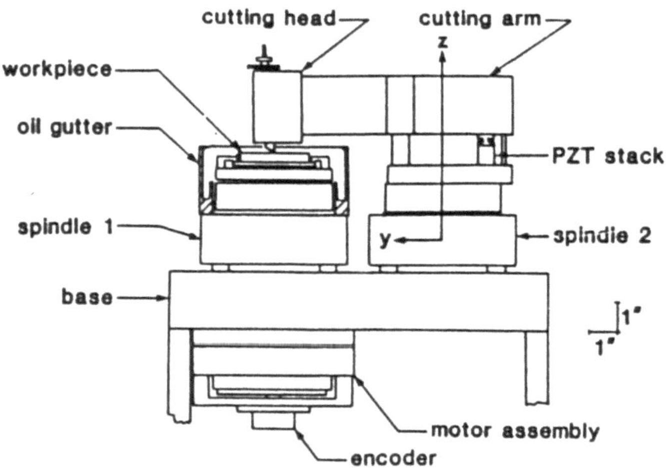

a) Original Machine Geometry

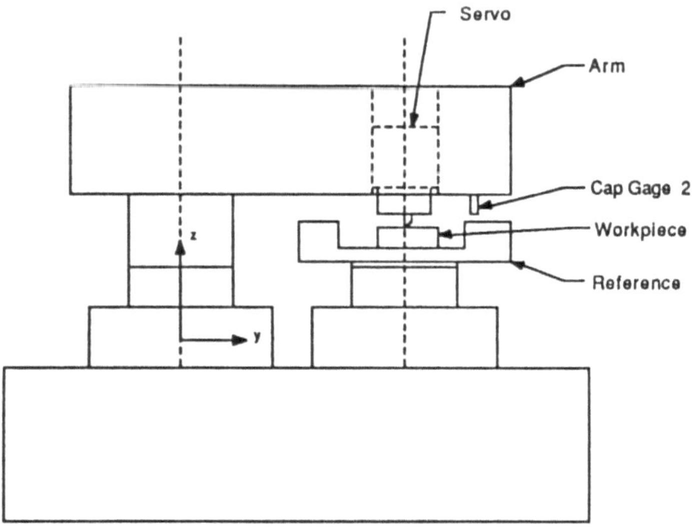

b) Machine Geometry Modified for Fast Tool Servo

Figure 1: Original and Modified PAUL Machines

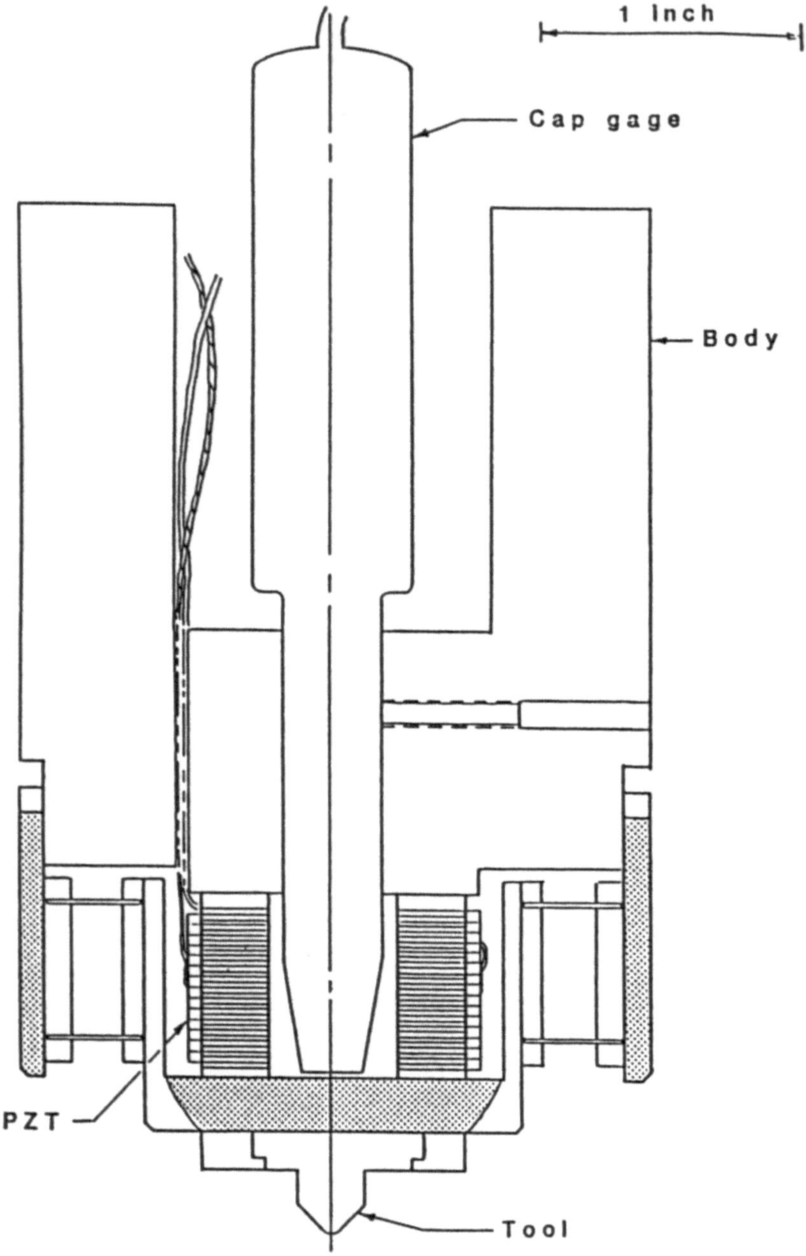

Figure 2: Tool Servo – FLATS

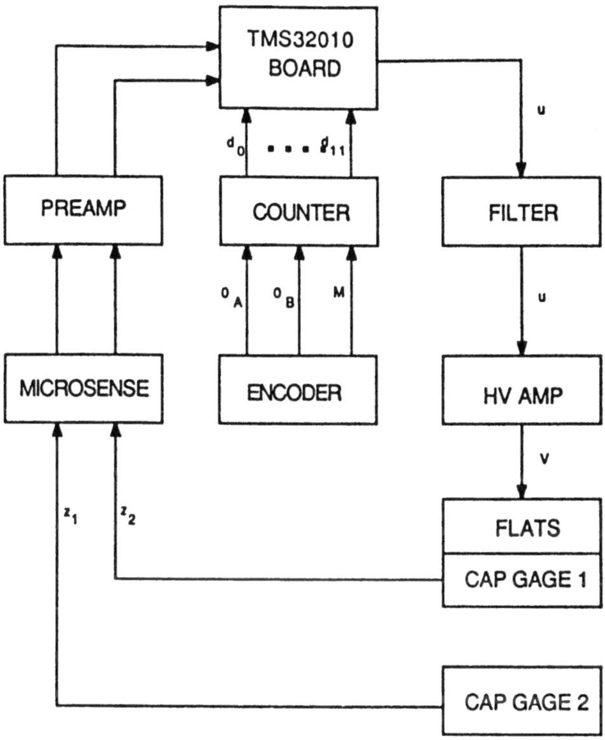

Figure 3: Flow Chart of Control Signals

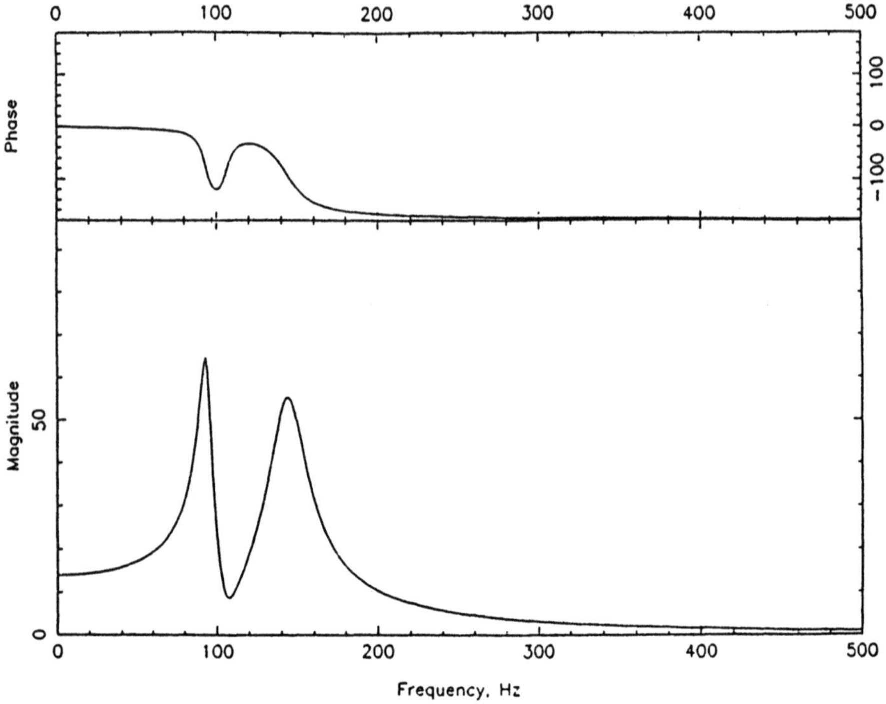

Figure 4: Frequency Response Function of Structure

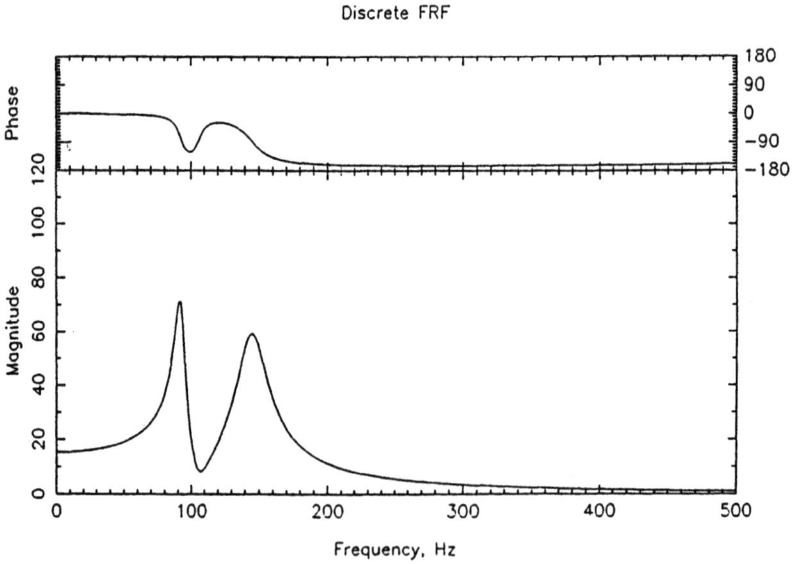

Figure 5: Modeled Frequency Response Function of Structure

Figure 6: Detail of Letter Machined into Brass

The Design and Construction of a Sub-Microinch Resolution Lathe

D.H. YOUDEN

Rank Pneumo Inc.
Precision Park
Keene, N.H. 03431 USA

Summary

In an effort to further improve the performance of our diamond turning and grinding machines, Rank Pneumo has been pursuing the development of machine slides which are capable of extreme accuracy. Our success with these slides has led us to the design of a lathe with feedback resolution of 0.05 microinches (1.27 nm). The design of this machine represents a substantial departure from the design of earlier diamond turning equipment produced by Rank Pneumo.

Reasoning

Over the past several years we have received several requests for quotations for the construction of machines with sub micro inch resolution. The usual reason cited for needing this resolution is that it allows the compensation of machine and process errors in increments which are too small to be visible in the finish machined surface. Increased resolution reduces the amplitude of the servo "hunting" that goes on when an axis is maintaining a fixed position, and therefore it will make it possible to machine better flat surfaces on two axis machines. Users of diamond turning machines are serious about sub micro inch resolution equipment and someone is going to fill their requirements. Increased resolution, coupled with our improved hydrostatic slides and newly designed spindle will help make the next generation diamond turning machines capable of day to day performance at a level that can only be achieved with patience and luck using today's machines.

Resolution and Accuracy

Increasing the resolution of a machine's servo systems will improve its repeatability provided that the system dynamics are good enough to allow it to achieve its commanded position within a few "counts" of its feedback device. Increasing resolution does not improve the accuracy of the system. A machinists scale that is a millimeter too short is not accurate no matter how fine the graduations. Accuracy is achieved by controlling, or compensating for, the machine geometry and environment (temperature, pressure, and humidity) and accurately knowing the length of our "yardstick" which is the wavelength of the light used in the interferometers. Sub micro inch resolution does allow us to command the machine to move in increments which are so small as to be insignificant on the surface of the work piece.

Machine Design

Figure 1 illustrates the general design of the new lathe. The spindle is located on the X axis which is mounted on risers so that it spans the Z axis. This arrangement allows the use of hard, non-telescoping way covers which do not influence the slide accuracy and are capable of withstanding a copious flood of coolant. The differing heights of the slide tops provides space for increased swing capacity or a rotary third axis. A large granite block mounted on a self leveling air suspension system forms the supporting structure beneath the slides.

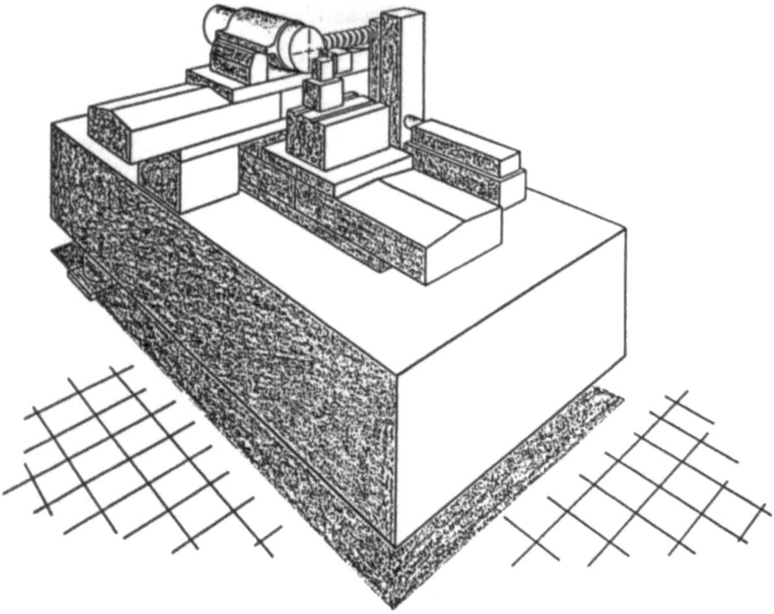

Figure 1

The geometry of our sub micro inch resolution lathe has been improved beyond our previous capability by using modular hydrostatic slides which can be ground to attain straightness of travel better than 0.5 parts per million without the need for tedious hand lapping. This is twice as good as the specifications on our current machines. Similar improvements have been made in roll, pitch, and yaw. The use of a ball bearing intermediate slide, hydrostatically coupled to the main slide, has allowed us to increase the slide longtitudinal stiffness by almost a factor of four and, at the same time, introduce substantial damping into the system. The increased stiffness and damping are responsible for an improvement in machine dynamics which, in turn, improve the servo performance.

The squareness of the axes of the new machine is adjustable after assembly is complete and it can be set to within a few tenths of an arc second. As the metrology for the Z axis is located at some distance from the tool tip in the vertical plane, we measure Z axis pitch and correct the axis position for this error in real time.

Slides

The slides are modular units which can be employed in various configurations and sizes to construct different types of machines. They utilize hydrostatic bearings in a fully constrained, box way design.

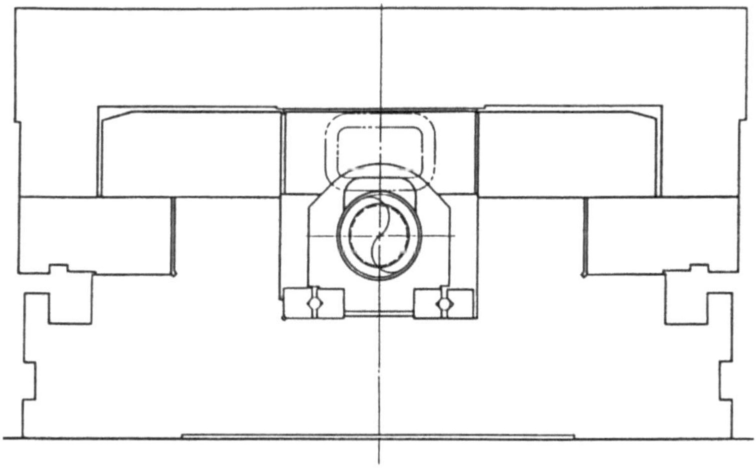

Figure 2

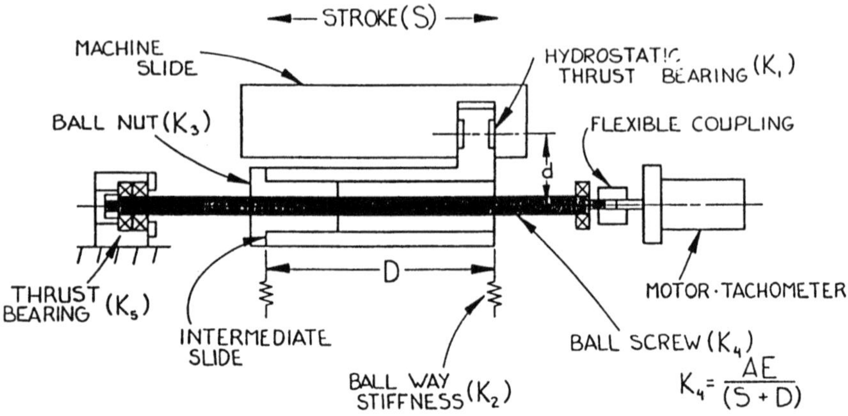

$$K_{SYSTEM} = \frac{1}{\frac{1}{K_1} + \frac{2d^2}{D^2 K_2} + \frac{1}{K_3} + \frac{(S+D)}{AE} + \frac{1}{K_5}}$$

Figure 3

The slides are isolated from the lead screws by an intermediate slide which translates on ball ways. This intermediate slide carries the ball nut and a hydrostatic thrust bearing that transfers the thrust force from the ball nut to the main slide. Figure 2 is a cross section through one of the slides and figure 3 shows how the intermediate slide is arranged. By careful design of the thrust bearing it is possible to achieve just the combination of stiffness and damping required to obtain a responsive, yet highly damped system. D.C. servomotors are used to drive the ball screws. The motors are, in turn, driven by linear power amplifiers. Limit switches are non-contacting proximity switches selected to avoid the influence on slide straightness caused by the force needed to actuate conventional switches.

Spindle

A new air bearing, hydrostatic spindle was designed for use

on the machine. The new spindle is motorized with the rotor of the motor affixed permanently to the spindle. The spindle thrust bearing is located directly behind the front journal bearing to minimize axial drifting of the spindle nose. The span of the journal bearings has been increased over earlier designs and the rotational moment of inertia has been reduced to provide better acceleration characteristics. The spindle housing has its mounting points at the spindle centerline to prevent the centerline from rising as the speed is increased. All of this results in a spindle which is stiffer, runs faster, and is more stable than ever before. Figure 4 is a section through the spindle and illustrates the features of the design.

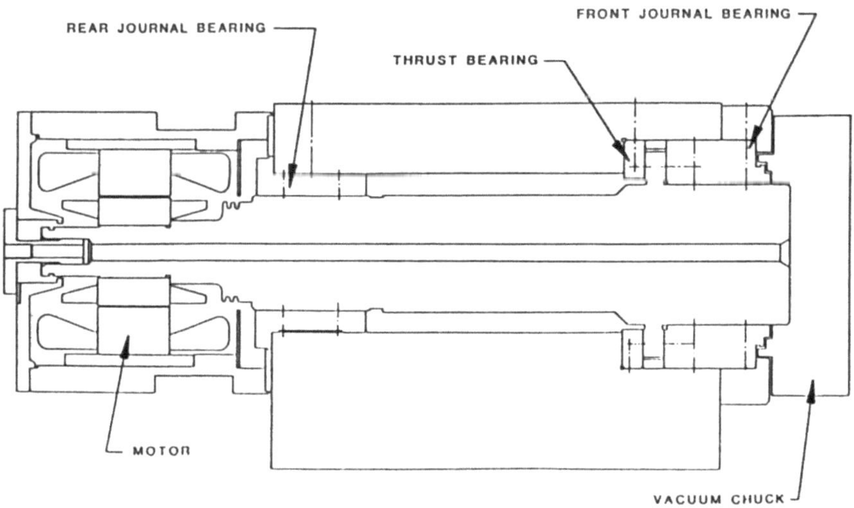

Figure 4

Environment

While the environment is not strictly part of the machine tool in the usual sense, it controls the ability of the machine to produce quality parts. Any improvement in the environment results in improved machine accuracy. In order to be able to even document the geometry of a sub micro inch resolution lathe, we will need an environment where the temperature is controlled to within a few tenths of a degree Fahrenheit and the relative humidity is held within a few percentage points. We are currently preparing a small (16'X16') room to enclose the prototype machine.

Wavelength of Light

The velocity of light and thus its wavelength is dependent on the medium through which it is passing. While it is possible to enclose all of the interferometry in a vacuum, it is expensive and entails maintaining the vacuum pump and its accessories. The metal bellows usually used to enclose the variable length portion of the light path exert substantial force upon the machine slides. This force and its effects can be avoided. Rather than use a vacuum, our design employs a refractometer to measure the refractive index of the air and correct, in real time, for the errors induced by locally changing conditions. The refractometer is located behind the Z axis slide and close to the surface of the granite machine base.

Interferometry

With the expiration of the patents surrounding the Hewlett Packard laser technology there has been a recent flurry of activity resulting in several new products which have come to market within the last several months. Among these is an offering by Zygo Corporation which uses an acousto-optic modulator to generate a dual frequency laser beam with a 20 Mhz. frequency split. This is an order of magnitude greater than that used by HP and others who manufacture similar products. The result is that the Zygo double pass interferometers have a resolution of 0.05 micro inches and will function at up to 0.5 meters per second. The output data is in the form of a 32 bit parallel word which can be delivered at up to 2 Mhz. The cost of the Zygo equipment is comparable to the cost of similar HP equipment. Figure 5 shows the layout of the interferometry on the machine. It can be seen that the X axis beam path is at spindle centerline height and located immediately to the rear of the vacuum chuck.

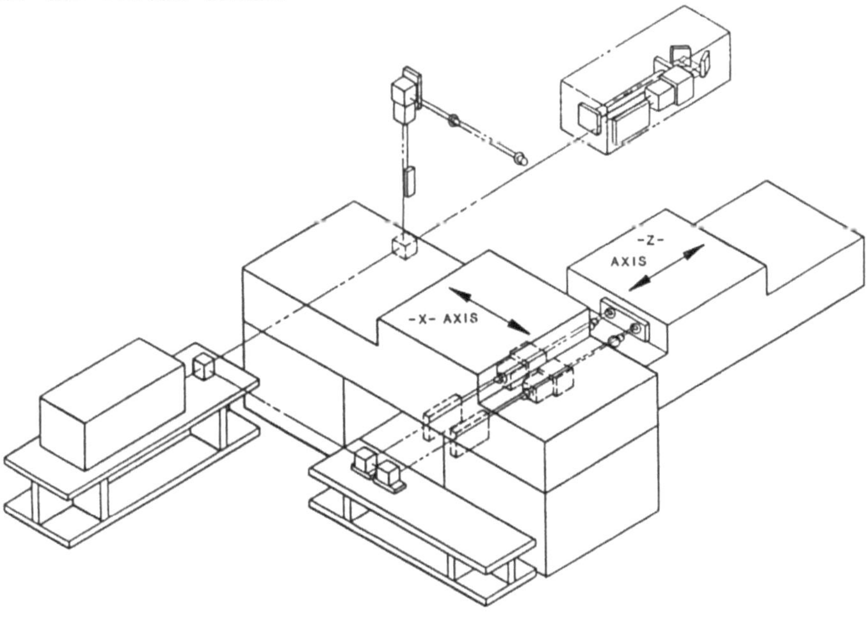

Figure 5

The Z axis has two interferometers, one to measure the displacement of the slide and the other to measure the pitch. This allows us to correct for Abbe errors introduced because the beam paths are below the slide way covers. The location of the interferometry on the Z axis was chosen to prevent the measuring system from interfering with workpieces and tooling.

Closing the Servo Loops

To complete the machine, there must be a controller which can accept the 32 bit parallel data from the interferometers, process this data at a high rate, and close the servo loops. What is required is more of a computer than the typical CNC machine controller. Ideally, this same computer will be able to generate the commands necessary to machine an optical surface given only the equation which describes that surface as input. We have located such a device. It is manufactured by a division of CUPE called Cranfield Precision Systems (CPS). The CUPROC-16 controller is currently used by several other machine tool manufacturers, both in the US and the UK. The cost of the CUPROC-16 is competitive with the cost of our current Allen-Bradley controller.

Future Work

There are a number of accessories that can be made available in the near future. A fast tool servo has been designed and constructed which will make it possible to machine parts which are not rotationally symmetric. This device is capable of moving the tool at up to 5 KHz. with an amplitude approaching 25 micrometers. Additional testing must be done before the device is ready for the market but we are getting closer. A second spindle is contemplated for the machine. This spindle will have a lower speed range but will be significantly more powerful and will have oil hydrostatic bearings with higher stiffness and damping than its air bearing counterpart.

Conclusion

Increasing the feedback resolution is only one factor which contributes to making a more capable machine. Attention must be paid to the overall machine design, the environment in which it must operate, and the peripheral electronics which control it. For the first time, lasers and controllers are available to combine with our unique machine design background to build a truly superior diamond turning machine.

M400 and P400 – A Pair of Machines for Computer-Controlled Fine Correction of Optical Surfaces

K. Becker and K. Beckstette

Carl Zeiss
D - 7082 Oberkochen

1 Abstract

This article presents a process for fine correction of optical surfaces in which a coordinate measuring machine with a resolution of 10 nm is used to measure the departures of the components from the nominal shape. The data measured serve to determine the work function of a computer-controlled polishing machine. The dwell time of the polishing tool at different locations of the component is controlled as a function of the magnitude of the residual errors. In iterative processes, a fine correction of aspheric optical components with as low a residual error as 0.7 arcsec has been achieved.

2 Fabrication of Optical Surfaces

The conventional fabrication of optical surfaces includes 3 processing steps:

1. Rough grinding
2. Fine grinding
3. Polishing

Starting with blanks and using a two-step grinding process, the components are approximated to their final form; the grinding tools operate in line contact.

In the subsequent polishing process, the final form and the structure of the surface is generated. This process not only smoothes the surface but also

- o eliminates the surface and subsurface damage caused by the grinding process o reduces the surface roughness

o reduces the departure from the nominal shape.

Polishing is therefore a figuring operation.

While spherical surfaces can be generated in a one-step polishing process, this is not possible with non-spherical (aspheric) surfaces (1). An additional correcting operation is necessary in which small tools are used to process particularly those zones of the component which still display a positive departure from the nominal shape. To be able to perform this procedure, it is necessary to know the exact actual shape of the component.

It is the objective of the procedure described here to automate the fine-correction process by using computer-aided evaluation methods and a CNC polishing machine. As it is not possible to fully predict the result of the operation, an iterative process consisting of processing and measuring is necessary.

In the following, we will discuss some basic questions regarding computer-controlled polishing. After that, there will be a detailed description of the machines developed and a presentation of the results obtained up to now.

3 Computer-Controlled Polishing

3.1 Basics

To be able to influence the shape of a component, it is necessary to check the different material removal rate across the surface. This rate is determined by a large number of variables. For a wide range of values, however, the removal rate can be expressed by a linear equation:

$$a(x,y) = \frac{dz(x,y)}{dt} = K * p(x,y,t) * v(x,y,t) \qquad (1)$$

where

a = removal rate

z = surface height

K = constant (Preston's constant)

p = processing pressure

v = relative speed between component and tool

t = processing time

To predict the removal rate, exact knowledge of the specific tool removal rate function R(x,y) is required. This function can be obtained experimentally or arithmetically.

Using the dwell function D which indicates how long the polishing tool remains at a specific location (x,y) of the component, it is now possible to calculate the total removal rate for a polishing cycle:

$$a(x,y) = R(x,y) ** D(x,y) \qquad (2)$$

The removal rate function a can be determined on the basis of measurements. The dwell function which is at the same time the control function of the polishing tool can therefore be calculated by a deconvolution of equation (2):

$$D(x,y) = a(x,y) \; (**) \; R(x,y) \qquad (3)$$

4 Processing Steps

The main steps of the computer-controlled polishing process are:

- o calculation of the residual removal rate

  Using the data obtained when measuring the component, the departure of the component from the nominal shape of the finished component is calculated. Based on this data, it is determined how much material must be removed from which locations in order to reach the final state.

- o comparison of the nominal and actual removal

  In an iterative process like this, the results of a preceding step are normally available. In this step, the result of the latest processing is compared with the theoretical result expected to draw conclusions as to the quality of the processing.

- o calculation of the dwell function

  In this step, the dwell function is calculated from the desired removal rate, the process parameters selected (tool), polishing agent, number of revolutions etc.) and the tool function which is determined either experimentally or theoretically (see equation (3)).

- calculation of the control data

  On completion of all calculations, the path data of the machine are edited in NC data sets for control of the polishing machine.

- processing

  The final step is the processing of the component on the machine.

5 Process Control Software

A software package has been developed for the above calculations and the generation of the NC control data for the polishing machine. This package is described in (3) and (4). A coupling module has also been written for direct loading of the NC data into the control system of the polishing machine.

6 M400 Measuring Machine

To obtain the data required for fine correction, a measuring machine has been developed and built (2) (5) with the following characteristics:

- measuring resolution of 10 nm
- measuring volume 600x400x50 mm
- high thermal and mechanical rigidity
- results are shown as (nominal-actual) difference
- output of the results in a form compatible with the subsequent, numerically controlled polishing process.

This measuring machine is shown in Fig. 1; the sketch in Fig. 2 illustrates the individual assemblies.

6.1 Design of the M400 Measuring Machine

To be able to obtain the high measuring resolution necessary for the fabrication of optical components with extremely small tangential errors, a solution has been striven for which would make the measuring system independent of the structure of the measuring machine.

We have chosen the conventional mechanical probing method for probing of the object. During the measurements, the computer guides the probe along the surface so that the probe deflection remains in the region of a few µm. During the measurements, three mirrors perpendicular to each other (space normal) are used for interferometric measurement of the probe position. The interferometric measuring system always provides the correct spatial coordinates of the surface point probed, regardless of the position of the machine carriage.

The 5 measuring axes are arranged so that it is possible to measure the position of the probe relative to the plane mirrors of the space normal. Tilt errors can be determined and taken into account in the evaluation process.

An independent measuring system (grating scales) with a resolution of 0.1 µm is used for machine control. Due to this separation of machine control and measuring system, the accuracy of the machine guideways and bearings is only of minor importance. Comprehensive precautions have been taken regarding the thermal stability of the setup to ensure that the temperature remains constant between +/- 0.02 K during the measurement. Deviations from the nominal values are measured and corrected in the evaluation process.

6.2 Data Evaluation Software

Two different evaluation procedures have been implemented:

- o a rapid procedure for profile measurements
- o a procedure with across-the-surface evaluation taking into account all 6 measuring systems including compensation of the drifts caused by environmental effects.

The first procedure substracts the tilt and nominal shape data from the data measured. Fourier analysis methods and digital filtering can be applied to the data obtained in this way. This makes it possible to gain information on the errors contained in the different spatial frequency regions. The output of this programme includes a plot of the surface and the statistical data of the tangential errors.

The second procedure enables the complete evaluation of all five spatial measuring axes and the compensation of the temperature drift.

The possible individual processing steps are:

1. correction of systematic errors (temperature drift, Abbe error, correction of the spherical radius etc.)
2. geometrical best fitting of the theoretical surface to the data measured using tilt and translation
3. calculation of the best-fitting parameters for second-order surfaces
4. calculation of the difference between the data measured resulting from 2. or 3. and the theoretical data
5. output of the statistical values of the surface errors and of the tangential errors
6. graphical output of the results in the form of profils or 3D representations
7. output of the data measured in a format that is suitable for further processing by the software of the P400.

6.3 Achieved Characteristics

The accuracy of the M400 measuring machine is given by

$$\Delta z\,[\mu m] = 0.02 + (0.5 + 2.4 + 1.5 * \Delta T + 0.2 * \Delta p) * l \qquad (4)$$

where

- z = measuring error in μm
- ΔT = temperature in degrees centigrade
- Δp = pressure in hPa
- l = measuring length in mm

The resolution is 10 nm.

Important terms in (4) are given by the constant 2.4 which results from the misalignment of the reference mirrors, and by the pressure influence p. The effects of the temperature drifts can be neglected in view of the good temperature control ($\Delta T <$ +/- 0.025 K) in the measuring room.

In this way, the measuring achieves an absolute accuracy in the region of micrometers. This accuracy is sufficient for practical purposes as the primary task is the detection of minimal height differences which requires a high resolution and low noise. The absolute dimensions of the components to be processed are less important than the local departures from the nominal shape.

The reproducibility of the tangential error measured is only 0.1 arcsec RMS up to a cutoff wavelength of 2 mm, measured on a component with an error of 0.5 arcsec RMS.

The reproducibility of the results obtained in surface measurements is 5 nm RMS and 30 nm P/V for a surface with approx. 100 nm RMS.

7 P400 Polishing Machine

7.1 Objectives

The design of the P400 polishing machine is experimental, permitting testing of different types of processing at a low expenditure.

The spectrum of components for the machine has been determined taking into account the specifications for synchrotron radiation optics.

7.2 Machine

7.2.1 Machine Design

Fig. 3 and 4 show a schematic and a photograph of the polishing machine respectively. The polishing head is adjustable in 5 axes (3 translational and 2 rotational axes) by path control. The translational X, Y, Z adjustment is performed in the cartesian coordinate system. The rotational axes D and E enable the continuous adaption of the surface normals of the tool and the component to each other. The angular adjustment is performed about a pivot "0" which is located at the intersection of the axes of the polishing head and the surface of the component. In this way, the angle coordinates are separated from the 3 translational coordinates.

7.2.2 Processing Modes

A machine with different processing modes has been designed to satisfy the demand for a flexible experimental setup for computer-assisted fine processing of optical surfaces made of different materials and dimensions. These modes enable both processing with small tools, as usual in computer-controlled polishing, and the performance of the large-scale movements of a normal optical polishing machine.

7.2.3 Basic Machine

The exacting demands imposed on the manufacture of optical mirrors had to be taken into account in the design of the machine. Of particular importance have been the static and dynamic rigidity. No strict requirements need be placed on tool positioning as relatively large tools are used. The lateral extension of the rate function is in the magnitude of centimeters. This means that the tools need not be set any more exactly than to a few percent of their radius.

7.2.4 Generation of the Polishing Force

With small to medium-size tools, the polishing force is generated by pneumatic cylinders; with large tools, however, it is generated direct by the weight of the Z-carriage and the head.

7.3 CNC Control System

The control system must satisfy the following requirements:

- o control of 9 NC axes (X,Y,Z,D,E,F,G,W)
- o movement of up to 6 axes in the path mode with linear interpolation
- o switchover of individual axes between path mode, positioning mode and mode with a given number of revolutions
- o storage capacity up to 10,000 vectors for the 6-axis path mode
- o implementation of high path speeds and short cycle times of the NC data sets
- o DNC path mode
- o programming facility on machine control

After a market analysis, we chose a General-Electric Mark-Century 2000 MC control system.

For computation of the path data, a VAX 11/750 computer is used. Owing to the complexity of the data computation and the large number of generated NC data sets, it is not possible to change the path data within the control system.

The process software was developed in cooperation with the WZL in Aachen.

Summary

Computer-controlled fine correction is a procedure which has been known for several years. To develop this procedure further, Carl Zeiss has designed a pair of machines, one of which is a high-resolution three-coordinate measuring machine and the other a CNC-polishing machine. A special process and evaluation software permits the processing of different components.

The illustrations 5 and 6 show the processing of a plano-ellipsoidal mirror. Starting from a state which has already been processed very well, with maximum peak-to-valley departures of 2.6 μm across the entire surface, and 0.6 μm across the useful area, we have been able to achieve a maximum departure of 0.2 μm across the useful area. Tangential errors of 0.7 arcsec have been attained. The individual data can be taken from the illustrations. The use of this new processing technique enables illustrations. The use of this new processing technique enables Carl Zeiss for the first time to fabricate strongly aspheric surfaces with tangential errors below one second of arc.

Literature

1) E. Heynacher, "Asphärische Optik - warum und wie sie hergestellt wird", Zeiss Information 24 (1979) 19

2) Preston, F.W., "The Theory and Design of Plate Glass Finishing Machines", Journal of the Soc. of Glass Tech., Vol. IX, No. 42, 1927

3) Beckstette, K., Becker, K., Baumgärtel, H., "Rechnergestützte Feinbearbeitung optischer Flächen"; Forschungsbericht des BMFT, Förderkennzeichen 13 N 5269

4) Dressen, H.C., "Entwicklung eines Algorithmus zur Realisierung vorgegebener Abtragsprofile für die rechnergesteuerte Feinbearbeitung asphärischer optischer Flächen", Diplomarbeit am WZL der TH-Aachen, 1985

5) Becker, K., Heynacher, E., "M400 - A Coordinate Measuring Machine with 10 nm Resolution", Proc. SPIE 802 (1987).

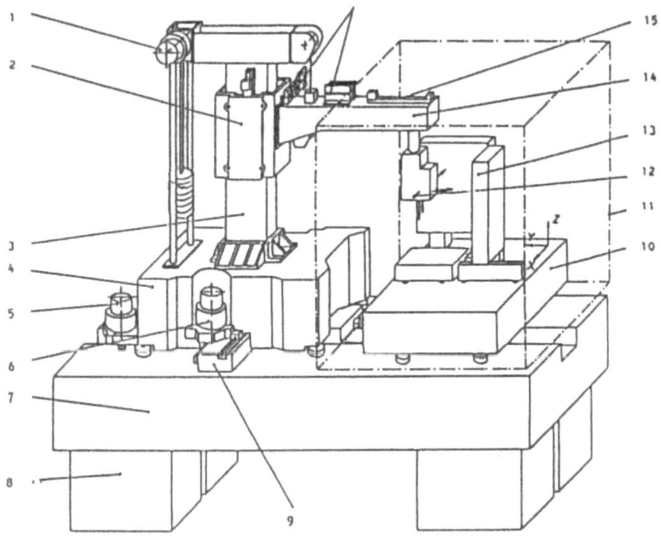

Fig. 1: Schematic of the high resolution coordinate measuring machine M400. Legend: 1: z-axis drive (tape); 2: z-slide; 3: granite column; 4: xy-slide; 5: capstan drive y-axis; 6: capstan drive x-axis; 7: granite machine bed; 8: vibration isolation element; 9: xy-guides; 10: baseplate (Zerodur); 11: temperature enclosure; 12: probe head with 5 measuring axes; 13: Zerodur reference mirrors (space normal); 14: measuring arm; 15: temperature reference axis on Zerodur base.

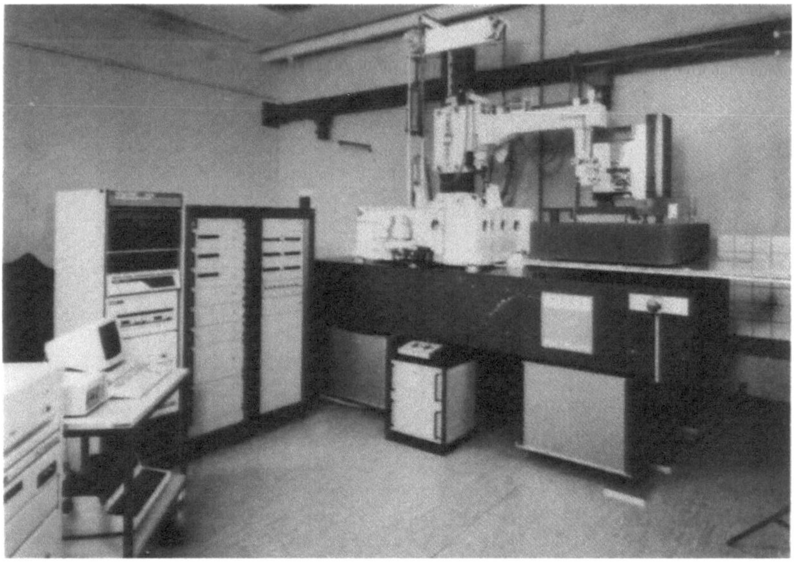

Fig. 2: Complete measuring system M400. Starting from the left one finds the computers for machine control and data collection followed by the servo- and data-collecting electronics. For details of the machine itself compare Fig. 1. The temperature enclosures are not yet mounted.

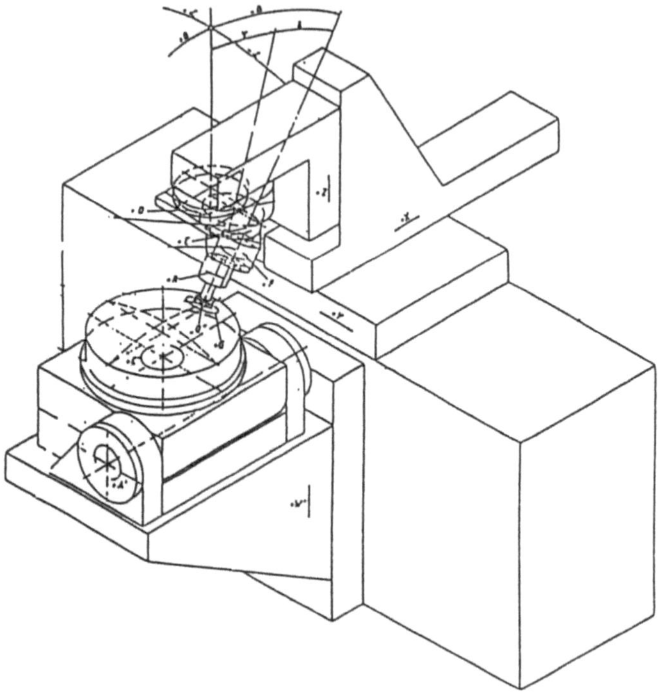

Fig. 3: Schematic of the computer controlled polishing machine P400. In total 9 axes can be moved by numerical control: These are the 6 axes of the polishing head (indicated as (D, E, F, G, R), 3 translational axes (x, y, z) and the rotation of the work table.

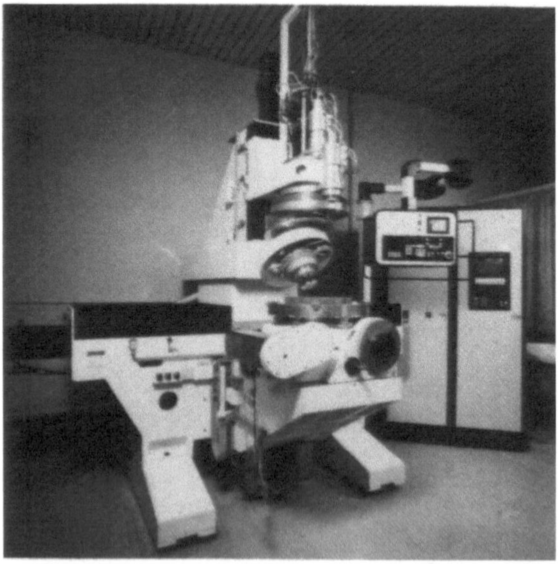

Fig. 4: Photograph of the P400 polishing machine. In the center one finds the polishing head with the work table below.

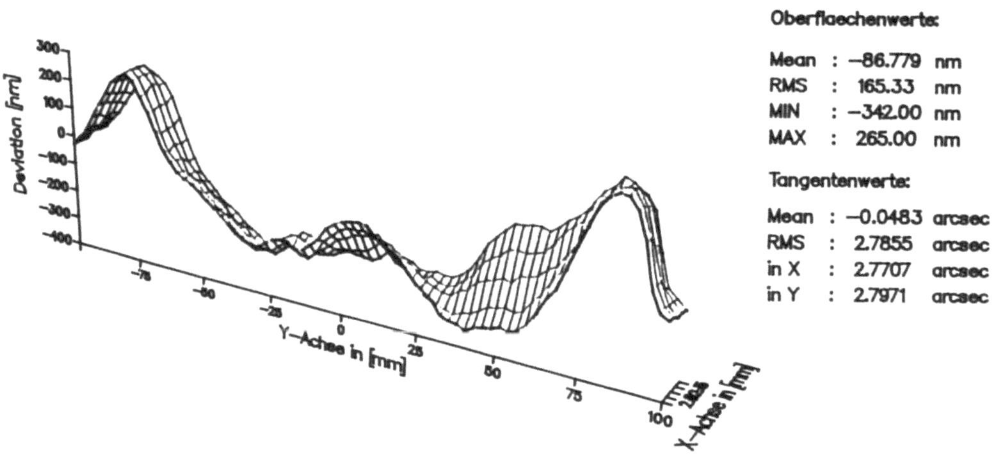

Fig. 5: Surface deviations of a plano-elliptical mirror at the start of the fine correction process. The peak-to-valley-value is about 0.6 µm in the useful area.

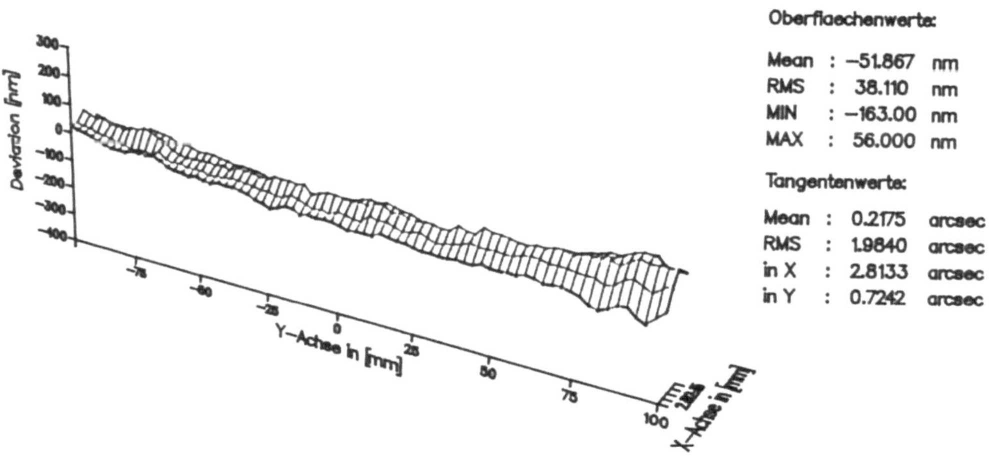

Fig. 6: Surface deviations of the mirror shown in Fig. 5 at the end of the fine correction process. The peak-to-valley-error is now below 0.2 µm.

New Design Features of a Fully Hydrostatic Ultra Precision CNC Turning Machine, Type Super-Microturn CNC

Pierre H. Knol, Denes Szepesi, Martin Huijbers

Hembrug B.V., P.O. Box 531, 2003 RM HAARLEM, the Netherlands

ABSTRACT

In the early seventies Hembrug started the design and production of Mikroturn ultra precision diamond lathes, originally for facing computer memory discs with a surface finish of Ra $\leq$ 0.01 um. These conventional facing lathes were equipped with a hydrostatic main spindle, developed by Hembrug and giving an axial and radial spindle run-out of 0.1 um.

The first generation of machines was followed by the advanced computer numerically controlled types for machining in contour of complex precision workpieces, requiring a high degree of surface finish, dimensional and shape accuracy. The accuracy required was maintained by the fully hydrostatic concept of both main spindle and slide assemblies as well as by the integrated computer numerical control and linear measuring feed back systems with a resolution of 0.1 um.

The company strategy for continuous improvement of accuracy and productivity recently resulted in new design features.

* higher spindle speeds up to 5,000 rpm [1]
* internal cooling system for hydrostatic headstock
* improved resolution of 0.05 um for CNC and measuring system
* built-in system of vibration dampers between machine base and pedestal
* double coolant arrangement, i.e. mist coolant and wet coolant, including special guarding for slides and measuring system and splash guard housing for working area
* improved ergonomic aspects, viz. better accessibility for loading and unloading workpiece, tool set-up and measuring inspection
* increased flexibility in tool arrangements, such as single tool posts, linear tooling systems, 8- or 12-station vertical tool turrets including rotary tools (C-axis) for premachining operations in combination with fix-type finishing tools.
* natural granite machine base
* numerically controlled high precision hydrostatic rotary table with special software for parabolic mirrors.

The paper will deal with the last two features and explain them in greater detail.

NATURAL GRANITE MACHINE BASE

Hembrug has been successfully using different kind of materials for machine base structures like cast iron and epoxy granite. After studying the physical data and possible applications of natural granite, we tested this new material and find it an excellent alternative for high precision turning machines.
The comparison between different materials for machine base structure is given in table 1.

	Modulus of elasticity N/mm²	Density kg/m³	Coefficient thermal expansion $°K^{-1} . 10^{-6}$	Thermal conductivity $Wm^{-1}\ °K^{-1}$	Tensile strength N/mm²	Compressive strength N/mm²
Cast iron (grade 17)	117,000	7,210	12	75	230	
Epoxy granite (granitan S100)	39,000	2,450	14	0.8	20	120
Polymer concrete (Motema)	42,000	2,300	17		11	120
Natural granite (Garbo/Impala)	125,000	2,900	5	3	14	250

Table 1. Physical data of different materials for machine base structure.

The use of natural granite as alternative material to cast iron and epoxy granite offers some significant advantages:

* high E-modulus
* low density
* long-time stability without shrinkage
* low coefficient of thermal expansion
* high degree of surface quality can be obtained
* unaffected by mineral oils, mineral emulsions and chemicals

Other advantages are also offered by design aspects:

* allows modular machine concept by machining the entire mounting surfaces
* no need of moulds, shorter lead time
* no need of corrosion protection by painting
* contributes to precision image by surfaces polished all over
* easily cleaned and maintained

The dynamic behaviour of natural granite was studied in our test room. By using a Fast Fourier Transform Analyser type HP 5423 A, we established reference values and compared the measurements of natural granite machine (figure 1) with those of cast iron (figure 2).

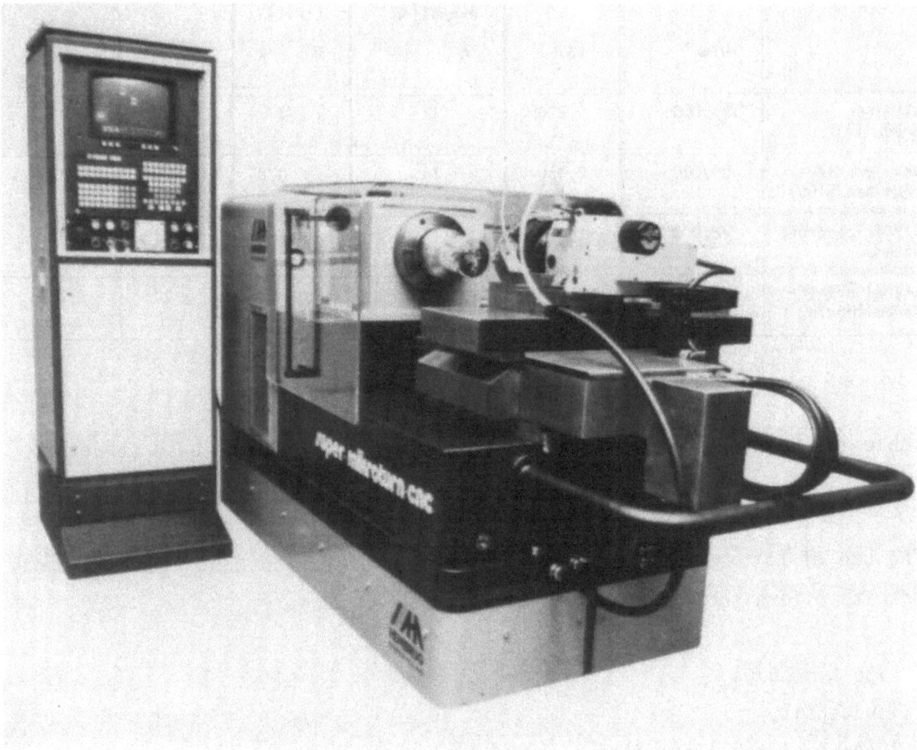

Figure 1. Natural granite base

Figure 2. Cast iron base

The frequency response measurements proved that the natural granite design increases the dynamic stiffness of the machine; some of the resonance peaks disappeared in the frequency spectrum and other peaks had a lower value.

The modal analysis of the cast iron bed (figure 3) shows typical deformations on the horizontal plane and these deformations are significantly reduced on the natural granite base (figure 4).

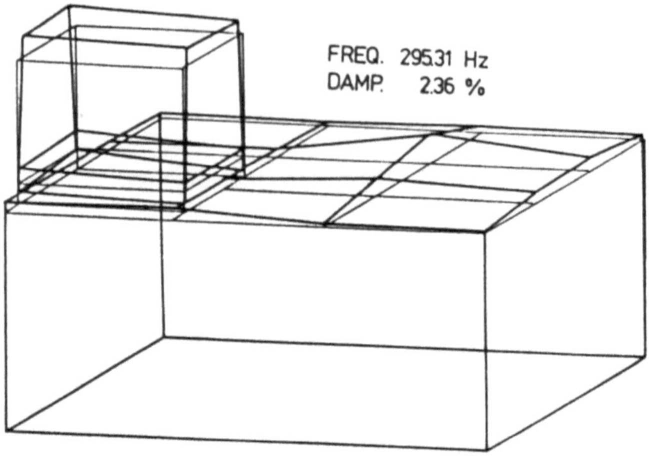

Figure 3. Modal analysis of cast iron bed

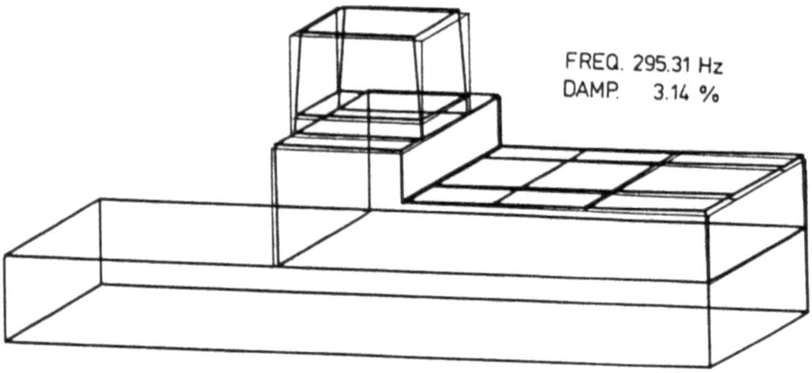

Figure 4. Modal analysis of natural granite bed

NUMERICALLY CONTROLLED HIGH PRECISION HYDROSTATIC ROTARY TABLE

Manufacture of aspheric optical components, often called "optomechanics", is described in a number of scientific publications. A lot of successful work has been done at Philips Research Laboratories, Eindhoven, the Netherlands [2]. This technology is mainly based on single point diamond turning on ultra precision CNC lathes, whereby the whole radius edge of the monocrystal diamond tool-tip is involved in the cutting process. It is obvious that the accuracy of the tool-tip radius plays an important role among the many factors which influence the aspheric quality.

In order to eliminate this influence, a new method was worked out by a team of scientists at the University of Bremen, Department of Production Technology, West Germany [3].
The method describes a high precision diamond CNC lathe equipped with a numerically controlled rotary table as a tool holder whereby only one point of the tool tip radius is involved in the cutting process. Figure 5 illustrates the method, also called "tool-normality":

The X-axis (cross-slide), Z-axis (longitudinal slide) and B-axis (rotary table) of the CNC lathe are controlled in such a way that only one point of the tool-tip accurately follows the programmed aspheric profile. The line between the cutting point and the center of tool rotation is constantly perpendicular to the tangent in the cutting point on the aspheric profile.

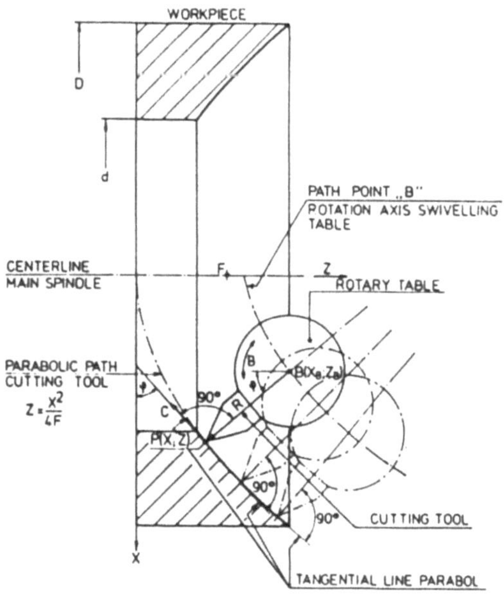

Figure 5. Tool-normality and 3rd axis interpolation with high precision hydrostatic NC rotary table

The method of tool-normality also offers another advantage, namely the equivalent tool-geometry during the whole cutting process which can hardly be achieved with high precision diamond tools and 2 axis interpolation.

The University of Bremen supported by BMFT and DFG placed an order at Hembrug to deliver a high precision diamond CNC lathe for production of aspheric components with the tool-normality method.

Performance data aspherics:

* parabolic mirrors (laser mirrors) $z = \frac{1}{4 \cdot F} \cdot x^2$

* focus F = 30-100 mm

* outside diameter mirrors: 100-600 mm

* surface roughness $Rz \leq 0.2$ um

* focus displacement ≤ 100 um

* profile continuity ≤ 0.2 um

Hembrug carried out the development of the new machine in 3 phases:

1. Arithmetical calculations of rotary table in combination with 3-axis interpolation.

2. Design of the NC controlled rotary table.

3. Development of software package for parabolic interpolation and 3rd axis control.

The arithmetical calculations resulted in the following equations:

X-coordinate of rotation point "B":

$$X_B = X - R \cdot \sin\left(\arctan\frac{X}{2F}\right)$$

Z-coordinate of rotation point "B":

$$Z_B = \frac{X^2}{4F} + R \cdot \cos\left(\arctan\frac{X}{2F}\right)$$

Rotation angle:

$$\varphi = \arctan\frac{X}{2F}$$

X-axis speed of rotation point "B":

$$\dot{X}_B = \left(1 - \frac{R}{2F} \cdot \cos^3\varphi\right) \cdot C \cdot \cos\varphi$$

Z-axis speed of rotation point "B":

$$\dot{Z}_B = \left(1 - \frac{R}{2F} \cdot \cos^3\varphi\right) \cdot C \cdot \sin\varphi$$

Angular speed "B" axis:

$$\dot{\varphi} = \frac{C}{2F} \cdot \cos^3\varphi$$

X-axis acceleration of rotation point "B":

$$\ddot{X}_B = 2 \cdot R \cdot q \cdot (F \cdot \sin p + X \cdot \cos p)$$

$$p = -\arctan\left(\frac{X}{2F}\right)$$

$$q = -\frac{2F}{(2F)^2 + X^2}$$

Z-axis acceleration of rotation point "B":

$$\ddot{Z}_B = \frac{1}{2F} - 2Rq(F \cdot \cos p - X \cdot \sin p)$$

X-axis speed of the tool point:

$$\dot{X} = \dot{X}_B + R \cdot \sin \varphi \cdot \dot{\varphi}$$

Z-axis speed of the tool point:

$$\dot{Z} = \dot{X}_Z + R \cdot \cos \varphi \cdot \dot{\varphi}$$

Speed of tool point (contour feed):

$$C = \sqrt{\dot{X}^2 + \dot{Z}^2}$$

According to the above equations, the movement of rotation point "B" is strongly dependent on the tool offset radius "R", shown on figure 6.

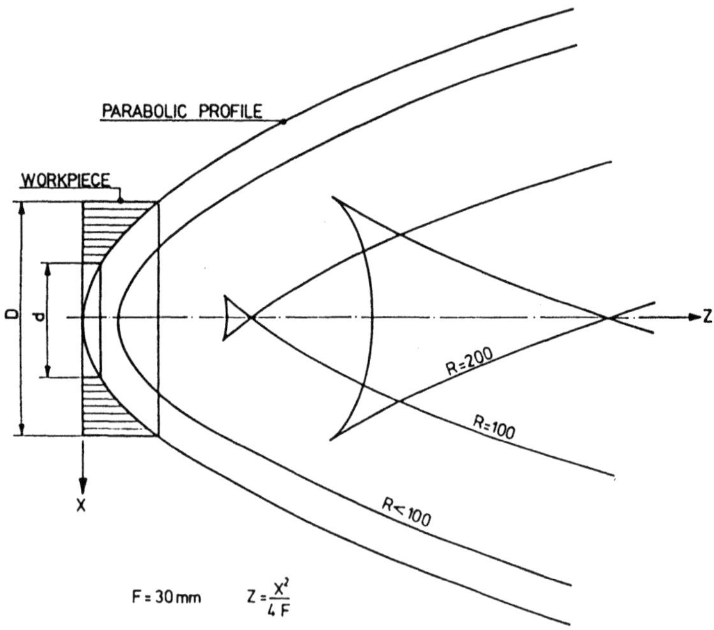

Figure 6. Movement of rotation point "B"

To develop the high precision rotary table, the required design parameters first had to be determined:

Maximum swivelling angle	200 degrees
Tool length (distance between rotation center "B" and tool-point)	150-200 mm
Minimum contour feed	5 mm/min
Minimum rotation speed	0.004 rpm
Rapid rotation speed	7 rpm
Maximum feed force	100 N
Measuring system resolution	0.36"
Positioning accuracy	2"/180°
Repeatability ± 2 Sigma	0.7"/180°
Spindle run-out	0.0001 mm

In the interests of high accuracy, Hembrug designed the rotary table with a hydrostatic radial and axial spindle bearing with an integrated incremental feed-back measuring system. The high response DC servo-motor controls the rotary movement via 1:400 precision gearing. The toolpost construction allows the mounting of 3 different types of diamond toolholders. One of them has an integrated piezo transducer for 3 dimensional in-process measurement of cutting forces (figure 7).
The diamond tools are adjustable both in radial direction for the desired tool-point offset and in height.

Figure 7.
Numerically controlled high precision hydrostatic rotary table

The NUM 760 T computer numerical control makes use of a multi-microprocessor system.
This technique allows very fast internal communications, which is important as the control works with small resolutions of 0.0001 mm or 0.0001 degree. The main NC processor is a Motorola 68000 chip using a 16-bit memory address with 32-bit architecture.
The parabolic interpolation and the 3rd axis rotary table control is performed by the dynamic operator of the CNC (figure 8). A short sample time of 10 ms allows contour errors of less than 0.1 um.

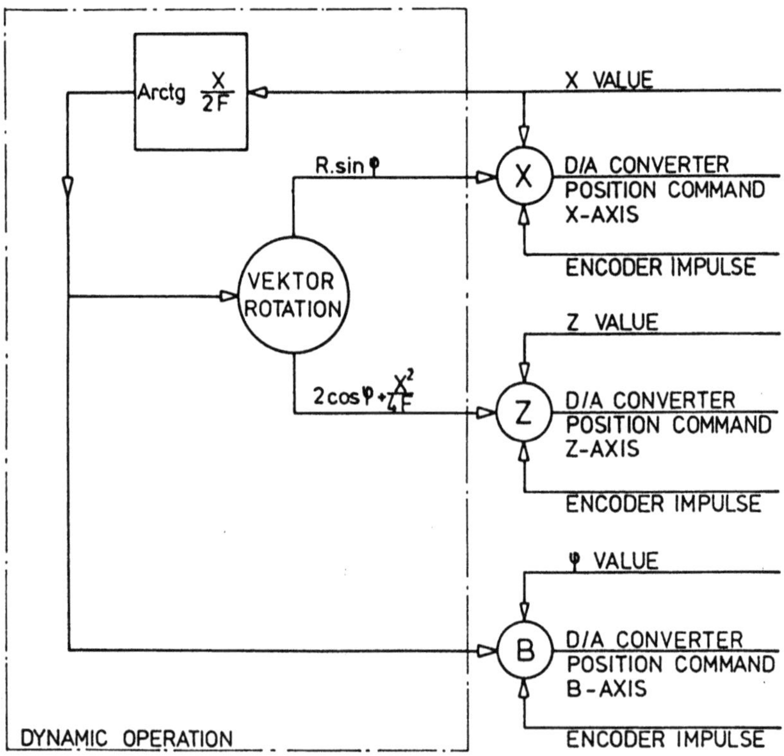

Figure 8. NUM 760 T dynamic operator for parabolic interpolation and 3rd axis NC rotary table control.

The workpiece NC program of an aspheric component includes the following variable parameters:

D	external diameter, parabol	mm
d	internal diameter, parabol	mm
F	focus, parabol	mm
C	contour feed	mm/min
R	tool-point offset	mm
n	rapid rotation speed	rpm

Since the contour feed "C" is constant, the X-axis speed $\dot{X}_B$ of rotation point "B" is corrected within a programmable interval (figure 9).

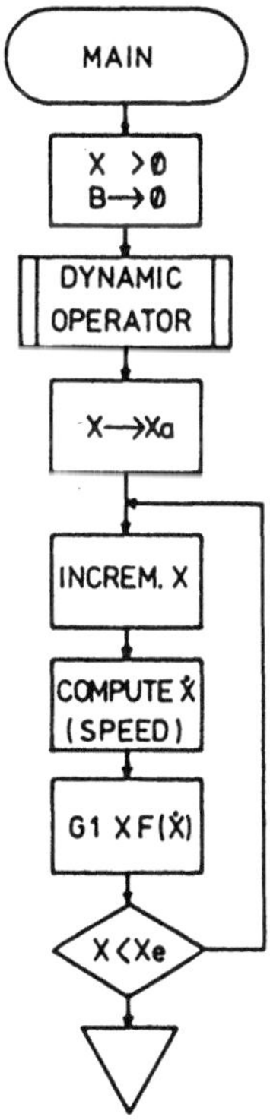

Figure 9. Correction of $\dot{X}_B$ speed to result a constant contour speed C.

SUMMARY

The recent developments of the Super-Mikroturn CNC ultra precision lathe now include interesting design features.
The natural granite base construction increases the overall dynamic stiffness of the machine. A numerically controlled high precision rotary table together with a dedicated software package allows the "tool-normality" method for production of precision aspheric components.

References:

1. P.H. Knol, D. Szepesi, J. Deurwaarder: "Air bearings machined on an ultra precision, hydrostatic CNC-lathe". SPIE Vol. 803, 1987.

2. J. Haisma, T. Gijsbers: "Optomechanics, an ultra-high-precision machining technique". Philips Tech. Rev. 41. 1983/84 No. 10.

3. G. Werner, W. Knappert, E. Minke: "Entwicklung von Werkzeugen und Steuerungseinrichtungen zur spanenden Feinstbearbeitung asphärischer Metallwerkstücke". Report to DFG. University of Bremen, Dept. of Production Technology, 1984.

A Personal Prospective on the Quest for Higher Precision Diamond Turning Lathes

PHILLIP R. HANNAH

Manager, Diamond Turning
Moore Special Tool Co.,Inc.
800 Union Ave.
Bridgeport, CT USA

As I wrote this talk, I reflected on when I worked for The Perkin-Elmer Corp. I recalled many discussions we had centering around a general attitude that then prevailed in the optical industry, and only recently has abated. Maybe you who are, or have been, involved in this industry also thought the same way.

The optics community and it's ability to make and measure was more accurate than any other, and as far as we at Perkin-Elmer were concerned, the machine tool industry's accuracy stopped at 2.5 micrometers. Why, we worked to less than a micrometer! Never in our wildest dreams did we ever imagine a machine tool capable of making a finished optical element accurate to a fraction of a wavelength of light. Yet, it happened, and today we would be surprised at what a high percentage of optical components are either directly machined or moulded on machined moulds that use the single point diamond turning technique.

For the next few minutes we will take a look back at some early machine data and the significant related areas that have contributed to bring the industry to where it is today, as well as some thoughts for the future.

Surprisingly, over the last 20 years, the period of adolescence for the single point diamond turning; the machine's accuracy has changed little. However, techniques, materials, peripheral equipment, and components have improved.

We would like to make clear that we are talking about today's commercial machines and components. There are machines like the Large Optics Diamond Turning Machine (LODTM) at Lawrence Livermore Laboratory (LLNL), as well as many one-of kind machines working quietly in industry that are more accurate than the commercially available machines. Let's look at some early reports which should give us a feel for what capabilities existed in the past.

Twenty-six years ago (1962) DuPont produced a 100 millimeter diameter hemisphere, round to 0.05 micrometers, with geometric (form) accuracy to 0.125 micrometers, and the part was specularly reflective (unfortunately no data is given). This was reported by Arnold and Steger, then with the Union Carbide Corporation, Oak Ridge, Tennessee, in the SME paper "Diamond Turning on Advanced Machine Tool Prototypes."

They also reported 2 machines which were built at Oak Ridge, an R Theta design, and a machine utilizing 2-Axis Moore Measuring Base. Both machines were reported to produce a surface finish on a single axis (flat) disc of approximately 0.01 micrometers Peak to Valley for the Moore machine, and 0.1 to 0.15 micrometers Peak to Valley for the R Theta machine.

In 1974 at Perkin-Elmer we were diamond turning optical quality plastic lenses as reported in the SME paper "Diamond Machining of Plastic & Metal Surfaces" written by the speaker. At that time, we were making 50 millimeter diameter spherical lenses in PMMA, commonly called Perspex or Plexiglas to a contour accuracy of 0.3 micrometers, and with a measured surface irregularity height or roughness of 0.175 micrometers. At that time, we were trying to produce low scatter surfaces. Typical measured total intergrated scatter values for these parts were 0.02%. A recticule or super polish on fused quartz typically exhibits a 0.005% T.I.S.

Only 12 years ago Ted Saito, then with the U.S. Air Force, presented the following as being typical industry wide. This information was presented at an Optical Society Workshop. The paper was titled "Scheming and Machining of Optics."

	Shape Accuracy (Peak-Valley Deviation)	Surface Smoothness (RMS Å)
Flats	λ/4 (1500 Å)	40
High f/Number Spheres	λ/1.6(3800 Å)	50
Low f/Number Spheres	λ/1.3(4500 Å)	100

for a nominal diameter of 150 millimeters.

Tom Gijsbers of Phillips Eindhoven N.V. reported on the "Colath" Lathe's results:

Form Error	0.5 micrometers
Suface Roughness	20 nanometers Peak to Valley

Impressive data from the past.

Let's now go back to that early DuPont machine. Interestingly, DuPont was not interested in making optics of any kind! Instead wanted a low force, low wear, machining method that would give them a "window" allowing them to look at the machine's performance. The results of the work came to our attention at Perkin-Elmer, and from the PMMA sample they provided, we realized that they were not only achieving their goal, but were producing Optical Quality Surfaces!

We should mention that at that point in time, we were not aware that people like Rank in England were using a single point technique on glass. Or, that the Swiss watch industry had been diamond turning gold and silver dial numerals for watches on a Posilux machine since the 1920's. Nor did we realize that N.V. Phillips and Polaroid, during World War 2, were also diamond turning.

We were so taken with the DuPont sample that we arranged a visit to learn more. They were more than willing to show us what they were doing. At once we realized that the machine incorporated some unique features:

. Single Crystal Diamond Tools
. Air Bearings
. A Massive Base
. Smooth, Accurate Slideways
. Good Machine Geometry
. Vibration Isolation

I remember well coming home from that trip, and immediately obtaining a microtome (a tissue sectioning device) and some of the single crystal diamond metalurgical knives that DuPont was then manufacturing. We wanted to see if we could observe things like chip formation, the effect of rake, and maybe some optimum machining parameters for PMMA. We had an old Moore Jig Borer Base with two linear axes, hydrodynamic slides, and lead screws. This Base provided accurate (0.5 micrometers full travel straightness) straight line geometry and smooth travel. By adding an air bearing where the vertical column would normally sit, we had an experimental machine. I want to point out again that these developments at Perkin-Elmer were not unique. Unknown to us, Jim Bryan at LLNL and the group at Oak Ridge were also hot on the trail.

In looking again at part accuracy over the last 15 or 20 years, it doesn't seem, in light of today's typical part accuracy, that much has improved. We still make 1500 Å parts with 30 Å surfaces, but if we dig a bit deeper we find there have been significant contributions made to the technology of single point diamond turning:

. Precision Machine Tool Specification
. Part Inspection Technique
. Tool Setting
. The Thermal Environment
. Part Mounting (chucking)

- Materials
- The Tool
- The N/C Control

This list is not meant to be all inclusive. I am sure that many of you could add to it, but for the sake of time we will hit some high points.

Precision Machine Tool Specification
Gradually, as various companies became interested in the process and purchasing a machine, they had a problem. Optical companies simply wanted to make accurate parts to a specified accuracy, and didn't really care how the machine's individual components or, for that matter, how the whole machine performed.

The machine manufacturer, on the other hand, wanted to know how straight, how square, how fast, how etc., this machine should be.

Some credit must go to the potential user as many machine purchasers learned how to make error budgets, understand programming resolution, feedback systems, servoloops, and the like, which led to their fuller understanding of a precision machine tool's capabilities. Today, most purchasers have very definite requirements that dictate such things as a preferred accuracy or machining area, or volume, in which most of their machine use will occur. They also are much more sophisticated in their understanding of the effects of the environment. Over the years these machines have left the laboratory and are now commercially traded; another factor in the development of "specifications", the method of communication between builder and user. Certainly, credit also must be given to ANSI, CERP, and other committees that have developed a uniform language and definition for many precision components, like the error motions of the air bearing spindle These facts have done much to cause the precision machine manufacturer and the precision part or optical component manufacturer to grow toward each other.

Part Inspection Techniques

The very early machines at DuPont, Oak Ridge and Perkin-Elmer were sphere and flat makers, and checking geometric accuracy was relatively simple. A Twyman-Green or Fizeau interferometer yielded excellent results. The machine tool builder had to be educated in the interpretation of interference fringes, but basically there were few problems. We measured surface roughness with a Rank Taylor Talystep, Brush, or Perthen Stylus device. With the advent of aspherical surfaces, the problems compounded; especially from the machine manufacturer's point of view. Holographic interferometers were just too expensive, so we at Moore decided the way to measure form was with a stylus or LVDT (Linear Variable Displacement Transformer) Device, taking many measurements or data points and then creating a contour map from the data. The basic problem with this was that in trying to do this on the machine, not removing the part from the chuck, we never see the chucking or mounting distortion.

Several companies, including Phillips, have built very accurate and stable measuring machines which can measure part geometry to a few hundreths of a micrometer. Machines made of thermally stable material, and set up in a temperature controlled environment exhibit repeatability to a few hundredths also. I could never understand why someone hasn't made a device like this commercially available; a device capable of measuring Radical Aspheres or High Order Polynomial Surfaces. It's got to be less expensive than interfermetric techniques.

In looking toward the future, what we would really like to do is in realtime (just as the cut is being made) measure, and then feedback making almost instantaneous tool path corrections. The high speed fast tool servo reported by Donaldson & Magrab (Precision Engineering 1983), as well as more recent work by Falter & Dow at N.C. State, certainly provides a major step in this direction.

This basic measure, program, and recut technique (not in real-time) has been used successfully. Jim Bryan, a good many years ago, was using this technique, making correction tapes, and re-cutting the part on the X-Ray microscope program. After several iterations part contour accuracy went from 0.175 micrometers to a final accuracy of 0.050 micrometers. Of course, lead screw error compensation programs have been available for years, but are not generally used, as the data spacing is usually too coarse for precision machines where you would like to correct only a few hundredths of a micrometer.

Tool Setting

One of the significant advances or improvements is the use of the tool setting station. Back when we built our first sphere/ flat machine at Perkin-Elmer in 1970, the biggest problem, and the most time consuming, was putting the tool on center; this was especially true at that time, as we were using two air bearings (the work spindle and the tool holder). By rotating the bearings, with respect to each other, make either flats or increasingly shorter radii spheres.

We didn't initially realize the importance of the alignment of these two bearings' axes with respect to each other. We found out the hard way that slight bearing misalignments produced the most interesting torics instead of flats or spheres.

On today's multi-axis machines with a numerical control system, precise alignment of the tool with respect to the work spindle axis, rotary table axis, and the N/C program is critical. A tool not finishing on axis will produce either an oblate or prolate spheroid. In the early days we used to cut a small cylinder, measure its diameter, and then move the tool to the part's other side and cut again. By measuring the difference between the two cuts, and dividing by two, we could find our spindle axis; a time consuming method, but a method used by many even today.

Another technique reported by Union-Carbide, Oak Ridge, permits the user to interferometrically measure the cone error (decentering) and thereby find the spindle axis or part's center. Unfortunately, it usually means demounting the part, and suffering the potential errors of remounting.

Today's use of two LVDT's, one vertical and one horizontal, provides a rapid means of setting the tool once the spindle center datum is established. The vertical probe sets the tool height. We have found that this method will yield centering accuracies of 0.25 to 0.05 micrometers.

The production of accurate parts depends on many factors; however, tool positioning, or the lack thereof, can mean the difference between acceptable or rejected parts.

The Thermal Environment

I think most of you are familiar with the work done at LLNL by Jim Bryan and company in providing a method of thermally stablizing the precision turning machine's temperature.

An interesting paper entitled "An Order of Magnitude Improvement in Thermal Stability with the use of Liquid Shower on a General Purpose Measuring Machine" by Bryan, Carter, Clouser, and Hamilton concludes with the following statement:

"Accuracy and repeatability errors in a measuring machine should be significantly less than the errors in the part being measured. An order of magnitude is the goal in many areas of manufacturing. At LLNL, for the past 12 years, the temperature controlled diamond turning lathes have produced parts which had errors less than the measuring machines used for inspection. Now for the first time in 12 years we have a general purpose inspection machine more accurate than our diamond turning machines. It is primarily the result of thermal stability.

There has been, and will be, psychological resistance to oil showered inspection machines. We believe that this initial resistance will be overcome by the need to progress, and the lack of practical alternatives."

We would like to add that on our (Moore's) first oil shower machine (Corning Glass), we achieved the kind of repeatability just mentioned. Ten electroless nickel plated parts, fifteen millimeters in diameter, were machined consecutively. The total deviation in form (geometric error) for all ten parts was less than 0.03 of a micrometer. Again, the importance placed on any precision machine's thermal environment cannot be over emphasized!

We haven't mentioned the method most used by diamond turning practioners, the air shower; certainly a successful thermal control system with many designs in use. The air system has evolved over the years where the oil system was a major advance, a step function if you will, and a system that we feel is superior.

Part Mounting (Chucking)

The literature is full of problems associated with part holding or mounting. We could spend the next hour just talking on this subject alone. But, for the sake of time, let me relate only two; two cases where the solution might not have occurred to the listener.

The first relates to the turning of a relatively soft, not too stiff, Polly Crystalline, infra-red material, Zinc Sulfide. This material is widely used for missile noses and serves as a window and protects the internal optics.

The near hemisphere is about 200 millimeters in diameter and 80 millimeters deep. The material diamond turns relatively easily, but due to its low modulus, it is hard to chuck or hold without distortion. A spherical diamond turned vacuum chuck with a near perfect matching radii you would think would be the answer, but it wasn't. Using a spherical test plate master and monochromatic light source, it showed several waves (1.25 micrometers) of distortion.

Only when the vacuum was reduced from a full atmosphere to about a fifth of an atmosphere were we able to achieve distortion free chucking. The fact that the part was axially symmetric, and the tool forces were low made this technique possible.

The second example comes from work done at LLNL for the Space Sciences Lab. at the University of California, Berkeley. Ten years of fabricating grazing incidence, cylindrical metal (Aluminum) mirrors are reported.

The significant performance improvement was not that the machine was improved, but that the process was improved with each iteration due to changing chucking (mounting) techniques.

Jim Bryan reported that through the use of accurate tool setting they were able to hold roundness and absolute size 0.125 micrometers or less, part to part; this on parts 450 millimeters in diameter.

Interestingly, these cylindrical optical elements were being held in a pot chuck with a rubber compound, but after several failures it was found that a simple mount, using three bolts 120 degrees apart, was the answer.

The final report in its conclusion states:
"Over the past ten years, we have developed four generations of grazing incidence metal mirrors. In the course of this effort, we have developed techniques which have improved the figure quality, the surface finish, and our ability to test and analyze the performance of these mirrors. In Figure 12 we show the throughput of these telescopes which dramatically emphasizes the effect of these improvements. In overall terms, we have been able to improve the surface finish to the point that the short wavelength cutoff closely follows the theroretical

limits. To illustrate the cost effectiveness of our fabrication process, we note that our mirrors exceed the performance of the Einstein mirrors with regard to FWHM imaging, yet cost two orders of magnitude less to produce."

How many times have you see astigmatism in your interferograms? Generally the machine can't produce astigmatic surfaces. Invariably the problem is thermal, or more likely a mounting or chucking problem.

Materials

We mention materials not because there are lots of new diamond turnable metals or ceramics, but because we have learned much about the ones we have, and how to specify and handle them.

I can recall when we thought stock 2024 Aluminum was the best available. We soon learned that the material was probably the worst choice. As we became more sophisticated precision machinists, we soon realized that most "as supplied" materials were potential problems. This was particularly true with metals which always tend to be full of stress, and needed heat treatment.

At first, a simple two hours at 180 degree C, and air cooling was enough. As the parts became larger and more complicated, we found that even this technique wasn't enough.

Going back to our U.V. telescope example, we find that the final stablization cycle for the 6061-T6 Aluminum required cycling in liquid nitrogen, then into boiling water several times before the material was stable. We should point out that the recommended procedure uses dry ice instead of liquid nitrogen. We must not overlook an often forgotten, but important, part of the diamond turning process.

The Tool

This is a subject dear to my heart, and there has been significant advances made in this industry.

New single crystal diamond tool manufacturers have made a much larger selection of tools available. The use of man-made diamond, like the Sumitomo diamond, has in our tests on crystalline material shown longer tool life, and for the 2-Axis machine more accurate tool tip radii is a real help. Radii to 0.5 micrometers or less are available.

One of my favorite subjects is crystal orientation and its effect on tool life. I believe that we, the diamond turning community, are limited to a degree in the materials we can turn by tools that are commercially available; tools which can be profitably manufactured. Crystal orientations that can be lapped in a reasonable time, keeping the tool cost within reason. There are crystal orientations that are difficult or impossible to lap or grind by conventional techniques. However, the newly emerging chemical polishing technique allows super tool orientation which may permit us to extend the range of nominally diamond turnable materials. Maybe you think this is ridiculous, but as Chris Evans of the U.S. National Bureau of Standards points out, Jessie Ramsden, the famous inventor, back in 1777 turned a steel lead screw using diamond, and there is considerable 19th century literature on the diamond turning of steel.

Most probably it was the tip of an Octahedral Crystal mounted like a wheel dresser or scribe diamond. But, it worked; wear (diamond failure) can't be all chemical.

There is much we need to do and understand in the tool area. Understanding the tool/work interaction will help our list of turnable materials to grow.

The N/C Control

Here is a major area of growth; in this area alone lies much of the future.

In looking beyond the Allen-Bradleys and G.E./Fanucs to some of the newer controls available today, but designed for the future, we can make some significant improvements in our machine's performance.

Let's look at some features available on the IBH and CUPE Controls available today.

Thanks to Pat McKeown of CUPE for his help here.

Our first (Moore's) experience in this area was due to a customer's insistance that our in-house lab machine just wasn't making good parts. But it wasn't the machine, it was the control system! Our customer wanted to produce small (12 millimeter) diameter aspheric waveguide lenses. These were turned into a Lithium Niobate Substrate, and were more like depressions in the material than lenses as we know them. The customer insisted that microscopic examination was showing tool marks at each program end of block. We were finally convinced that another control should be fitted to the machine. An IBH control was married to our M-18AG, and, Wonderbar, the problem went away!

After much analysis, we now feel that due to the relatively slow internal clock in the Allen-Bradley control, the machine's axes were not closing out at the same time. Further examination of the test data showed a marked improvement in the machine's overall axis's performance.

The IBH and CUPE control have a number of very real technological advances that anyone building or buying a machine should seriously consider.

Some unique features of the new CUPE Series 2000 CNC System are:

1. Velocity feedforward technique provides virtually zero following errors during multi-axis generation of complex profiles.
2. Interfaces available for the latest very high resolution laser interferometer systems (resolution 1 nanometer and beyond). On line refractive index compensation for the laser interferometer.
3. Servo control techniques allow the use of any position transducer resolution without imposing constraints on maximum axis velocity.
4. Software "Electronic Gearbox" allows multiple axes to be synchronized to provide complex profile generation with high accuracy even when servo drive limitations would otherwise lead to reduced performance.
5. Hardware "Electronic Gearbox" extends the profile generation capability to the highest levels such as those required for CNC cam grinders.
6. Axis error correction facility allows individual axis position transducers to be calibrated so that systematic position measurement errors can be removed.
7. Full 3D error correction capability.
8. Availability of polynomial interpolation in addition to linear and circular interpolation. The form of polynomial interpolation used guarantees continuity of slope and position at each point (without instability) and allows forms such as cam profiles and aspherics to be generated with high accuracy from a minimum amount of input data.

These are only a few major improvements awaiting the precision machine tool user/builder.

We have glimpsed changes over the years. Not so much changes in accuracy, as changes in productivity. We purposely spent little time on commercially available machines as they have not changed significantly over the years. However, a point worth mentioning before we close takes us into the area of precision grinding, specifically ductile grinding. There have been several reports that have addressed the area of precision diamond turning and grinding machine loop stiffness. A recent article in the Journal of Precision Engineering by Dow and Co. at N.C. State reports that the loop stiffness on two commercially available diamond turning machines was between 12 and 17 newtons per micrometer. Keep these values in mind as we look at the new area of ductile grinding. There are a number of people working in this area in Japan, England, Europe, and the United States. Pat McKeown of CUPE reports that Zerodur parts ground by this method showed 8 nanometers Peak to Valley surface finish irregularity. (The Japanese report 2 nanometers Peak to Valley). Pat also says that they measured chip thicknesses of 100 nanometers. The machine's loop stiffness was approximately 150 newtons per micrometer.

With all the interest today in the production of precision ground specular surfaces, that is: finished parts, no post publishing; newer, stiffer machines are going to be needed as well as new techniques.

I hope this has challenged your thinking as you look forward to greater precision and productivity.

Thank you for your patience.

Session 3

Measurement Techniques and Equipments

Nanotechnology at the National Physical Laboratory

A. FRANKS

Division of Mechanical and Optical Metrology
National Physical Laboratory
Teddington, Middlesex, TW11 OLW, UK.

Summary

The development of nanotechnology at the National Physical Laboratory was stimulated by the very fine manufacturing tolerances that were required to produce x-ray optical components; the technology is now being applied more widely. Fundamental studies of the brittle-ductile fracture transition have led to a better understanding of the mechanisms involved and, together with the development of a very stiff machine tool, should enable brittle materials to be machined without introducing sub-surface damage. Amplitude-wavelength mapping is a powerful technique that has been developed to provide a quantitative and graphical description of the performance of measuring instruments, which now have measuring capabilities down to 0.05 nm. X-ray interferometry is used for calibration at this level of accuracy. The work at NPL has resulted in the launch of the UK National Initiative on Nanotechnology.

Introduction

We define nanotechnology as the technology of manufacture to dimensions or tolerances in the range from 0.1 - 100 nanometres, (from the size of the atom to the wavelength of light).

Areas in which nanotechnology has important contributions to make are continually expanding, particularly with the continuing drive to miniaturisation and higher precision. Examples of present and potential applications of nano-technology include micromechanisms, optical systems, magnetic and optical data storage, optoelectronics, microelectronics, particularly low dimensional structures, scanning tunnelling microscopy applications (ie scanning tunnelling engineering), x-ray interferometry and holography using multilayer-coated mirrors. These applications are discussed in some detail in [1].

At NPL, the stimulus for entering the field arose both from the need to produce x-ray optical mirrors, the surface tolerances of which could not be met by the traditional craftsman's approach to optical manufacture, and from the challenge to widen the scope of engineering to encompass optical manufacture. The range of the work at NPL has now expanded and includes the finishing of optical components such as laser gyroscope mirrors, semi-conductor substrates, the development of precision slideways for measuring machines, and precision machine tools.

The production of very precise components goes hand-in-hand with the development of the necessary metrology, and a wide range of measuring instruments has been devised to cater for the evaluation of surfaces and structures down to the 0.05 nm level. An important advance has been made in evaluating and comparing the performance characteristics of measuring instruments, in terms of their amplitude and surface wavelength measuring capabilities, in which complex data is presented succinctly on Stedman charts [2,3].

Manufacturing tolerances

(i) X-ray optics

X-rays incident at a small grazing angle to a polished surface will be reflected from that surface and this phenomenon of *total external reflection* forms the basis of x-ray optics. The principles and applicatons of x-ray optics are discussed in [4]. Mirrors for focusing x-rays are usually spherical, cylindrical, toroidal or conicoidal in form. X-ray telescopes consisting of parabolic and hyperbolic figures of revolution have been made, with diameters and lengths of about 1 m, while x-ray microscopes which also consist of conicoidal surfaces have corresponding dimensions in the 10-40 mm range.

It is not possible to make generalised statements about the specifications of x-ray optical components, except to state that the trend is towards arc second or better resolution in telescopes and to sub-micrometre resolution in microscopes, which translates to sub-arc second resolution in angular terms.

Taking the AXAF telescope [5] as a specific example, the axial slope tolerance of its 0.84 m long mirrors is 0.05 arcsec. Hence the tolerance on the amplitude deviation for a surface wavelength of 500 mm is 0.12 μm. For shorter surface wavelengths the amplitude deviations are reduced proportionally, and are 25 nm, 12 nm, 6 nm and 2.5 nm for surface wavelengths of 100 mm, 50 mm, 25 mm and 10 mm respectively. The rms surface roughness must be in the sub-nanometre region: a surface roughness of 1 nm will scatter over 50% of 0.15 nm wavelength x-rays outside a 1 arcsec diameter circle in the focal plane, and will thus seriously impair the target resolution of 0.5 arcsec. Another example of an x-ray component which must be made to extremely fine tolerances is an x-ray microscope now being developed at NPL [6]. Its internal diameter is 30 mm and both its mirrors are 30 mm long. To work to its specification, the profile error should not exceed 2 nm and the surface roughness should be in the sub-nanometre region.

Summarising, quite severe specifications have to be met to satisfy the demands of the more critical x-ray optical components: surface finish must be in the sub-nanometre region, while the tolerance on surface figure of pocket-sized components is about a nanometre and that of metre-sized components less than a micrometre.

(ii) Other applications

In information technology, the trend towards miniaturisation and higher memory densities has stimulated manufacture to nanometre tolerances. Surface finishes of 3 nm are now being achieved on magnetic memory discs, subnanometre surface finishes are required on some semiconductor substrates, and Winchester disc reading heads and spindles for video cassette recorders fall in the manufacturing tolerance band of 5 nm to 50 nm.

In optics, the manufacturing tolerances for laser gyroscope and free electron laser mirrors are as severe as those for the most precise x-ray mirrors.

Conventional manufacturing principles

The concepts discussed in this section do not refer specifically to x-ray optics; they are more widely applicable.

(i) Choice of material

Favoured materials for high precision applications include silica, low expansion glass ceramics (eg Zerodur), strongly bound ceramics such as the refractory carbides (eg silicon carbide), electroless nickel coated metals and non-age hardening aluminium alloys. These materials exhibit some or all of the following properties: good polishability (because of their fine microstructure), good mechanical stability (low internal stress), good thermal properties (favourable combination of expansion coefficient, thermal conductivity and specific heat).

(ii) Pre-machining

Pre-machining covers all the machining processes up to the final lapping and polishing stages. Lapping attenuates the remaining errors in surface figure and finish, so that the objective of the pre-machining stage is to produce a component as close as possible to the required specification in order to minimise the time-consuming lapping process. The final pre-machining stage should therefore be undertaken with a stiff, precision machine tool. The design principles of such machines are well understood and many precision machine tools are now in use, particularly for diamond machining.

(iii) Lapping and polishing

Some current machine tools are capable of machining to accuracies of a few tens of nanometres or better. Lapping techniques, in which the workpiece and lapping tool are worked together without reference to external machine axes, offer the possibilities of achieving the required tolerances of a few nanometres in figure, and sub-nanometre surface roughness. Lapping principles are well-established: an important feature of the work at NPL has been the development of a series of lapping and polishing machines designed specifically for x-ray optical mirrors [7]. Flat samples can now be lapped and polished to about 0.05 nm rms [8], which is at the limit of measurement of the Rank Taylor Hobson 'Talystep' machine.

New manufacturing principles

(i) The brittle-ductile fracture transition

Reference has already been made to the fact that ceramics and glasses are favoured materials for precision applications. These are brittle materials, and when they are machined with conventional machine tools, cracks are formed in the surface which both weaken the component and make the surface unstable. It would be of significant technological importance if these materials could be machined in a ductile mode, without the formation of sub-surface damage, and with the ultimate aim of eliminating or minimising the need for finishing by lapping and polishing.

It is known that there is a transition from brittle to ductile behaviour which is a result of the non-geometrical scaling laws of brittle fracture [9]. Thus below a critical linear dimension d_c of the stressed volume of material, yielding rather than cracking occurs, ie when

$$d_c = \alpha ER/\sigma_y^2 \qquad (1)$$

where E is Young's modulus, σ_y is the yield stress for plastic flow, R is the specific work per unit area to propagate a crack, and α is a numerical parameter whose magnitude is a function of the type of test. This equation has been verified for tests in compression [10] and in indentation [11].

A small, stiff, single-point diamond turning facing machine was constructed for the purpose of undertaking fundamental studies of the role played by the brittle-ductile transition in the machining process. It was anticipated that below a critical size of cut it would be possible to remove material without initiating residual cracks in the surface. This indeed proved to be the case; however, the removal mechanisms were complex, but are amenable to quantitative interpretation [12].

Experiments were carried out on two materials: soda-lime glass and 'Spectrosil', the latter is harder and is a synthetic form of silica glass; some results obtained with this material are discussed here. The diamond tool was a standard Vickers indenter, which is pyramidal in form and has a maximum included angle of 136°. For Spectrosil, crack-free cutting occurred when the depth of cut was less than 0.25 μm. SEM micrographs of the surface produced with this depth of cut and with a traverse speed of 120 μm/min are shown in Figs. 1-3. Fig. 1 shows a general view where the swarf can be seen to spring from the cut in the form of tightly bound spirals. Under carefully controlled conditions, such swarf may be detached from a groove as a continuous ribbon of indefinite length. Fig. 2 shows that the ribbons do not necessarily curl up from one end, but may remain as arched loops at either end of the groove. Fig. 3 demonstrates that neighbouring ribbons may peel off in different directions.

These and other results indicate that micromachining of glasses is accomplished by at least two processes, groove formation and subsequent swarf removal. The fact that the two are not concurrent is particularly well demonstrated in the micrographs showing ribbons of material in spiral and loop form adhering to the groove in the process of peeling off, sometimes in opposite directions. The spiral swarf in Figs. 1-3 also indicates that plastic flow is involved in groove formation: the material removed is in a state of residual stress due to partial yielding in the surface layers of the swarf.

These and other observations of delamination and subsequent distortion tend to the conclusion that the material in the machined grooves is in a state of longitudinal compressive stress due to plastic flow around the tool. It is postulated that the first stage in the detachment of the swarf occurs by shear fracture along a path within the stressed region parallel to the machined surface. When this fracture has attained a certain length, the operation of the compressive stresses on the end of the free segment may cause it to bend or buckle. It is striking how stable the crack path is parallel to the surface: this is due to the well-defined region of the compressive stress traced by the path of the tool. The coils are produced by fracture of one or other end of the buckled strip under the combined stress due to the original longitudinal compression and a bending moment associated with the curvature of the buckled strip.

(ii) Development of a stiff machine tool

Theoretical considerations indicate that the conditions for machining brittle materials in the ductile mode require that the depth of cut should not appreciably exceed 100 nm and that the normal force should be in the range of 0.1-0.01 N. These conditions can be met only by machine tools that are much stiffer than conventional ones.

One approach to the design of a stiff machine tool, now being developed at NPL, is to construct it in the form of a tetrahedron [13] (Fig. 4). This is an intrinsically stiff configuration, and the first version made of mild steel has a static loop stiffness of about 1 nm/kg (10 000 000 lb/in), which is an order of magnitude greater than most diamond turning machines, as is its resonant frequency which exceeds 1 kHz.

Metrology

The quest for higher precision has led to the development of a number of measuring machines having measuring capabilities down to sub-atomic dimensions. The application of the concept of amplitude-wavelength space has been valuable in clarifying the meaning of these measurements and, more generally, in achieving a better understanding of the performance and characteristics of measuring instruments.

(i) Amplitude-wavelength space

The basic concepts of amplitude-wavelength space (AW space) are described in [2,3]. Measuring instruments differ widely in performance, both in their vertical range and resolution measured normal to the surface (amplitude), and in the range of surface wavelengths that are detectable. The performance of instruments is limited mainly by the specifications of a number of critical components. A range of parameters specifying, for example, probe geometry, transducer sensitivity, accuracies of the ways, scan length, and scale resolution are needed to define the capabilities and limitations of the instrument. The response of the instrument to sinusoidal perturbations of varying amplitude and wavelength can conveniently be modelled. Mapping the limit lines in AW space shows the effect of each parameter, and thus the effective working range of the instrument can be delineated as the polygonal area shown in Fig. 5 for a stylus instrument, where R_v and R_h are the ranges of the vertical and horizontal transducers respectively, and r_v, r_h are their corresponding resolutions, *a* is the slope angle of the flanks of the stylus, R_s is the radius of its tip, and *b* is the slope error of the measuring instrument's reference surface.

(ii) Measuring instruments

A wide range of instruments has been developed at NPL for the metrology of surface topography. Particularly noteworthy are the Nanosurf 2 [13], the polarising interferometer [14], the stylus profilometer [15], the laser profilometer [16], the Nanorond 1 [17], and the x-ray interferometer [18]. This powerful array of instruments provides a measuring capability that ranges from 50 pm (picometres) to 15 mm in surface amplitude and from 50 nm to 250 mm in surface wavelength, techniques for roundness measurement to 1 nm, and displacement calibration to 10 pm traceable to the national standards of length. The performance of these and some other measuring instruments has been mapped on Stedman charts in AW space [2,3]. The limits of performance of the Talystep (TS) and the NPL Nanosurf 2 (NS 2) are shown in Fig. 6. For the TS, the lines *a* and *b* show the limits for the diamond chisel and the standard styli respectively. The plotted zone represents the overal capability of the instrument. The performance of the NS2, which is based on the TS transducer head and chisel stylus, extends that of the TS through the increased range and quality of its horizontal motion. This is illustrated in Figs. 7 and 8 which demonstrate that the reproducibility over a traverse length of 40 mm is approximately 1.5 nm rms, and over 5 mm it is better than 0.25 nm rms.

The increasingly widespread availability of measuring instruments having nanometre and sub-nanometre measuring sensitivities has focused attention on the need to calibrate these instruments to this level of accuracy. This is done at NPL by means of x-ray interferometry. The principle of the method has previously been described [18], and the interferometer is shown in Fig. 9. The essence of the technique is that the transducer that is to be calibrated is attached to the moving member of the single crystal silicon interferometer. Motion of the latter by one lattice spacing produces an x-ray fringe that can easily be observed with the aid of an x-ray detector. The (111) silicon planes used in this work have a spacing of 0.3135626 ± 0.0000016 nm; it is known to an accuracy of 1 part in 10^6 and is traceable to the primary standard of length [19]. The x-ray fringes can be sub-divided to give a traceable measurement to 20 pm.

The UK National Initiative on Nanotechnology (NION)

The first five programmes under LINK, the UK Government's initiative to encourage collaboration between the scientific community and industry were announced in February 1988.

The initiative aims to boost industrial R&D and to make scientists and the business community more aware of the potential for profitable exploitation of science and technology.

Up to £83m will be spent on the five programmes, of which the government will provide up to half. Over the next five years further programmes are expected to take the total government and industry spending on LINK to at least £420m.

The LINK Nanotechnology programme is being co-ordinated by the Department of Trade and Industry through NPL as part of the National Initiative on Nanotechnology (NION), which is seeking to raise awareness and encourage the adoption of nanotechnology techniques by UK industry. NPL is assisted in the management of NION by a strategy committee consisting of experts from industry and academe nominated by DTI and the Science and Engineering Research Council, together with officials from these two organisations.

References

1. Franks, A.: Nanotechnology. J. Phys. E. Sci: Instrum. 20 (1987) 1442-1451.

2. Stedman, M.: Mapping the performance of surface-measuring instruments. Proc. SPIE. 803 (1987) 138-142.

3. Stedman, M.: Basis for the comparison of surface-measuring machines. Prec. Eng. 9 (1987) 149-152.

4. Franks, A.: X-ray optics. Science Prog. 64 (1977) 371-422.

5. Zombeck, M.V.: Advanced x-ray astrophysics facility (AXAF). Opt. Eng. 20 (1981) 297-309.

6. Franks, A. and Gale, B.: X-ray microscopy. Schmahl, G. and Rudolph, D. (eds.) Grazing incidence optics for x-ray microscopy. Springer Ser. in Opt. Sci. 43 (1984) 129-138.

7. Franks, A.: Präzisionsmechanik im röntgenoptischen Gerätebau. Feingerätetech. 35 (1986) 531-535.

8. Lindsey, K.: The attainable limits in x-ray mirror fabrication. Nucl. Instrum. Methods Phys. Res. 221 (1984) 14-19.

9. Puttick, K.E.: The mechanics of indentation fracture in polymethylmethacrylate. J. Phys. D: Appl. Phys. 11 (1978) 595-603.

10. Kendall, K.: The impossibility of comminuting small particles by compression. Nature 272 (1978) 710-711.

11. Puttick, K.E. and Yousif, R.H.: Indentation fracture transitions in polymethylmethacrylate. J. Phys. D: Appl. Phys. 16 (1983) 621-633.

12. Puttick, K.E., Rudman, M.R., Smith, K.J., Franks, A. and Lindsey, K.: Single point diamond machining of glasses. (1988). To be published.

13. Lindsey, K., Smith, S.T. and Robbie, C.R.: Sub-nanometre surface texture and profile measurement with Nanosurf 2. Ann. CIRP (1988) in the press.

14. Downs, M.J., McGivern, W.H. and Ferguson, H.J.: Optical system for measuring the profiles of super-smooth surfaces. Prec.Eng. 7 (1985) 211-215.

15. Stedman, M. and Stanley, V.W.: Machine for the rapid and accurate measurement of profile. Proc. SPIE 163 (1979) 99-102.

16. Ennos, A.E. and Virdee, M.: Precision measurement of surface form by laser autocollimation. Proc. SPIE 398 (1983) 252-257.

17. Stedman, M.: The precision measurement of roundness. Wear 109 (1986) 367-373.

18. Bowen, D.K., Chetwynd, D.G. and Davies, S.T.: Calibration of surface roughness transducers at Angstrom levels, using x-ray interferometry. Proc. SPIE 563 (1985) 412-419.

19. Becker, P., Dorenwendt, K., Ebeling, G., Lauer, R., Lucas, W., Probst, R., Rademacher, H-J., Reim, G., Seyfried, P. and Siegert, H.: Absolute measurement of the (220) lattice plane spacing in a silicon crystal. Phys. Rev. Lett. 46 (1981) 1540-1543.

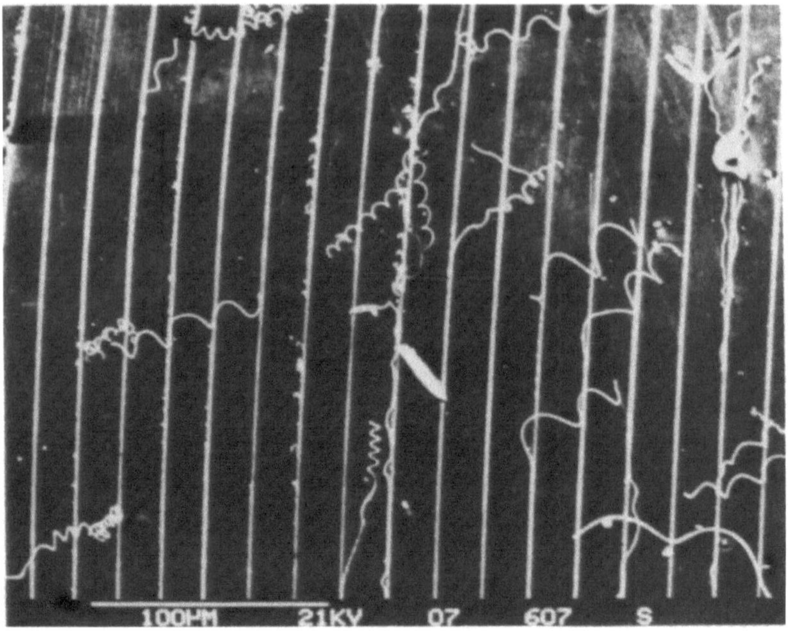

Fig.1. Diamond-turned Spectrosil

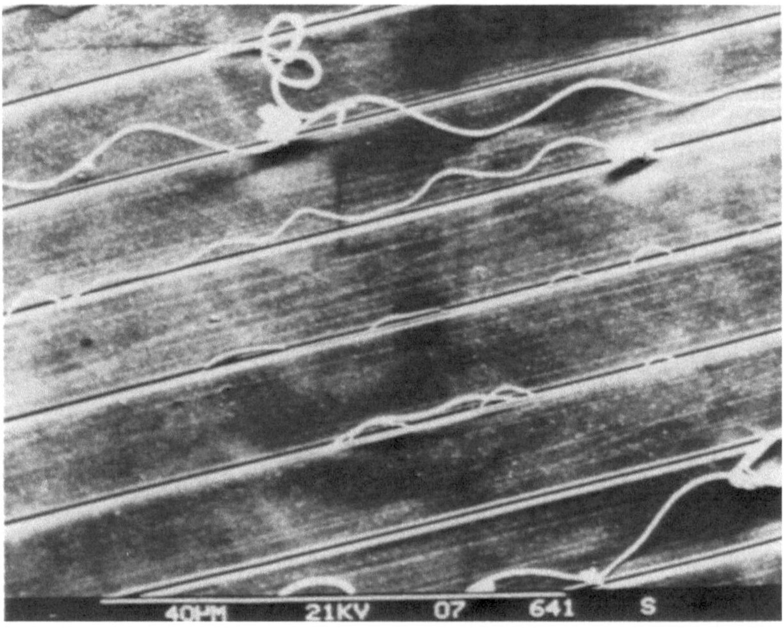

Fig.2. Spectrosil ribbons adhering to substrate

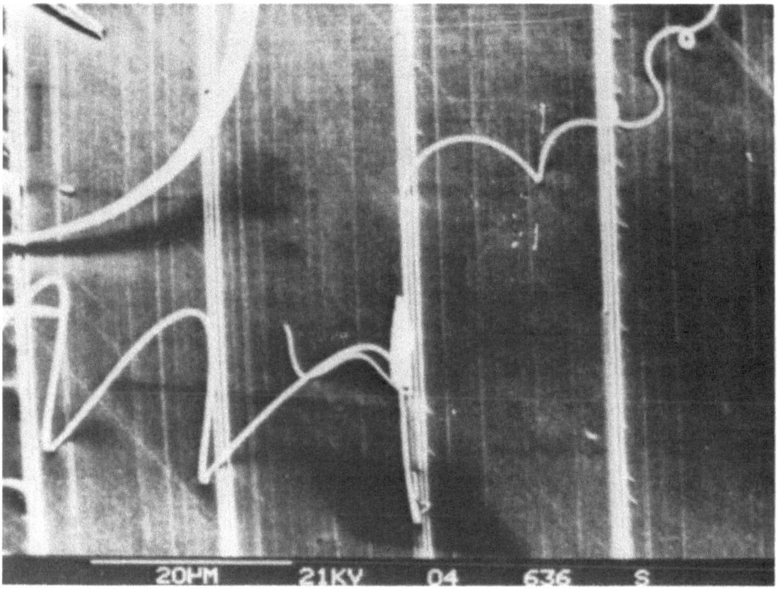

Fig.3. Spectrosil ribbons peeling off in opposite directions

Fig.4. Tetrahedral machine tool - the Tetraform

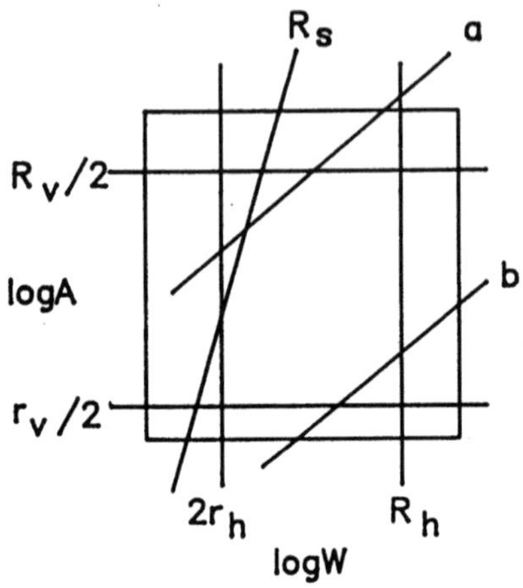

Fig.5. Parametric limits in AW space

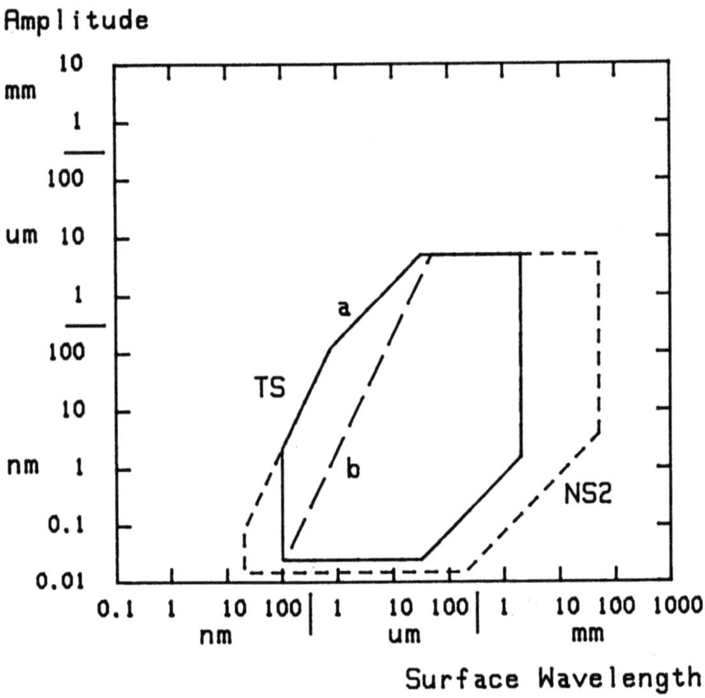

Fig.6. AW plots of the Talystep and Nanosurf 2

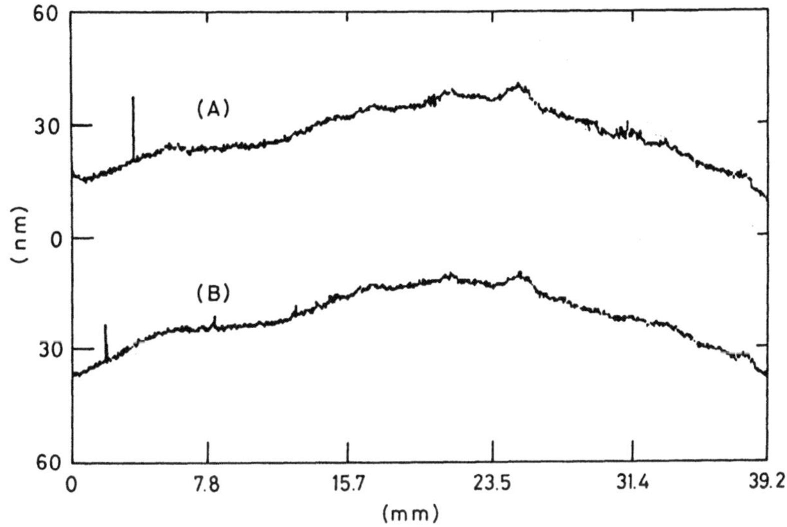

Fig.7. Successive Nanosurf 2 traces showing reproducibility of 1.5 nm rms over a 40 mm scan

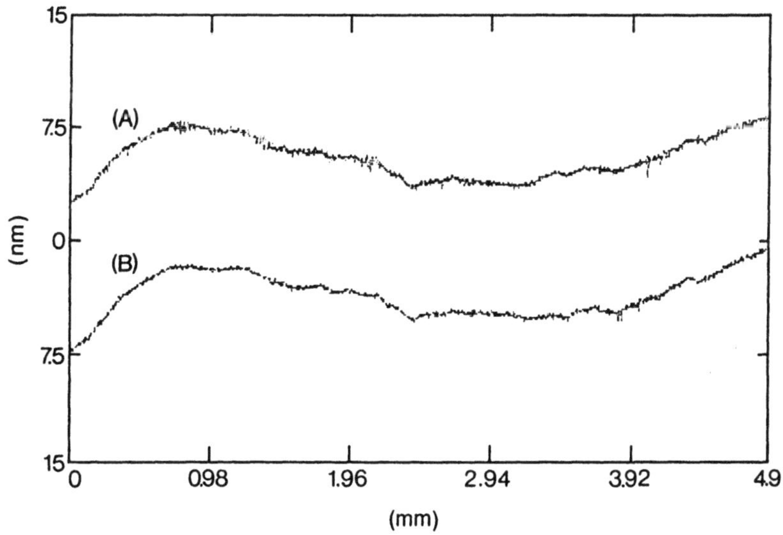

Fig.8. Successive Nanosurf 2 traces showing reproducibility of 0.25 nm rms over a 5 mm scan

Fig.9. X-ray interferometer. The silicon diffraction blades are shown at A

Micro-Indentation Assessment of Near Surface Material Properties and Sub-Surface Damage in Precision Machining

Robert S. Polvani and Chris Evans

Precision Engineering Division
National Bureau of Standards
Gaithersburg, MD 20899

Abstract: A technique is being developed for nondestructively determining the mechanical properties of material within surface layers. With it the effects of processing on the properties of coated or surface treated materials can be evaluated. The approach is based on hardness testing, but with two important innovations. First is to use instrumented, dynamic microindentation. This instrument performs dynamic loadings and characterizes sample indentations by continuous measurements of the applied load and penetration depth. Second is adaption of Sneddon's analytical solutions for indentation with pointed tools. The approach will be described in detail, its application to detection of the subsurface damage caused by single point diamond turning will be outlined, and its use to measure apparent elastic moduli reported.

Introduction: Ranging from advanced materials to surface treated and fabricated materials, the applications needing a nondestructive and in-situ means of mechanical testing are broadly varied. Composites, ion implanted surfaces and now microlaminates, use refined and complex microstructure to achieve extraordinary mechanical performance. Advanced manufacturing techniques, such as single point diamond turning and high precision grinding, are being developed to achieve workpieces with surface finishes of < 10 nm R_a and a minimally effected subsurface. With all of these, mechanical properties from micrometer or smaller sized regions or features are the key factor. Beyond the scale of a region or feature can be the complication of a singular sensitivity to processing

Official contribution of the National Bureau of Standards, and not subject to copyright in the United States.

conditions. An example is the tendency of single point diamond turned copper to recrystallize either during turning or shortly afterwards [1].

A new and nondestructive approach to measuring near surface mechanical properties is being developed. The approach is derived from conventional hardness testing, but with two important innovations. Indentations are described by continuous measurements of applied load and penetration depth, and indentations can be made over a wide range of loading times, hours to milliseconds. Hardness testing, measuring the flow resistance of materials to pointed tool indentation, is a convenient and widely accepted means of performing mechanical properties testing. Implicit in the extensive use of hardness testing is a need to obtain properties either from surface layers, within very small areas of a sample, or from very small quantities of sample stock. The belief that hardness numbers are measurements relatable to the yield or ultimate strengths of a sample serves as an additional incentive. Shortcomings of hardness testing are values which are, actually, relative rankings of one material with respect to others, and depend on test procedure. An exception is copper. For work hardened copper, an analytical basis for an equivalence of the two exists and this has been empirically demonstrated by Tabor [2]. Although empirical correlations between microhardness and bulk mechanical properties are available for many materials, a direct equivalence has not been established [3-4].

Originally, two goals underlied this technique: to develop an instrument, and to make more effective use of conventional hardness testing practice [5]. Hardness-depth profiling is one outgrowth of this perspective. We use profiling to determine the near surface alterations caused by precision machining processes. A second is using indentations as a basis for obtaining conventional mechanical properties, Young's Modulus and where possible yield and ultimate strengths. Both goals have required appropriately utilizing the theoretical works of Hertz, Love, Lawn and especially Tabor and Sneddon [6-11]. Now our emphasis is to determine and understand the microstructural effects of precision machining processes.

__Experimental__: Development and use of a load-displacement measuring indentation instrument for measuring mechanical properties is not unique to this effort [12-16]. The tester described here, see Figure 1, has several features that are different from the others. First, it is able to operate over a range of loading rates covering five decades of time. The shape of the loading waveform can also be selected over a wide range. Waveform programming permits maintaining peak load for times long enough to gauge creep behavior. Second, deformation energy can be measured and separated into elastic, plastic, and anelastic omponents. Lastly, the method of applying a load differs from conventional microhardness testing practice. The other instruments, start with the indenter well above the sample surface. It then continuously moves into, and finally away from the sample. Pretravel of the indenter introduces two problems: (1) indenters may impact the surface; and (2) surface location must be defined under dynamic conditions. This instrument also starts with the indenter well above the surface; but it then incrementally and very slowly advances the indenter to the surface where all motion stops. Load application follows as a separate event. Further, withdrawal is made in stages. The indentor is not entirely unloaded, one percent or less of maximum loading is retained. This feature maintains contact between the indenter and impression surface, while the amount of post indentation anelastic recovery is gauged.

The tester has four major functional subassemblies; these are an HP-85 computer/data logger*, electro-mechanical drive

---

* Specific commercial items are identified only to provide a complete description of the test system and does not imply an endorsement of the products.

assembly, main test frame, and electrical instrumentation package. The computer controls a test, generating necessary loading waveforms, controlling the different electrical subassemblies, and processing data. Data processing includes statistical analyses, plotting, and tabulating final results.

The shape of the loading waveform is a programmable electrical signal that can come from one of two sources. For testing at nearly static loading rates, seconds to hours, the computer does the task digitally. For dynamic loadings, seconds to fractions of milliseconds, a Krohn-Hite Model 5200* electrical function generator is used. An electro-magnetic driver (vibration shaker) is the means of developing driving forces. The Bruel and Kjaer Model 4810* shaker is a low force, high frequency unit; it can push or pull with a peak force of 10 N from an RMS input current of 1.8 amperes. Its first axial resonance is at 18 khz. The necessary currents for the driver are supplied by a power amplifier designed here. The amplifier has three subsections: (1) a biasing power supply used to initially position the indenter, (2) a dual input instrumentation amplifier based on an AD521 chip, and (3) a power amplifier supplying 1 ampere output current from a A791 chip.

The main frame is a modified stamping press die set. Samples are mounted on an X-Y translation stage which is position keyed to a metallographic microscope; the combination of the two gives an ability to target and then indent a feature with a working precision of $\mp$ 1μm. Specimens are directly mounted onto the translation stage table with a cellulose cement or low temperature mounting wax. Samples as large as 10 cm in diameter and to 2 cm high can be tested. The electrical instrumentation package permits observation of test events through indications of both the applied force and indenter displacement. Load is measured two ways. At static loading rates, the current through the electromagnetic drive coil is used. Measurement of coil current is a well established means of electrically determining forces. At dynamic rates, a piezoelectric load cell is used. The piezoelectric force cell is a Kistler Model 912* feeding a 904* charge amplifier. Eddy-current type transducers are used to measure the displacement and are arranged in a full 4-arm bridge configuration. The transducers are the Kaman Scientific type KD-5100X*. Bridge resolution is at least 0.01 μm. Equally important is the width of the frequency band pass of the displacement transducers; this is direct current to 22 kHz at the -3 db points. Both the load and displacement transducers are directly coupled to the samples. This feature eliminates a need to correct for the compliance of the load frame. A Nicolet Model 2090A transient digital oscilloscope is used to record the test signals. Use of a digital type oscilloscope increases the detailing of the test event. Potentially, 2056 individual data points are available on each of the two channels for describing the test event.

Knoop and Vicker's Microhardnesses, and a 800 μm diameter steel ball are the three indenters routinely used. A 200 μm diameter diamond ball based on a modification of a Rockwell C is being developed. For determining surface layer properties ball indenters are used. The advantages are: (1) a tested volume which maximizes surface area and minimizes depth of penetration, (2) relatively large loading can be applied, and (3) availability of analytical and finite element solutions for ball indentation.

Figures 2a and b show results for a 800 μm diameter ball microindentation of an NBS Microhardness Standard Reference Material, copper with a grain size of about 1 μm. Fig. 2a shows the load and corresponding depth versus time curves. This case involved using a square wave loading of 16 seconds duration. Figure 2b is the matching load versus depth curve. This curve combines the results of loading, dwelling and unloading a sample. The line superimposed onto the unloading portion of the curve is an initial tangent. It is fitted by means of linear regression and it is used in the evaluation of Young's Modulus. Fig. 3 illustrates how the character of a material and relative mix of elastic and plastic energies affect these curves. Three different situations are considered. First is the differences between indentations in hard or glassy, and soft metallic materials. Second is the observation of differences in extent of elastic recovery; for example, lead would exhibit no recovery while a hardened steel exhibits extensive recovery. Lastly, areas under the curves, permit partitioning the deformation energy into components and their quantitative evaluation.

Equations 1, 2 and 3 were derived from the usual expressions for Vickers, Knoop and Brinell microhardness numbers, but modified to give hardness in units of MPa:

$$HV = (304.72)P/h^2 \qquad [1]$$

$$HK = (149.31)P/h^2 \qquad [2]$$

$$HB = (4.014)P/h \qquad [3]$$

P is the applied force in Newtons, and h is the maximum indentation depth in micrometers. Because an evaluation is based on indentation depth under load rather than the diagonal length of the residual impression, the constants in these expressions are modified from the usual geometric factors.

<u>Results</u>: Hardness-Depth Profiling is an outgrowth of trying to make more effective use of hardness testing, and needing to determine the subsurface damage resulting from precision machining ductile materials. For brittle materials, the ability to determine local values of Young's Modulus from an effected surface layer better serves that purpose. Hardness-Depth Profiling uses the instrument's ability to perform and then report a continuous hardness curve. Consider an example; single point diamond turning has a number of advantages. One is a precisely machined surface, and another is the potential

for a workpiece with a nearly damage free surface. Compared with conventional machine tool cutting, the damage is nonexistent. The damage can be thought of as having two characteristics, a depth and an intensity. With ductile materials, the physical form of the damage is likely to be typical of work hardened materials. Damage in brittle materials such as glass, germanium and gallium arsenide takes a different form. These suffer from a tendency to microfracture or craze.

A series of fully annealed OFHC copper and 6061 aluminum parts were single point diamond turned. The samples were longitudinally split cylinders 25 mm in diameter by 12.5 mm long. The longitudinal face was was only "rough" machined. The radial face was treated as an optical surface and received a series of progressively finer "finishing" cuts. The final cuts were < 0.5 μm deep, made at 100 rev/min. and the slowest feed rate achievable. A third face was made by chemically polishing the cylindrical surface. The hardness profiles for the three faces are strikingly different; see Figure 4a. OHFC and 6061 exhibited similar performance. The near surface region of the "finished face for both materials has a much higher hardness. This elevation can be attributed to workhardening from the plasticity inherent in the shear process and possibly local smearing of the workpiece surface due to less than ideal cutting conditions. For both materials the penetration depth is estimated to be $\approx$ 1 μm. This result was independently confirmed with metallographic sections. Taper sections through the OFHC copper parts showed the bulk comprised large, about 50 μm, grains; the surface showed substantial plastic work and crystal refinement. The worked layer extended 1 to 1.5 μm into the material.

Young's Modulus Measurements: For brittle materials, the ability to determine values of Young's Modulus from within an effected surface layer can be a means of gauging subsurface damage. Experimentally, we find a depression of the modulus with increasing damage. The effects of grinding for three conditions of an optical glass shown in Figure 5. Abrasion removes material by shearing the surface and crazing the subsurface region. For this reason the coarse ground material load-depth curve is elongated with respect to the HF dip polished control. Further, the slope of the unloading portion of the curve is about a half. Using methods to be described apparent elastic moduli can be obtained from unloading portion of the curve. The simplest application of the apparent moduli values is a comparison of values from cut samples with either handbook values or preferably reference "bulk" material. Implicit in this strategy are two assumptions: brittle materials craze; and modulus values are a weighted responses from a mix of the bulk and crazed regions. The method of extracting values of Young's Modulus from sample indentations stems from Sneddon's attempts to analytically solve Boussinesq's problem, penetration of a pointed tools into an elastic-plastic material [10, 11]. The solution was first used by Tabor and more recently by Loubet to obtain values of Young's Modulus with Vicker's indenters [11,16]. In the 1948 paper, Sneddon analytically solved the problem of elastic-plastic indentation for a "cone point" indenter, and in the 1965 paper for several other axisymmetric shapes, notably a flat circular punch and ball.

Vickers Indentation: Loubet's method for extracting moduli depends on having load-depth curves available, and on making the following three assumptions. Penetration results from combined but separable elastic and plastic deformation components. Second, the initial unloading behavior is entirely elastic, and so the slope of the unloading portion of the curve is a direct measure of Young's Modulus. Third, Vicker's and "circular flat punch" indenters" with the same cross sectional areas are equivalent. The relation between applied load, P, and Young's Modulus of a material, E, is given by:

$$P = 2\delta hE/(1-\nu^2) \qquad [4]$$

where δ is the diameter of an equivalent "flat circular punch", maximum plastic penetration depth is h, and ν is

Poisson's ratio for the sample. A standard value of 0.33 is typically used for ν unless special accuracy is needed.

$$dP/dh = 2\delta E/(1-\nu^2) \qquad [5]$$

For a Vickers indenter the equivalent diameter expressed in terms of depth is given by the expression:

$$\delta^2 = 98h^2/\pi \qquad [6]$$

and so:

$$E = 1.14dP/7hdh \qquad [7]$$

Experimentally, dP/dh is the initial tangent of the unloading curve, and will typically include the only upper half. Depth, h, is the maximum plastic deformation depth, and not the unrecovered or total depth of penetration. This value is given by the intercept of the initial tangent to the unloading curve and the depth axis of the load-depth curve. See Figure 2b for an example.

MATERIAL	ELASTIC MODULUS (GPa)		
	Handbook	Vickers	800 μm Ball
Lead	<21	25	25
Aluminum	65	74	68
Copper			
poly-xtal	130	140	130
<100> xtal	68	74	x
<110> xtal	120	120	x
<111> xtal	220	120*	x
Ge	126	x	120
Glass	60	109	73
Steel	204	190	219
Nickel	207	220	x

Table 1. A comparison of the Young's Modulus using Handbook and the result of indentation determinations. * Subsequent examination with SEM channelling patterns revealed reversion to polycrystalline material.

Ball Indentation: Sneddon's solution can be used directly for the ball indenter geometry with no additional assumptions being necessary. The relation between applied force, P, and unrecovered or maximum impression diameter, δ is given by:

$$P = E((\delta^2+R^2)\ln R_d-\delta R)/2(1-\nu^2) \qquad [8]$$

R is the indenter radius, and R_d equals $(R+\delta)/(R-\delta)$. Differentiating with respect to δ gives the result:

$$dP/d\delta = E\{\delta \ln R_d - R/2 + R((R^2+\delta^2)/(R^2-\delta^2))\}/(1-\nu^2) \quad [9]$$

The exact expression for maximum penetration depth is given by:

$$h_{max} = \delta \ln R_d/2 \quad [11]$$

Rather than solving Equation 11 for δ in terms of h_{max} there is a much simpler way. The relation is approximated with a Taylor series and used for the very small values of h_{max}/R which are typically measured. By doing this values of δ are given by the relation:

$$\delta = \sqrt{Rh_{max}} \quad [12]$$

Young's Modulus from a ball indentation when Poisson's ratio is equal to 0.33 is given by:

$$E = 1.78(dP/dh_{max})(\delta/R)/[\delta \ln R_d - R/2 + R(R^2+\delta^2)/(R^2-\delta^2)] \quad [13]$$

Table 1. compares preliminary results from indentation measurements of Young's Modulus for a variety of materials against Handbook values. Each indentation value is an average of at least five (5) separate determinations. These results are offered to illustrate the potential utility of the technique.

Yield and Ultimate Strengths: Yield and ultimate strengths for surface effected materials should also be obtainable from indentation testing. Consider three more of Tabor's results. First was demonstrating an equivalence between hardness values and tensile test stress-strain curves for ball indented copper. Second, was verifying a theory of Ishlinsky [17]. For a ball indenter penetrating an ideally elastic-plastic material, yield strength and hardness are simply related:

$$\text{Yield Stress} = P_m/2.8 \quad [14]$$

Mean flow pressure, Pm, is the applied load divided by the projected impression area. Work hardened metals are good approximations of the ideally elastic-plastic metal. For several hardened metals Tabor showed the equivalence. See Table 2.

METAL	Y_o (MPa)	P_m (MPa)	P_m/Y_o
Lead	21	60	2.9
Aluminum	121	338	2.8
Copper	304	863	2.8
Steel	638	1,864	2.9

Table 2. Tabor's comparison of indentation and tensile yield strengths.

Conclusions: Using the load-depth curve obtainable with an instrumented microindenter instrument the micromechanical properties of surface coated, treated and especially surface effected materials can be determined. For ductile materials, work hardening effects serves as the basis for detecting the local alternations. For brittle materials, the tendency to microfracture or craze and a resulting depression of the local Young's Modulus serves as a basis for quantifiying alterations. Although any of the conventional indenters can be used a small ball indenter proved the most useful.

Acknowledgements: Thanks to Dr. Donald G. Polvani of the Westinghouse Corp., Oceanic Division for his help with the mathematics necessary for the application of Sneddon's solution for a ball indenter.

References:

1. H. H. Hurt: A Metallurgical Investigation of the Diamond Turned Optical Surface, PhD Thesis, U. of Southern California, CA, 1984.

2. D. Tabor: The Hardness and Strength of Metals, Proc. Roy. Soc. [A], Vol. 192, 1948, p. 1-18.

3. J. H. Westbrook: Flow in Rock-Salt Structures, Report #58-RL-2033, G.E. Research Laboratory (August 1958).

4. D.J. Abson, O. Kosik, J.L. Uvira and J.J.Jonas: Correlations Between Room Temperature Hardness and Flow Stress in Hot-Worked Metals, from The Science of Hardness

Testing and Its Research Applications (Edited by J.H. Westbrook and H.Conrad), ASM, Metals Park, Ohio, 1973, p. 91-98.

5. R. S. Polvani, A. W. Ruff and E. R. Whitenton: A Dynamic Microindentation Apparatus For Materials Characterization, to appear in the ASTM Journal of Testing and Evaluation January 1988.

6. H. Hertz: Gesammelte Werke, Vol 1, Leipzig, 1895.

7. S. Timoshenko and J. N. Goodier: Theory of Elasticity, 2nd Edition, Engineering Societies Monograph, Mc Graw-Hill Book Co. 1951.

8. A. E. H. Love: Quart. J. Math. Vol 10 , 1939, p 161.

9. B. R. Lawn and V. R. Howes: Elastic Recovery at Hardness Indentations, J. Mat. Science, Vol 16, 1981, p. 2745.

10. I. N. Sneddon: Boussinesq's Problem for a Rigid Cone, Proc. Cambr. Phil. Soc. Vol. 44, 1948, p. 492.

11. I. N. Sneddon:The Relation Between Load and Penetration in The Axisymmetric Boussinesq Problem for a Punch of Arbitrary Profile, Int. J. Engng Sci., Vol. 3, 1965, pp.47-57.

12. F. Frolich, P. Grau and W. Gullmann: Performance and Analysis of Recording Microhardness Test, Phys. stat. solidi a 2, 1977, p. 79.

13. W. H. Robinson and S. D. Truman: Stress-Strain Curve for Aluminum from a Continuous Indentation Test, J. Mat. Science, Vol 12, 1977, p. 1961.

14. J. B. Pethica: Microhardness Tests with Penetration Depths Less Than Ion Implanted Layer Thickness in Ion

Implantation into Metals (V. Ashworth et al. editors), Pergamon Press, 1982, p. 157.

15. R. W. Armstrong and R. H. Robinson: Combined Elastic and Plastic Deformation Behavior from a Continuous Indentation Hardness Test, New Zealand J. Science, Vol. 17, 1974, p. 429.

16. J.L. Loubet, J. M. Georges and G. Meille: Vickers Indentation Curves of Elastoplastic Materials, in Microindentation Techniques in Materials Science and Engineering, ASTM STP 889, P. R. Blau and B. R. Lawn Eds.,American Society for Testing and Materials, Philadelphia, 1986, pp.72-89.

17. A. J. Ishlinsky, Journal of Applied Mechanics (U.S.S.R.) Volume 8 pp.233

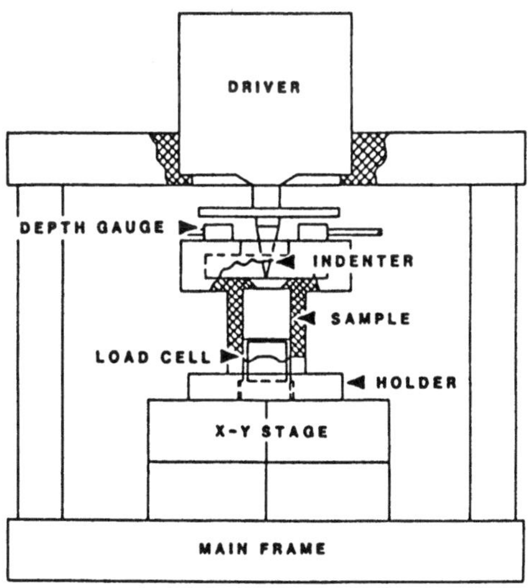

Figure 1. A schematic of the main test frame showing the placement of the sample and load and displacements gauges.

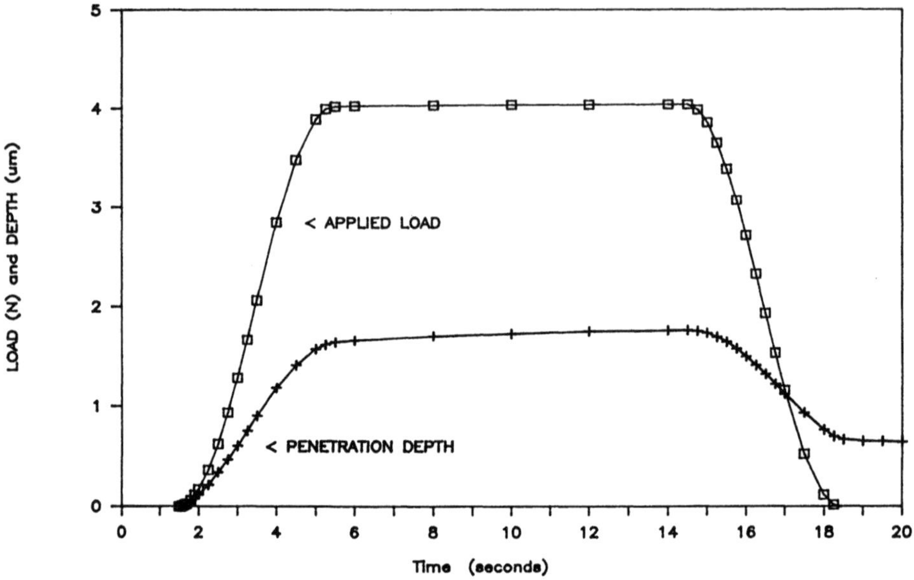

(a)

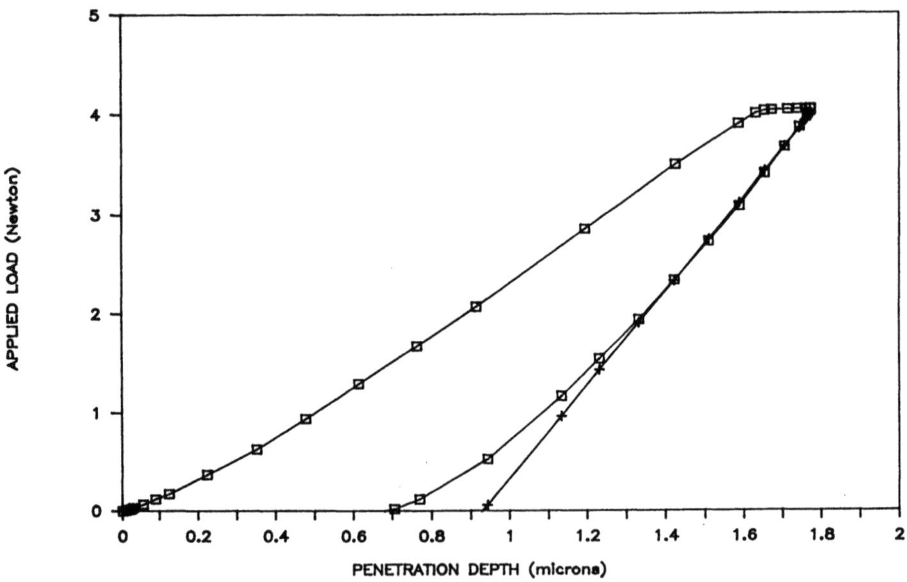

(b)

Figure 2. Ball microindentation of copper, (a) Applied load and penetration depth versus time, and (b) load-depth curve.

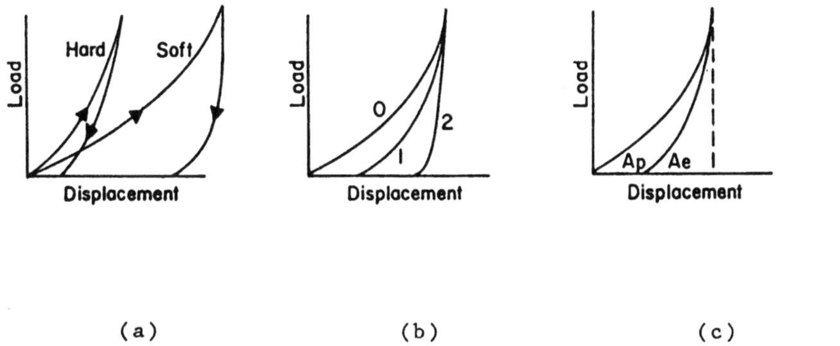

Figure 3. Schematic load-depth curves, (a) Hard versus soft materials. (b) largely elastic recovery upon unloading, curve #1, versus a mainly plastic recovery, curve #2, and (c) energy partitioning. A_e is the recoverable or elastic deformation energy and A_p is the plastic energy

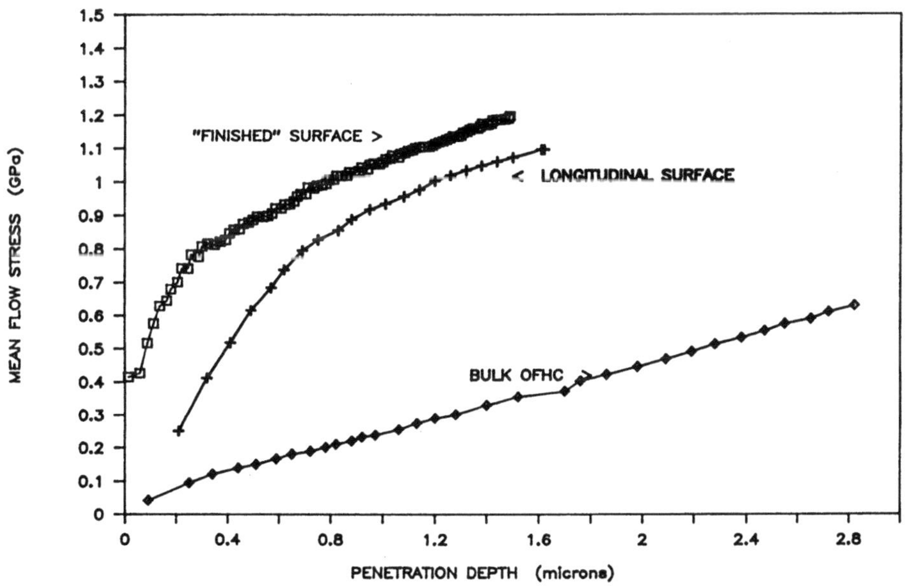

Figure 4. Flow Pressure-Depth Profile curves for single point diamond turned OFHC copper.

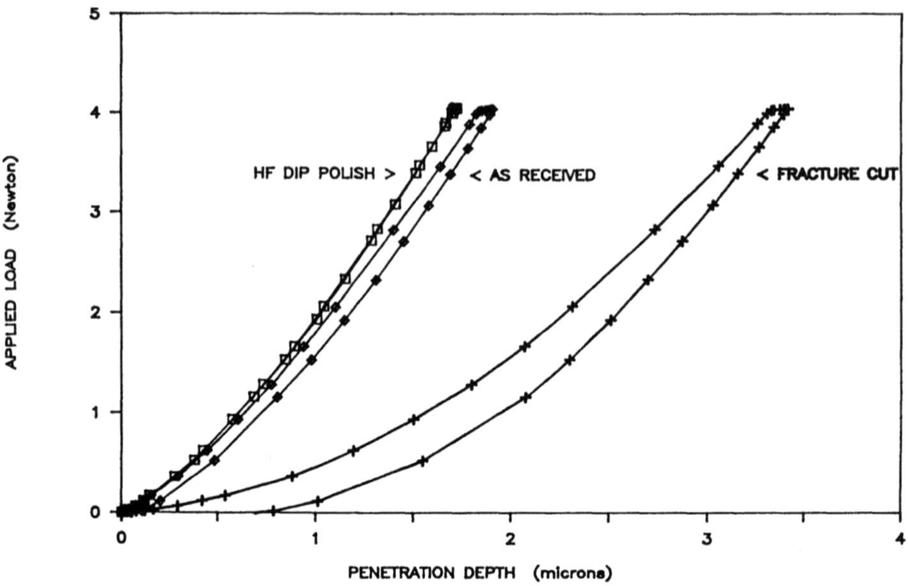

Figure 5. Load-Depth curves for three conditions of an optical glass: fracture cut, as finish ground and HF dip polished.

Measurement of Ultraprecision Components Using Non-Contact Interferometry Based Instrumentation

Katherine Creath and James C. Wyant

WYKO Corporation
1955 East Sixth Street
Tucson, Arizona 85719

Summary

WYKO Corporation produces interferometer systems for a number of metrology applications. Systems to be discussed include: TOPO, a microscope system for profiling surfaces such as mirrors, hard disks, magnetic tape, and paint; SIRIS, used to test optical components ranging in size from a millimeter to meters; and MAX, a projected fringe contouring system. All products use phase-measurement interferometry principles to obtain data. This paper gives the general operational principles of these systems and describes some of their applications and results.

Introduction

Optical metrology has been used for testing optical components such as lenses and mirrors for surface figure measurements, for determining the beam quality of lenses or optical systems in transmission, and for determining surface statistics. In addition, optical metrology has been applied to a large array of applications such as magnetic storage media, optical storage media, machined parts, and material properties. In the area of magnetic storage, it has been used to measure the surface roughness of hard disks, floppy disks, and magnetic tape. It also has been used to determine the flatness of a hard disk platter and to determine the profiles of magnetic read/write heads. In optical storage, the roughness of the substrate can be measured as well as the platter flatness or the pit profiles. Another application for optical storage is in measuring the wavefront quality of the read/write head assemblies. On a much rougher scale, machined parts are measured to determine their shape and surface finish quality. In terms of material properties, optical metrology can be used to measure object deformations, which can be related to strains in an object. For optical components, the homogeneity of a material may be measured as well as its refractive index. These are just some of the possible applications for optical metrology. They range from very small scale measurements to very large scale measurements, from surface microstructure to contouring automobile or aircraft bodies, and from measuring very precise geometries in microlithography to testing tires.

When using interferometric techniques in optical metrolgy, the quantity measured is a wavefront reflected off a surface or transmitted through the object. Most techniques involve the interference of two beams, one being a known reference beam, and the other being an unknown test beam.

To interferometrically test an object, we need to be able to interpret the interference fringes. Fringe interpretation can be split into two major categories: those using fringe-following techniques, and those using phase-shifting techniques. In fringe following, a single interferogram is analyzed by tracing either the light or dark fringe centers. Once the fringe centers are found, the data are fitted using a set of polynomials to find the wavefront, or some form of interpolation is used to determine the wavefront under test. Phase-measuring interferometry techniques modulate one beam in an interferometer while the interference fringes are detected. From multiple frames of interference data with different relative phase shifts between the object and reference beams, the wavefront phase can be directly calculated from the measured fringe intensities.

All of the systems to be described in this paper rely upon phase-measurement techniques except for WISP, which uses fringe following techniques. Emphasis will be placed on applications and new capabilities.

Phase-Measurement Interferometry

For the interferometer systems described in this paper, phase-measuring techniques are used to determine the phase of the wavefront from the test arm of the interferometer relative to the phase of the wavefront from the reference beam [1]. In all cases, the interference fringes are detected using an area detector array containing from 100 x 100 detector elements up to 256 x 256 detector elements. The phase of the reference beam relative to the test beam is shifted by either stepping a piezoelectric transducer (PZT) pushing a mirror or by ramping the PZT pushing the mirror as the detector array is integrating. Figure 1 shows both the phase-stepping and the integrating-bucket techniques. The only difference between them is that the integrating-bucket technique averages over a range of phase differences. If data can be taken without waiting for a settling time between exposures, then the integrating-bucket method is preferred because it is faster.

The intensity of a detected interference pattern at a detector positioned at point x,y can be written as [2]

$$I_i(x,y) = \frac{1}{\Delta} \int_{\alpha_i - \Delta/2}^{\alpha_i + \Delta/2} I_0(x,y) \{ 1 + \gamma_0 \cos [\phi(x,y) + \alpha(t)]\} \, d\alpha(t) \ , \qquad (1)$$

where I_0 is the dc intensity, γ_0 is the fringe modulation, $\alpha(t)$ is the phase shift with respect to time, Δ is the amount of phase shift over which the intensity is integrated, and $\phi(x,y)$ is the phase to be determined. Δ equals the amount of phase shift difference from one frame to the next for the integrating-bucket technique, and equals 0 for the phase-stepping technique. Evaluating the integral in Eq. (1) yields

$$I_i(x,y) = I_o(x,y) \left\{ 1 + \gamma_o \,\mathrm{sinc}\left[\frac{\Delta}{2}\right] \cos\left[\phi(x,y) + \alpha_i\right] \right\} \tag{2}$$

as the detected intensity where $\mathrm{sinc}(\Delta/2) = \sin(\Delta/2)/(\Delta/2)$, and α_i is the central value for the phase shift for that frame. For the phase-stepping technique, $\mathrm{sinc}(\Delta/2)=1$.

Because there are three unknowns I_o, γ_o, and $\phi(x,y)$, three intensity measurements are required to determine the phase $\phi(x,y)$. Using three measurements, I_1, I_2, and I_3 where $\alpha_1=\pi/4$, $\alpha_2=3\pi/4$, and $\alpha_3=5\pi/4$, the phase is given by

$$\phi(x,y) = \tan^{-1}\left[\frac{I_3(x,y) - I_2(x,y)}{I_1(x,y) - I_2(x,y)}\right] . \tag{3}$$

Another common phase-measurement technique uses four intensity measurements to calculate the phase. With intensity measurements taken at relative phase shifts of $\alpha_1=-3\pi/4$, $\alpha_2=-\pi/4$, $\alpha_3=\pi/4$, and $\alpha_4=3\pi/4$, the phase is calculated from

$$\phi(x,y) = \tan^{-1}\left[\frac{I_4(x,y) - I_2(x,y)}{I_1(x,y) - I_3(x,y)}\right] . \tag{4}$$

A phase-measurement technique that is independent of the value of the phase shift as long as it is constant and linear was devised by Carré where the four intensity measurements are assumed to have relative phase shifts of -3α, $-\alpha$, α, and 3α [3]. The phase can be calculated using

$$\phi = \tan^{-1}\frac{\sqrt{\{3(I_2 - I_3) - (I_1 - I_4)\}\{(I_2 - I_3) + (I_1 - I_4)\}}}{(I_2 + I_3) - (I_1 + I_4)} . \tag{5}$$

Once the phase is calculated modulo 2π from the above techniques, the phase ambiguities must be removed to reconstruct the wavefront. This is usually done by comparing adjacent pixels and adding or subtracting multiples of 2π until the difference in phase between adjacent pixels is less than π in phase. The best phase-measurement algorithm to use depends upon the application, the speed of the measurement, and errors that may be induced by the interferometer system itself. A detailed description and comparison of phase-measuring algorithms is given by Creath [4].

The advantages of using phase-measuring techniques over fringe-following techniques are many. First, the data are sampled over a uniform grid rather than only at the fringe centers. Second, fixed pattern noise in the detector array will cancel when the phase measurement equations are used. Third, a large number of data points are available using a detector array, and the data can be taken quickly and calculated in a short time. Finally, phase-measurement techniques easily have a repeatability of one-hundredth of a fringe, and it is not too hard to achieve one-thousandth of a fringe; whereas fringe following is only good to a tenth of a fringe.

There are limitations to the phase-measuring techniques. The most important is the sampling interval. The techniques mentioned require that the optical phase between adjacent detector elements not change by more than π (or 180°) in order to reconstruct the wavefront. This is because the phase measurements are modulo 2π owing to the arctangent calculation. Thus, two detector elements are needed for each fringe. It should be pointed out that the slope of the wavefront is more important than the actual number of fringes, because the area where the fringes are the most closely spaced will determine the limits of the measurement. This barrier due to the Nyquist sampling limit can be sidestepped in some instances as long as certain assumptions can be made about the wavefront under test as pointed out by Greivenkamp [5].

The other major limitation is a problem with all interferometry techniques. This problem arises because different materials shift the phase of the reflected beams by different amounts. Because interferometry is a relative measurement technique, the height of a step spanning an area containing two different materials may not be measured accurately. It should be noted that the measurement will be repeatable even though it will not be accurate. A constant to add to one side of the step can be determined from the complex index of refraction as derived by Church [6], but the value of the refractive index needs to be measured for the sample material in the form in which it is to be used. If the test surface has a transparent thin-film coating on it, the inhomogeneities in the film thickness and index of refraction will distort the measurement of the test surface.

TOPO

The TOPO system is an optical profiler for determining surface profiles, surface roughness, and statistical properties of surfaces [7-8]. Figure 2 shows a schematic of this system. It is essentially a modified microscope. An interference objective is used to generate the interference fringes, and a detector is placed at the conjugate plane to the surface. A phase shift is generated by a PZT which moves the reference surface of the interference objective. Data are passed through an electronic interface to a computer which controls the PZT and calculates the surface profile. The source consists of a narrowband interference filter in front of a white light source. This makes it simple to change the reference wavelength of the measurement. The detector array is either a linear 1024-element array (TOPO-2D), or a 256x256-element array (TOPO-3D). Measurements can be made with magnifications ranging from 1.5X to 200X.

A surface profiler can be used for a number of applications. Among them are the characterization of optical surfaces, as well as the measurement of magnetic and optical storage media. Virtually any smooth surface can be measured and its statistical properties determined. Examples of TOPO applications are shown in Figures 3-5. Figure 3 shows a measurement of a compact disk surface taken with a 200X objective. The pits in the surface can be seen very well. Figure 4 shows a two-dimensional measurement of a grating with 0.6 μm pitch. The center of a diamond- turned surface is shown in Figure 5. Note that the tooling marks are highly visible. Besides the profile of the surface, the user may view slope, curvatures, power spectra, and autocovariance information as a few of the available options. Errors due to the reference surface can be subtracted out.

The phase-measuring techniques for TOPO are limited to measuring steps which change by no more than $\lambda/4$ between adjacent detector elements. To increase the measurement range of the profiler, multiple wavelength techniques may be used [9]. By taking measurements at two different wavelengths, the phase ambiguities at a step boundary may be removed. A two-wavelength measurement consists of taking two separate measurements of a surface using two different interference filters. These two measurements are combined to yield a measurement that is equivalent to one taken at the beat wavelength between the two measurement wavelengths (see Figure 6). This means that the surface is effectively measured at a longer equivalent wavelength given by

$$\lambda_{eq} = \frac{\lambda_1 \lambda_2}{|\lambda_1 - \lambda_2|} \tag{6}$$

where λ_1 and λ_2 are the measurement wavelengths. The effective wavelength of the measurement is easily variable by changing the wavelengths. Steps of up to 20 μm in height can be measured using TOPO-2D with multiple wavelengths. Figure 7 shows a 1.3 μm amplitude grating which has been measured using wavelengths of 612 nm and 651 nm. The rms repeatability of a measurement is one-thousandth of the effective measurement wavelength, not the actual measurement wavelength. Thus the signal-to-noise ratio of the measurement is going to be determined by the equivalent wavelength used. The signal-to-noise ratio of the measurement can be increased by using the two-wavelength measurement to correct the phase ambiguities in a single wavelength measurement. Figure 8 shows a corrected measurement of the grating in Figure 7 at 651 nm. Note that the repeatability of the two-wavelength measurement is 8 nm rms, whereas the repeatability of the corrected two-wavelength measurement is 0.35 nm rms yielding an improvement of 23 in signal-to-noise ratio. This technique can be further extended by using more than two wavelengths to remove the phase ambiguities.

SIRIS and WYKO 6000

For general optical component testing, the SIRIS interferometer uses a Twyman-Green configuration as shown in Figure 9. The source can either be a HeNe laser operating at 633 nm or a laser diode source operating at 830 nm. The beam size is 12 mm and is geared for testing small optics. Larger optics can be measured using a beam expander. Spherical surfaces and lenses can be tested by using the configuration shown in Figure 9 in the test arm of the interferometer. A computer is used to control the phase shifter, take data from the detector array, control the intensity, focus on the test piece, and change the zoom ratio.

The system is flexible enough for measuring very small optics as well as for testing an entire hard disk substrate to determine flatness. Other applications include the measurement of corner cubes, molded aspheric lenses in transmission, and diamond-turned surfaces. Figure 10 shows a measurement of a corner cube with angular errors as shown in Table 1. An example of a diamond turned mirror is shown in Figure 11. Besides determining surface figure, the user may fit Zernike polynomials to the surface, calculate slopes, point spread function, and modulation transfer function as well as analyze any sub-portion of the data and compare it to another sub-portion of the data. Interference fringes are viewed live on a computer monitor, and the detector image may be masked to any desired area. Absolute measurements of spherical surfaces can be made using an algorithm first introduced by Jenson [10-11], and aberrations due to the interferometer may be subtracted from the measurements.

Recently, WYKO introduced a phase-measuring Fizeau interferometer with a 150 mm output beam. This instrument is useful for measuring surface contours of flat, spherical. and some aspheric surfaces.

Table 1. Angular Errors of the Corner Cube Shown in Figure 11.

Subareas	Dihedral Angle Error (degrees)	(seconds)	Subarea rms (waves)	
(1,2)(4.5)	0.00149	5.4	1: 0.007	4: 0.008
(3,4)(6,1)	0.00131	4.7	2: 0.005	5: 0.004
(5,6)(2,3)	0.00163	5.9	3: 0.007	6: 0.008

Wavelength: 632.8 nm; Prism refractive index: 1.500; Maximum detector view (inches): 0.500

MAX

The MAX system is designed for contouring objects such as machined and molded parts using fringe projection techniques [12-14]. Interference fringes are created by a fringe projector (see Figure 12) and projected onto the object. These fringes are then viewed with the camera from a different direction. Using the phase-shifting techniques mentioned earlier, the object surface profile is calculated from the interference fringe data. Virtually any surface can be contoured as long as the camera sees the fringes. Areas in shadows are not measurable. The fringe spacing as viewed by the camera determines the contour spacing for one fringe. This can be determined by measuring a known step to calibrate the system. The sensitivity of the measurement depends upon the contour spacing, and the measurement repeatability is roughly $\lambda/50$ to $\lambda/100$ rms. Examples of this technique are shown in Figures 13-14 where a live hand and a can have been contoured. Other applications include the measurement of feet for orthotics, the measurement of machined parts, and the measurement of car windshields or airplane parts.

Summary of Applications for WYKO Products

In summary, WYKO produces optical metrology instrumentation for a large variety of applications. All WYKO instruments use phase-shifting techniques to acquire data. For surface roughness measurements, TOPO-2D and TOPO-3D provide special surface analysis software. Surface figure of optical components is provided by SIRIS in the visible and near infrared. Finally, machined and molded parts can be contoured using fringe projection techniques by MAX.

Future Directions

As the optical metrology industry becomes used more widely in different fields, new instruments will be created to fulfill the measurement criteria. Most recent changes in the industry are related to the software that analyzes the data. Specialized software packages are necessary for many applications, and new measurement techniques using more sophisticated software are constantly being developed. Besides software, other areas of interest include the use of larger detector arrays, an increase in the interferometric measuring range both laterally and vertically, and an improve-ment in the accuracies of measurements using lens design software. To create more accurate and precise measurements, standard surfaces must be generated and measured to keep up with improved measurement capabilities.

References

[1] K. Creath, "Phase-measurement interferometry," to be published in *Progress in Optics, Vol. 26*, E. Wolf, ed., North Holland (1988).
[2] J. E. Greivenkamp, "Generalized data reduction for holographic interferometry," Opt. Eng. **23**, 350 (1984).

[3] P. Carré, "Installation et utilization du comparateur photoelectrique et interferentiel du Bureau International des Poids et Mesures," Metrologia **2**, 13-23 (1966).
[4] K. Creath, "A comparison of phase-measurement algorithms," Proc, SPIE **680**, 19-28 (1986).
[5] J. E. Greivenkamp, "Sub-Nyquist interferometry," to be published in Appl. Opt. (1987).
[6] E. L. Church, "Structure effects in optical surface metrology," Proc. SPIE **680**, 124-130 (1986).
[7] J. C. Wyant, C. Koliopoulos, B. Bhushan, and O. E. George,"An optical profilometer for surface characterization of magnetic media," ASLE Trans. **108**, 1-8 (1984).
[8] B. Bhushan, J. C. Wyant, and C. L. Koliopoulos, "Measurement of surface topography of magnetic tapes by Mirau interferometry," Appl. Opt. **24**, 1489-1497 (1985).
[9] K. Creath, "Step height measurement using two-wavelength phase-shifting interferometry," Appl. Opt.**26**, 2810-2818 (1987).
[10] A. E. Jensen, "Absolute calibration method for laser Twyman-Green wavefront testing interferometers," (abstract only), J. Opt. Soc. Am. **63**, 1313 (1973).
[11] J. H. Bruning, "Fringe scanning interferometers," in *Optical Shop Testing*, D. Malacara, ed., Wiley and Sons, New York (1978).
[12] G. Indebetouw, "Profile measurement using projection of running fringes," Appl. Opt. **17**, 2930 (1978).
[13] V. Srinivasan, H. C. Liu, and M. Halioua, "Automated phase-measuring profilometry: a phase mapping approach," Appl. Opt. **24**, 185 (1985).
[14] B. W. Bell, "Digital heterodyne topography," Dissertation, Univ. of Arizona (1985).

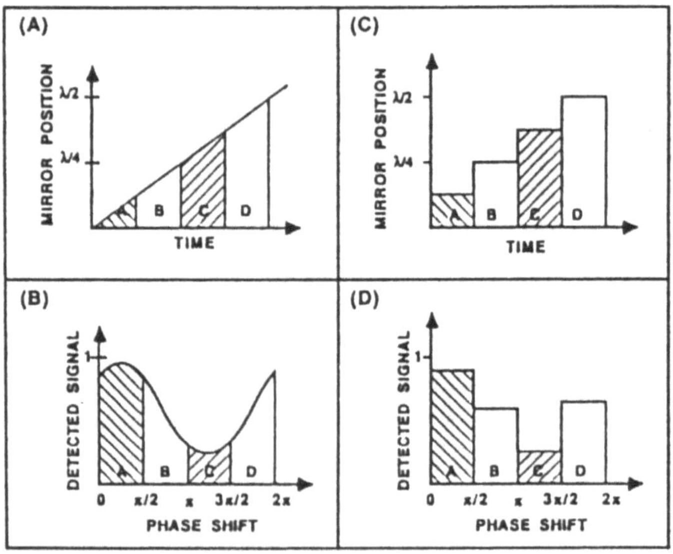

Figure 1. (A) Mirror position versus time, and (B) detected signal versus phase shift for the integrating bucket technique. (C) Mirror position versus time, and (D) detected signal versus phase shift for the phase stepping technique.

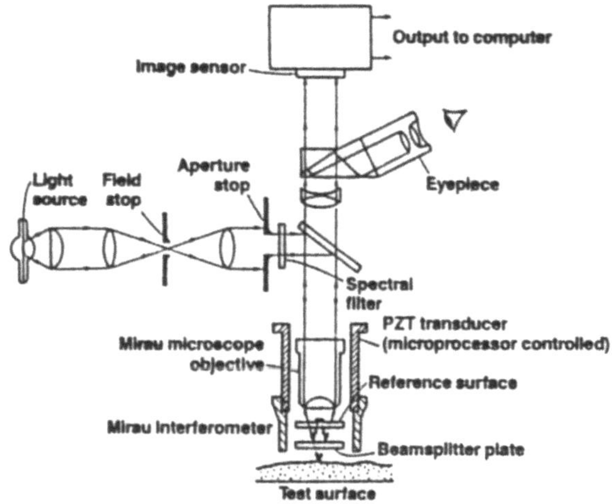

Figure 2. Optical layout of the TOPO optical profiler.

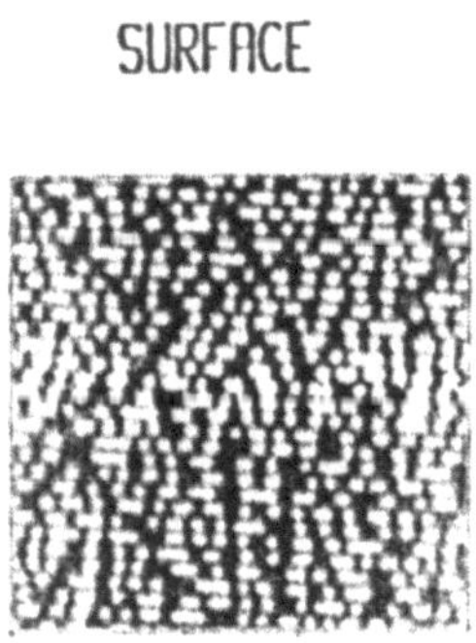

Figure 3. Surface plot of a compact disk taken with TOPO-3D using a 200X objective.

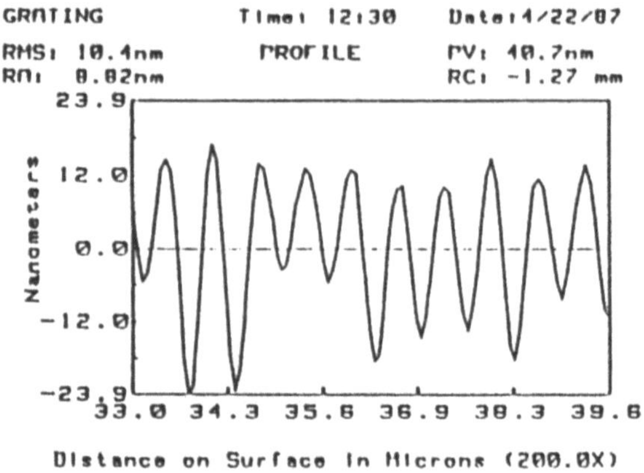

Figure 4. **Two-dimensional profile of a grating with a 0.6 μm period taken using TOPO-2D with a 200X objective.**

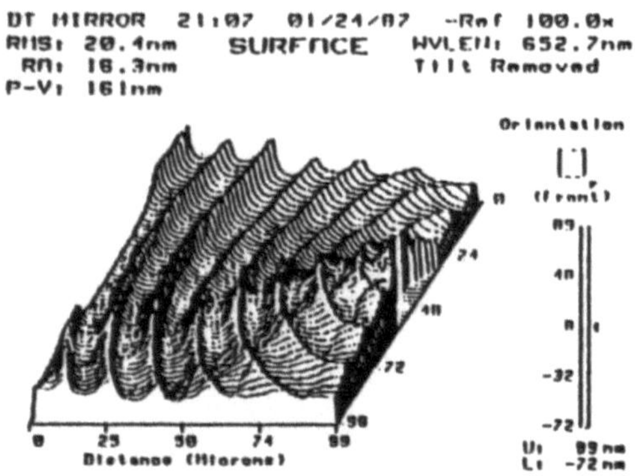

Figure 5. **Isometric contour plot of a diamond-turned mirror using TOPO-3D with a 100X objective.**

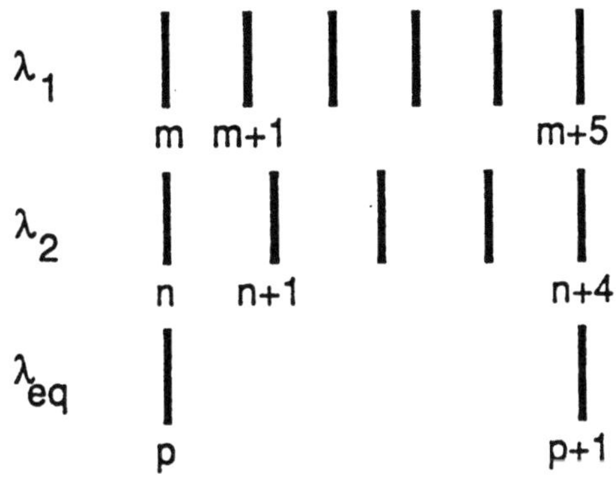

Figure 6. Creation of an equivalent wavelength at the beat frequency between two shorter wavelengths.

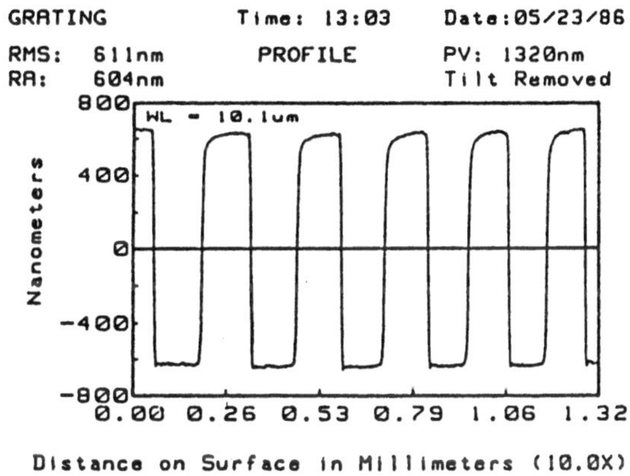

(A)

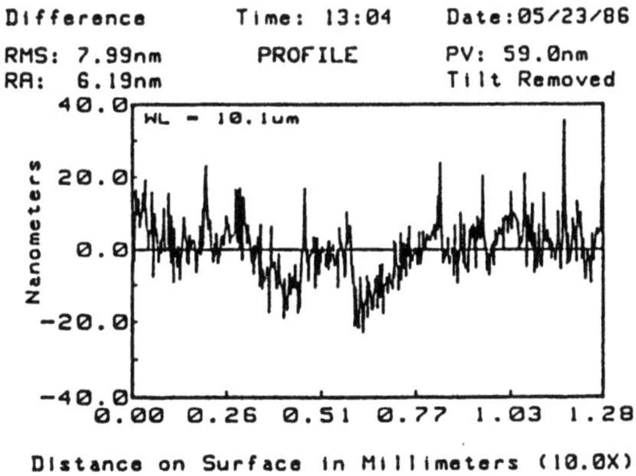

(B)

Figure 7. (A) Plot of a 1.3 μm amplitude grating using wavelengths of 612 nm and 651 nm to create an equivalent wavelength measurement at 10.1 μm. (B) Repeatability of the measurement shown in (A).

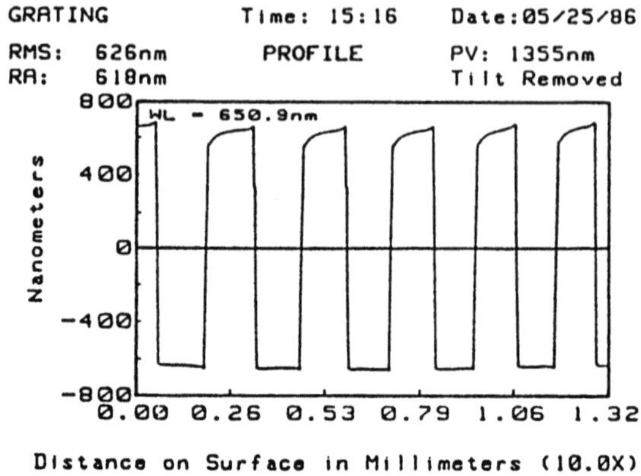

(A)

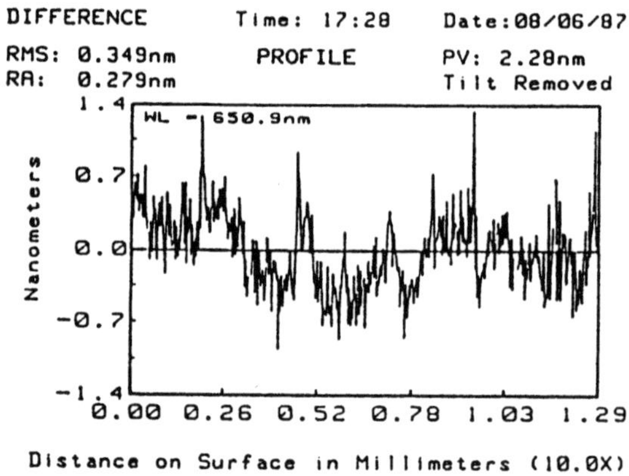

(B)

Figure 8. (A) Corrected plot at 651 nm for the grating in Figure 8 using wavelengths of 612 nm and 651 nm. (B) Repeatability of the measurement shown in (A).

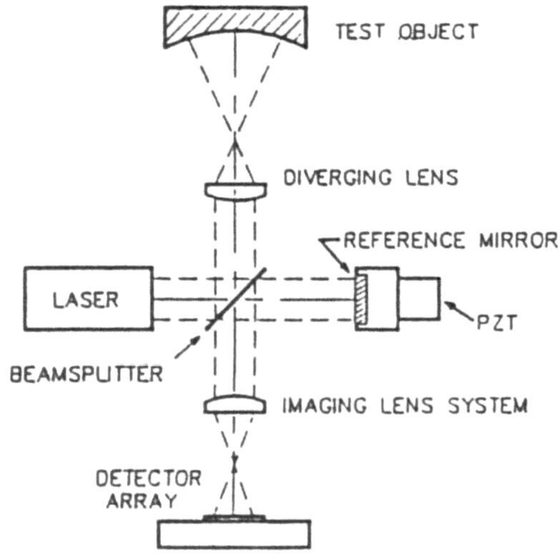

Figure 9. Simplified layout of the SIRIS interferometer.

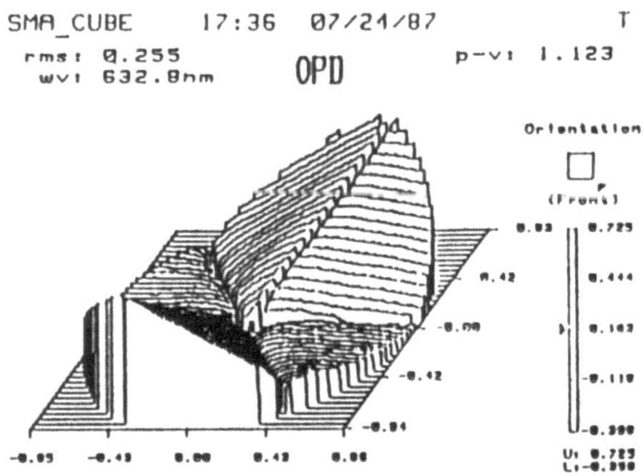

Figure 10. Isometric contour plot of a small corner cube with angular errors taken using SIRIS.

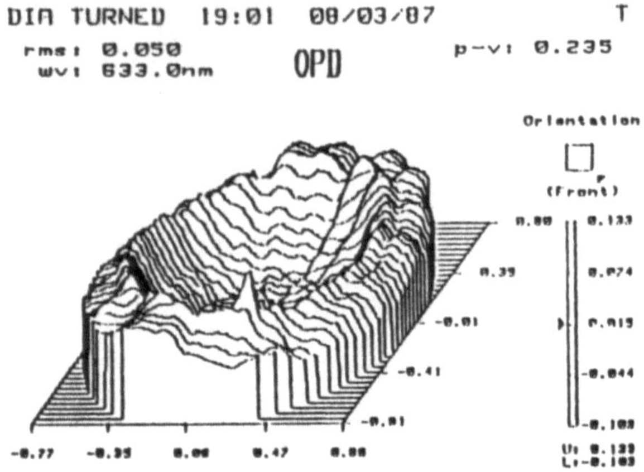

Figure 11. Isometric contour plot of a diamond-turned mirror taken with SIRIS.

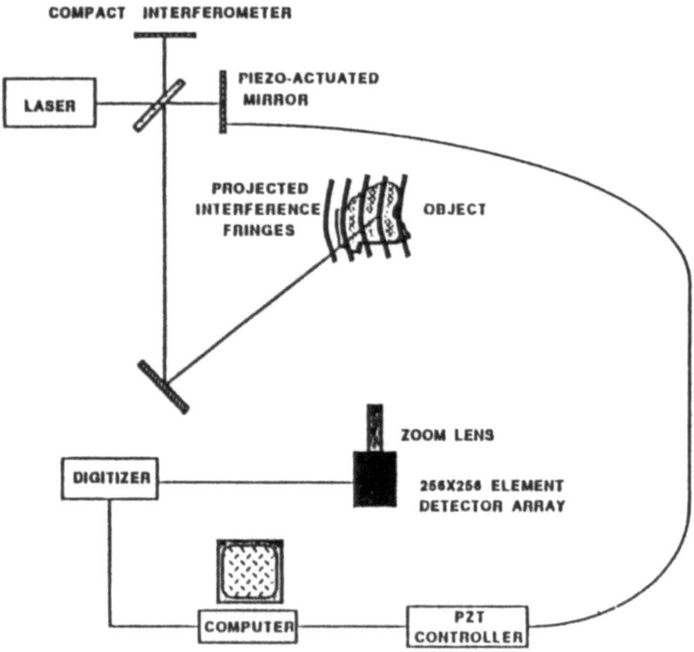

Figure 12. Schematic of layout to contour an object using the MAX system.

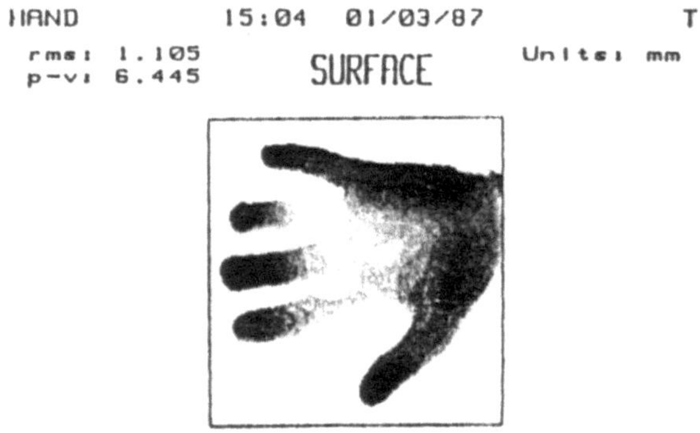

(A)

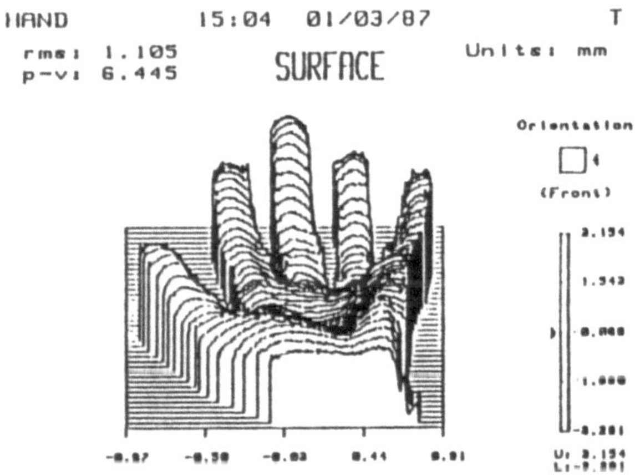

(B)

Figure 13. **(A) Contour plot and (B) isometric contour plot of a live human hand taken using MAX.**

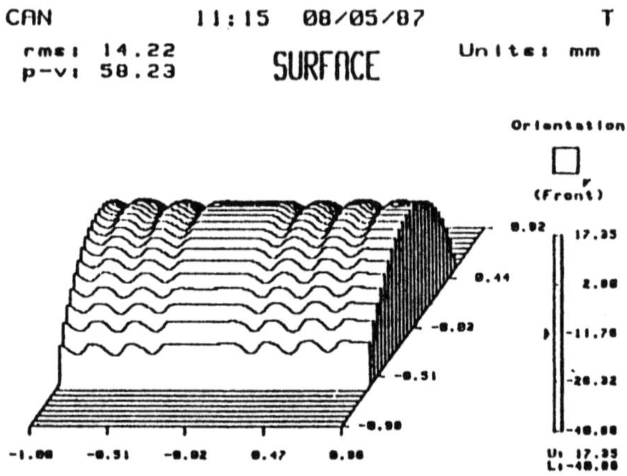

Figure 14. Isometric contour plot of a soup can taken using MAX.

In-Process Surface Finish Measurement of High Quality Components

J.H. RAKELS

Centre for Micro-Engineering and Metrology
Department of Engineering
University of Warwick
Coventry, CV4 7AL, U.K.

Summary

To avoid damage during the measurement of the surface finish of high quality components, non-contacting methods are required. The most promising techniques are optical ones, which is quite appropriate since most high quality components are used as optical elements. Optical techniques can be divided into optical profilometry, which includes interferometric and optical stylus methods, and optical parametric techniques. The latter have the advantage that they can be applied in-process, and therefore offer the possibility of being incorporated into the machine-tool control system.

Most optical parametric techniques are based upon diffraction or light scattering. The various systems differ essentially in the number of photo-detectors present in the instrument. For instance, Total Integrated Scatter (TIS) can be applied by using 2 photo-detectors while Angular Resolved Scatter (ARS) instruments can contain more than 500 photo-detectors. An instrument marketed by Rodenstock contains 20 photo-detectors which can be classified as Partial Integrated Scatter (PIS), it is a compromise between ARS and TIS. Obviously, the speed with which results are obtained is inversely proportional to the number of detectors. However, ARS methods can give more than surface finish only. It can provide machine-tool condition information and indicate how the non-specular light is distributed by the optical component, in other words it is a functional test.

In this contribution a comparison will be made between the PIS, ARS and TIS methods, showing their relative merits for surface finish analysis and for machine-tool condition monitoring.

Introduction.

The use of light scattering or diffraction for the determination of the surface roughness of machined metal components has been investigated by many researchers. As a result, a number of theoretical formulations of the rough surface scattering problem have been derived for the cases of surfaces with random and periodic surface height distributions. Most of these formulations relate the r.m.s. surface roughness parameter R_q to the specular reflectance as in Total Integrated Scatter (TIS) [1,2] for random and as in Angular Resolved Scatter (ARS) [3,4] for periodic surface profiles. These quantitative relationships are valid for surface roughness values R_q up to approximately 0.1λ, assuming normal incident light.

Experimental results for rougher surfaces have been reported and from them the individual researchers [5,6,7,8] have derived empirical relationships between the standard deviation σ of the angularly scattered light distribution and the r.m.s. surface slopes Δ_q. Each of these relationships are claimed to be valid for a certain class of surfaces. As a result of this, commercial optical surface finish sensors, based upon light scattering have been produced. These instruments are a compromise between TIS and ARS techniques, since they contain a limited number of photo diodes, they can be classified as Partial Integrated Scatter (PIS) instruments. The manufacturers of these instruments have defined some new surface finish parameters like S_n [9] and R_{op}, thereby contributing to the so-called "parameter rash" [10]. However, a formal relationship between σ and Δ_q can be derived [11], based upon scalar diffraction theory, thereby obviating the need of introducing new optical surface finish parameters.

Definition of the parameters.

The surface finish parameters R_q and Δ_q for respectively the surface height and slope are defined by

$$R_q = \sqrt{\langle h^2(x) \rangle} \tag{1}$$

and

$$\Delta_q = \sqrt{\langle (\frac{d}{dx} h(x))^2 \rangle} \tag{2}$$

where h(x) is the surface profile. The expression $\langle f(x) \rangle$ denotes the averaging process $\lim_{L\to\infty} 1/2L \int_L f(x)dx$.

The standard deviation σ of the optical diffraction pattern I(X) is defined by

$$\sigma = \sqrt{\int_{-\infty}^{\infty} X^2 I(X)dX \Big/ \int_{-\infty}^{\infty} I(X)dX} = \sqrt{M_2/M_0} \tag{3}$$

where M_0 and M_2 respectively are the zeroth and second moment of the diffraction pattern. M_0 also represents the total amount of scattered light.

Theory

The operation of the sensors under discussion is based on the same principle, they only differ in the acquisition and subsequent evaluation of the scattered light distribution.

The situation considered is that of a uniform plane wavefront, with amplitude A, incident normally on an imperfectly smooth flat metal surface, whose surface deviations from the mean, z=0, is h(x,y). It is assumed that no polarization and multiple scattering occurs and that the surface has finite derivatives. Let the phase modulation at the metal surface be $\phi(x,y)$, then the phase of the reflected wavefront, very close to the surface, is related by geometrical optics to the surface height by (see figure 1):

$$\phi(x,y)=2kh(x,y), \tag{4}$$

where $k=2\pi/\lambda$ and λ is the wavelength of the illumination. Now if the wavefront, bounded by an aperture of area S, leaves the scattering surface in a general direction perpendicular to the plane z=0, and if the far-field pattern is formed by a lens of focal length f, then applying Kirchhhoff diffraction theory, the far-field complex amplitude is

$$U(X,Y)=\frac{A}{S}\iint_S \exp[i2kh(x,y)]\ \exp[-ik(Xx+Yy)/f]dxdy. \tag{5}$$

Since the surface finish parameters are defined for surface profiles, the diffraction integral will be reduced to a one-dimensional one. In reality this can be achieved by using a cylindrical lens and surfaces with a unidirectional lay, where the surface height is a function only from say the x-ordinate. Equation (5) can then be rewritten as

$$U(X)=C_1\langle\exp[i2kh(x)]\cdot\exp[-ikXx/f]\rangle, \tag{6}$$

where C_1 is a constant. The light intensity I(X) in the diffraction plane can be expressed by

$$I(X)=C_2\left|U(X)\right|^2, \tag{7}$$

which, by using the Wiener-Khintchine theorem can be rewritten as

$$I(X)=C_2\int_L(\langle\exp[i2kh(x+s)]\cdot\exp[-i2kh(x)]\rangle)\cdot\exp[-ikXs/f]ds \tag{8}$$

With the aid of the equations (6), (7) and (8) various relationships can be derived linking the light intensity of the diffraction pattern to the r.m.s. surface roughness R_q and slope Δ_q, whose results are listed in the following.

For a surface with a Gaussian surface height distribution the specular reflected intensity is

$$I_r(R_q)=I_o\exp[-(4\pi R_q/\lambda)^2] \tag{9}$$

where I_o can be obtained by integrating the light intensity over the total diffraction plane, see figure 2.

For a surface profile with a profile $h(x)=\sqrt{2}R_q/\lambda\cos(2\pi x/D)$, where D is the period length, the specular reflected light intensity $I_p(R_q)$ equals:

$$I_p(R_q)=I_oJ_0^2(4\pi\sqrt{2}R_q/\lambda) \tag{10}$$

where $J_0()$ is a Bessel function of the first kind of integral order 0.

The magnitude of the specular reflected beam has also been calculated for periodic surface profiles with triangular and parabolic cusp shapes [4]. It has been found that for all these periodic surface profiles, as long as $R_q<0.06\lambda$, the specular reflected light can still be expressed by (10).

Comparison of the expressions for the specular reflectance for periodic and random surface profiles, with $R_q<0.06\lambda$, shows that both are very close and that the surface roughness might as well be obtained by equation (9)

$$R_q/\lambda=\sqrt{\ln[I_0/I(R_q)]}/4\pi, \tag{11}$$

where $I(R_q)$ is the specular and I_0 the total reflected light intensity. Formula (11) holds also very well for periodic surface profiles with $R_q<0.10\lambda$, obtained by diamond turning.

The r.m.s. surface slope Δq is related to the standard deviation σ of the scattered light intensity distribution by:

$$\Delta q=\sigma/2f \tag{12}$$

Description of the three scattering techniques.

As mentioned, all three techniques are based upon light scattering. The principle differences between the sensors are the number of photo detectors used, which collect the scattered light in the far-field. It is assumed that the photo detectors span the entire angular light distribution. Figure 3 shows the principle lay-out of the sensors together with a sketch of a diffraction pattern produced by a periodic surface.

Total integrated Scatter (TIS).

This method was originally used by Bennett and Porteus [2] to study surfaces having $R_q<0.1\lambda$. It is based on the relation between the specular reflected $I(R_q)$ and R_q as given in equation (11). This instrument contains essentially two detectors, where D_1 and D_2 together collect the total reflected light and D_2 the diffusely scattered light. Their actual system includes an aluminized collecting Coblentz sphere, which focusses the diffusely scattered light onto detector D_2. Since there are only two detectors involved, this technique can be very fast with simple electronic circuitry to obtain R_q. The upper surface roughness limit of 0.1λ is caused by the fact that for rougher surfaces the specular reflected beam effectively disappears. The lower limit is about 0.001λ, which is governed by the electronic noise in the detectors. The instrument needs accurate setting up to prevent the specular beam from reaching detector D_2. Because of this, the instrument is not suitable for in-process measurement and has to be used in a laboratory environment. Surface slope information cannot be obtained since the information for evaluation of equation (12) is missing.

Angular Resolved Scatter (ARS).

In principle, the entire angular light distribution contains a lot more information about the surface topography than can be obtained from TIS. In addition, measurement of the angular light distribution can yield the r.m.s. slope parameter and machine tool condition information [12]. In order to obtain this information, a great number of photo detectors has to be placed in the far-field of the instrument lens. The Warwick system contains 512 photo detectors. The size of the detectors has to be smaller than the minimum line width of the diffraction orders, in order to be able to resolve the diffraction pattern. To keep the size of the instrument small, CCD photo arrays will have to be used. These have the disadvantage over photo diodes that their output is linear with the amount of received light, thereby reducing the dynamic range of the detectors. The upper surface roughness limit is again 0.1λ for the same reason as in ARS, and the lower limit is about 0.01λ, because of electronic noise and the limited dynamic range. However, the dynamic range can be increased by suitable software in the microprocessor which controls the instrument. Alignment is not a big problem since microprocessor software can be devised to detect the position of the specular reflected beam, and this information can actually be used to determine the form of the object [13]. Because of the number of photo detectors, the acquisition of information is much slower than in TIS.

Partial Integrated Scatter.(PIS).

This type of instrument contains less photo detectors than the

ARS based system. The Rodenstock sensor for instance, has 20 photo detectors. Since they have to span the complete angular light distribution, the photo diodes will not be able to resolve the angular distribution well enough to separate the specular light intensity. Therefore it is impossible to obtain the R_q value as described in equation (11), and only the Δ_q parameter can be evaluated. Again there are no alignment problems, and by monitoring the centre of gravity of the detected light distribution , form information of the object can be collected. Since this instrument contains only a limited number of detectors, photo diodes can be used with their related advantages. The acquisition speed is higher than for an ARS system because only a few diodes have to be sampled.

Comparison of the instruments.

The table in figure 4 lists the properties of the three techniques. A more detailed comparison between the performance of the ARS and PIS system is made since both are capable of in-process measurements and they can measure the Δ_q parameter, whereas the TIS system can only measure the R_q parameter, albeit with superior performance with respect to ARS.

The comparison is based on the capability to obtain the Δ_q parameter of sinusoidal surface profiles. The R_q value of these sinusoids ranges from $.05\lambda$ to 10λ, the period length D is 100um and the illumination wavelength is .8um as produced by laser diodes. Furthermore, it is assumed that the detectors span the full angular distribution of the theoretical diffraction pattern produced by a sinusoid with an amplitude of 2.26um ($R_q=2\lambda$). The ARS system contains 512 and the PIS system 20 photo detectors. The light intensity pattern I(X) can be expressed by:

$$I(X)= J_m^2(4\pi\sqrt{2}R_q/\lambda)\delta(X+f\cdot\tan(m\pi/D)) \qquad (13)$$

where f is the focal length of the instrument lens and $\delta()$ is the Dirac delta function. Some of the patterns as they are perceived by the instruments are shown in the figures 5,6,7 and 8. The graph in figure 9 shows their theoretically obtained Δ_q values versus the analytical value which equals $2\pi\sqrt{2}R_q/\pi D$. It shows clearly the inability of the PIS system to resolve the diffraction pattern well enough for surface profiles with $R_q<0.2\lambda$. It is also interesting to note what happens when the diffraction pattern is to wide for the detector plane ($R_q>2\lambda$). This shows that one has to be careful when using these instruments and that a graphical display of the light distribution is important when judging the validity of the results obtained with these instruments. The patterns in figure 10 are obtained from the ARS pattern by convolving these patterns with a box function which spans 9 diffraction orders. It is interesting to see that for surfaces with relative large amplitudes the convolved pattern resembles very closely the amplitude density function of the surface slopes. In optical terms this means that from then onwards geometrical optics can be used to predict the diffraction pattern.

Conclusions.

PIS instruments can produce reliably the Δ_q surface slope parameter within a limited range. A lower limit is set by the resolving power of the photo detectors. The upper limit is set by the length of the detector array. They can also produce waviness or form information. Their acquisition speed is fairly high with an inherent high dynamic range. It is important to obtain a graphical output of the light distribution to ensure the validity of the measurement. They are useful for comparitive measurements.

ARS instruments can produce the R_q surface height parameter for $0.01\lambda<R_q<0.1\lambda$ and the Δ_q surface slope parameter. Here the surface slope range is improved in the lower end due to the higher resolving power of the instrument. Waviness and machine tool condition information is available. It is also very much a functional test for cases where one is interested in the actual reflected light distribution like in laser gyro mirrors. Again a graphical display is desirable. Because of the number of photo detectors involved, the acquisition time will be slower than for the other two systems.

TIS instruments can produce the R_q surface height parameter with high accuracy for $0.001\lambda<R_q<0.1\lambda$. This instrument needs accurate adjustments. However, its acquisition speed is very high.

References

1 H. Davies, "The refelection of electromagnetic waves from a rough surface" Proc. Inst. Elec. Engrs. 101, 209-214, 1954.

2 H.E. Bennett and J.O. Porteus, "Relation between surf ace roughness and specular reflectance at normal incidence" J. Opt. Soc. Am. 51, 123-129, 1967.

3 E.L. Church, H.A. Jenkinson and J.M. Zavada, "Measurement of the finish of diamond-turned metal surfaces by differential light scattering" Opt. Eng. 16, 360-374, 1977.

4 J.H. Rakels, "The use of Bessel functions to extend the range of optical diffraction techniques for in-process surface finish measurements of high precision turned parts" J. Phys. E: Sci. Instrum. 19, 76-79, 1986.

5 J.A. Aas, "Reconstruction of surface profiles from their diffraction spectra" Appl. Opt. 11, 1579-1584, 1972.

6 M. Kamei and T. Inari, "On-line monitoring of surface roughness using laser" SPIE, 508, 56-63, 1984.

7 E.L. Church and P.Z. Takacs, "Survey of the finish of machined optical surfaces" Opt. Eng. 24, 396-403, 1985.

8 M. Allgäuer and R. Brodmann, "Vergleichende Untersuchung der Mikrotopographie von Sinusprofilen mit Hilfe der mechanischen und optischen Profilometrie sowie Streulichtmethode" Feingerätetechnik, 37, 9-12, 1988.

9 R. Brodmann, Th. Gast and G. Thurn, "An optical instrument for measuring the surface roughness in production control" Annals of the CIRP. 33/1/, 403-406, 1984.

10 D.J. Whitehouse, "The parameter rash- is there a cure?" Wear 83, 75-78, 1982.

11 J.H. Rakels, "Recognized surface finish parameters obtained by light scattering" to be published.

12 J.H. Rakels, "Diffraction, an old optical phenomenon, used as an advanced metrology tool" Int. Conf. "Modern Production and Production Metrology" T.U. Vienna, April, 79.3-96.3, 1986

13 R. Brodmann, H. Pasidzior, N. Rau and G. Hübner, "Optische Messung der Rauheit und Welligkeit feinbearbeiteter Oberflächen" Int. Conf. "Modern Production and Production Metrology" T.U. Vienna, April, 113.3-126.3, 1986

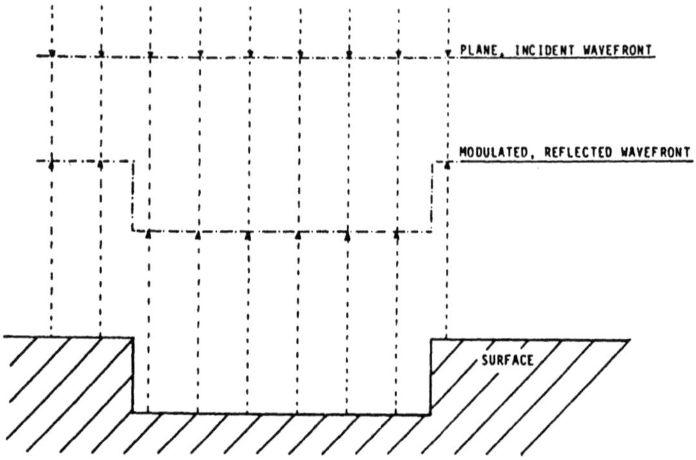

Fig.1. Phase modulation model near a metal reflecting surface.

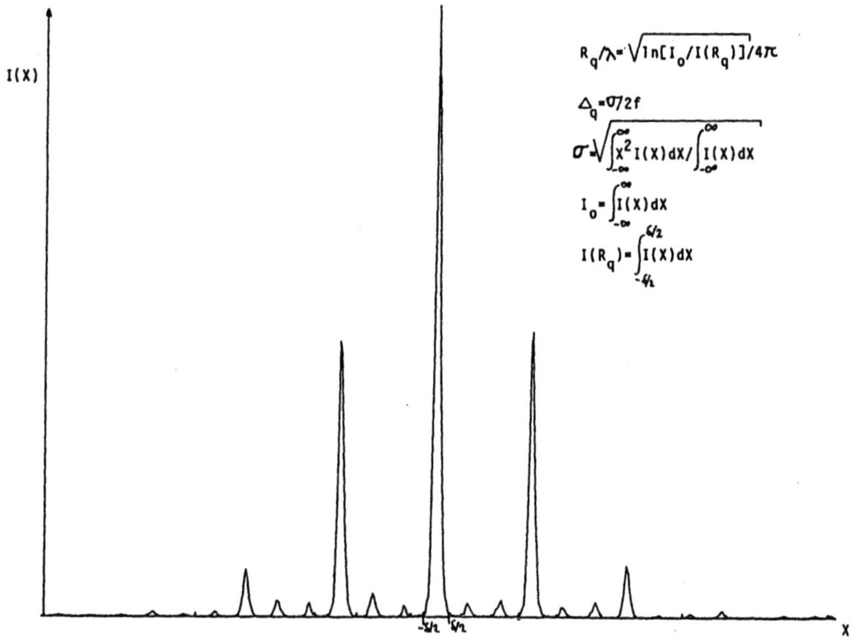

Fig.2. Relationships between the scattered light distribution and surface finish parameters.

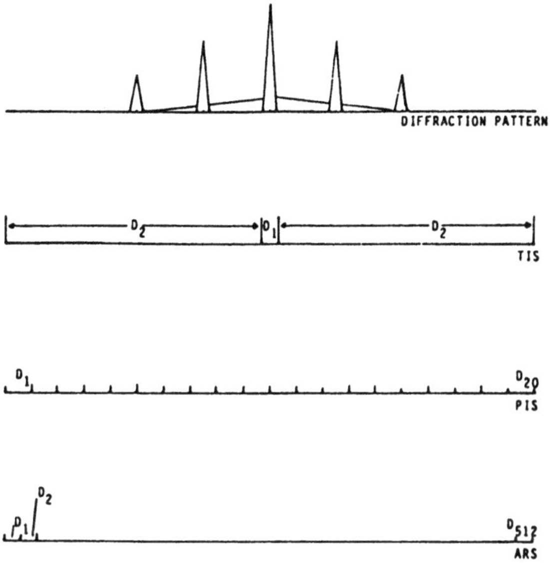

Fig.3. Principle arrangement of the photo detectors for the various systems.

METHOD	Δ_q	RELATIVE RANGE (1)	R_q	RANGE	RELATIVE SPEED (2)	IN-PROCESS	MACHINE TOOL CONDITION	WAVINESS and FORM
PIS	YES	MEDIUM	NO	-----------	HIGH	YES	NO	YES
ARS	YES	LARGE	YES	0.01λ - 0.1λ	MEDIUM	YES	YES	YES
TIS	NO	-----	YES	0.001λ - 0.1λ	VERY HIGH	NO	---	---

(1) LOWER LIMIT GOVERNED BY THE RESOLVING POWER OF THE DIODES.
UPPER LIMIT GOVERNED BY THE LENGTH OF THE DETECTOR ARRAY.

(2) SPEED IS IN GENERAL INVERSELY PROPORTIONAL TO THE NUMBER OF DETECTORS

Fig.4. Comparison table between the various scattering techniques.

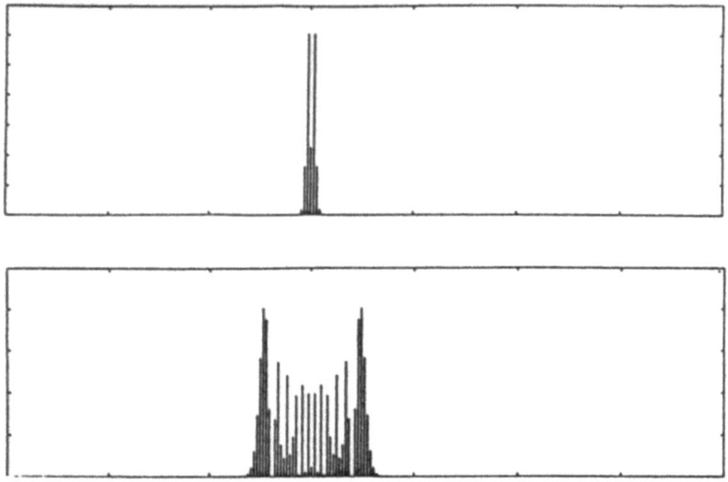

Fig.5. Diffractions patterns as perceived by an ARS instrument. Top for $R_q = .1\lambda$ and bottom for $R_q = 1\lambda$, Feed spacing is 100 µm.

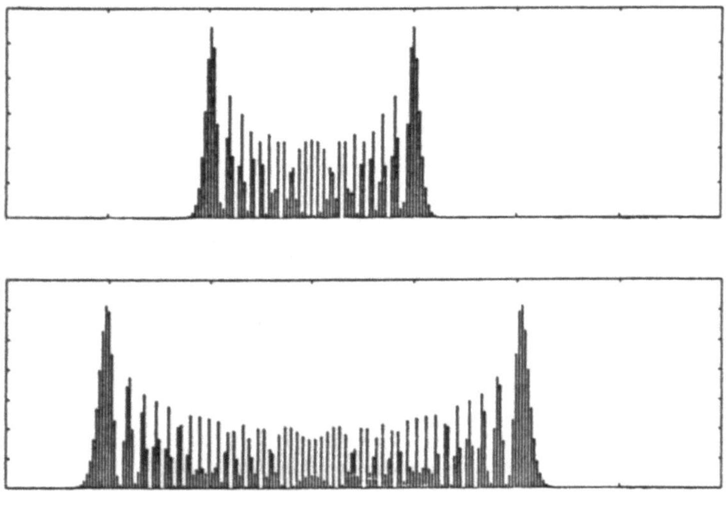

Fig.6. Diffractions patterns as perceived by an ARS instrument. Top for $R_q = 2\lambda$ and bottom for $R_q = 4\lambda$, Feed spacing is 100 µm.

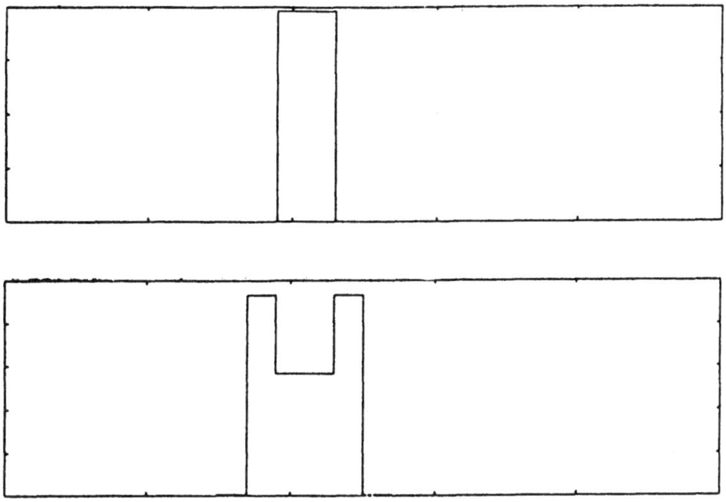

Fig.7. Diffractions patterns as perceived by a TIS instrument. Top for $R_q = .1\lambda$ and bottom for $R_q = 1\lambda$, Feed spacing is 100 µm.

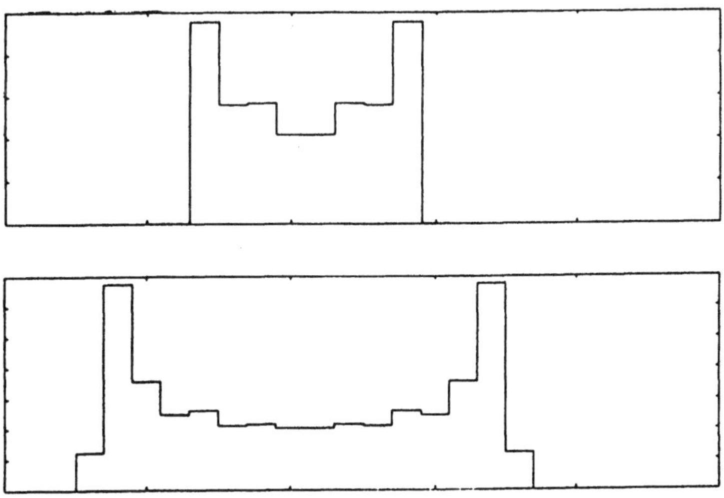

Fig.8. Diffracrions patterns as perceived by a TIS instrument. Top for $R_q = 2\lambda$ and bottom for $R_q = 4\lambda$, Feed spacing is 100 µm.

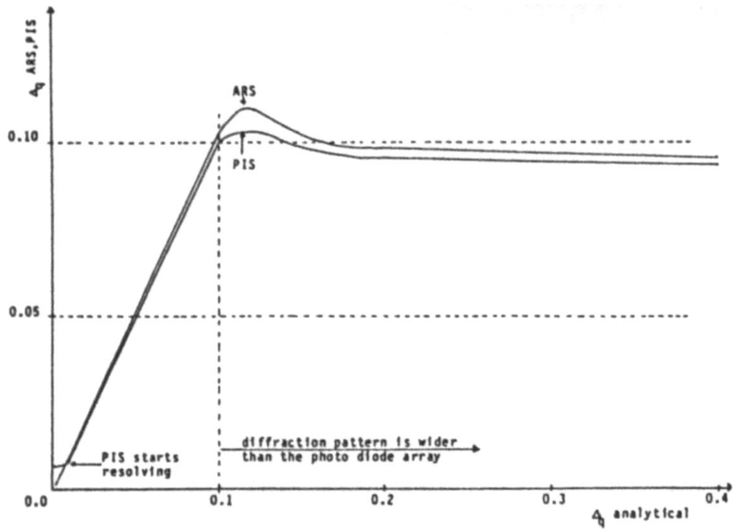

Fig.9. Comparison between Δq values obtained with ARS and PIS versus the analytically obtained Δq value.

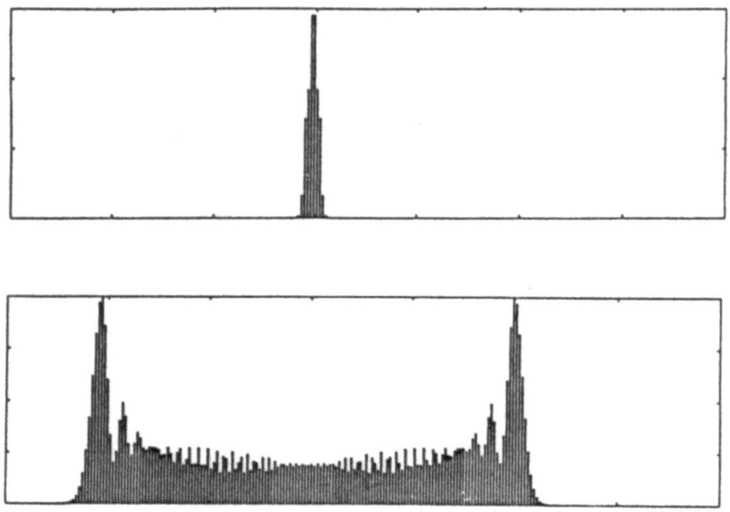

Fig.10. Convolved diffraction pattern for $R_q=0.1\lambda$ (top) and $R_q=4\lambda$. The bottom pattern aproaches a slope distribution for a sinus.

Measurement of Steep Aspheric Surfaces Using a Radial Shear Interferometer

G. Harbers, M. Struis

Philips Research Laboratories Eindhoven
The Netherlands

Summary
When measuring steep aspherical surfaces with standard (for example Twyman-Green) interferometers without using specially designed compensation systems or computer generated holograms, three major problems arise. These are the high fringe density, vignetting in the objective lens and aberrations in this lens. The limits of measuring aspheres with respect to these problems are derived. A radial shear interferometer has been built to overcome the problem of fringe density. The accuracies which can be reached and the measures which have to be taken to reach sub-micron accuracies are discussed.

1 Introduction.

In the last decade aspherical optics have been successfully employed in sophisticated optical systems like CD-players, zoom lenses for photographic systems and TV-projection systems. By applying aspherical optics, higher imaging quality can be reached using fewer components. Aspherical components can be fabricated using moulds produced on an ultra-high precision lathe. The use of high-precision test-methods is an essential feedback in the fabrication process of the moulds. In highly demanding applications, form accuracies down to 10 nm are needed and, in order to reach this accuracy, test methods should be at least as accurate.
Stylus profilometers as conventionally used for these measurements (Dil et al., 1978) have an inaccuracy of at least 0.1 μm. This limitation in accuracy and the potential damage of the surface make mechanical contact measurements for such aspheres not very suitable. Optical measurements of flats and spheres are nowadays performed with accuracies down to the nanometer region by using dynamic fringe analysis techniques. The expectation is that these optical measurement techniques can be extended to the surfaces as mentioned above.
In a project stimulated by the European Community (Brite), test methods are investigated which should reach these accuracies. In this project ('Development of

non-destructive high precision test methods for aspheric components and tools in optics') the University of Stuttgart (Institute for Technical Optics), Cerco (Paris) and the Philips Research Laboratories (Eindhoven) are cooperating. At the University of Stuttgart methods using computer generated holograms are developed, Cerco produces special optical devices needed and, at Philips, the conventional techniques like the quantitative knife-edge test and the classical quantitative interferometric tests are investigated.

In this paper some of the results of the work at the Philips Research Laboratories on the measurement of concave aspherical surfaces using standard measurement objectives in a radial shear interferometer are given. The problems encountered when interferometrically measuring steep aspherical surfaces are:

1) High fringe density.
   In a conventional interferometer the fringe density can be as high as 10.000 fringes/aperture and these cannot be resolved by ordinary electronic cameras as used in phase stepping interferometers. With radial shear interferometry the fringe density can, in principle, be decreased to such a degree that all fringes can be resolved.
2) Vignetting of rays within the objective lens.
   It is possible that the light from steep regions of the asphere is lost in the measuring lens (vignetting). By using super wide field objectives, this limitation can be shifted to higher degrees of asphericity.
3) Aberrations of the objective lens.
   Aberrations in these lenses limit the accuracy to at least several tenths of a micron depending on the numerical aperture of the lens. When demanding sub-micron accuracies, corrections for these aberrations have to be made. Experimental data on these aberrations will be given.

In the next paragraphs of this paper a further discussion of these problems is given.

2 Fringe Density.

A conventional set-up for measuring concave spheres interferometrically is shown in figure 1. A laser beam illuminates the sphere via an objective lens. The centre of curvature of the sphere coincides with the focal point (F) of the objective. If the lens is perfect the rays will strike the surface perpendicularly. In this case the reflected wavefront emerging from the objective will be flat. When this front is interferometrically compared with the front reflected from a mirror, straight

fringes (or no fringes at all) will be seen. In principle the same set-up can be used to measure aspheric surfaces. However, since the rays will no longer be incident normally at the surface of the asphere, the reflected wavefront will not converge to the focal point of the objective lens, and the front emerging from the objective lens will no longer be flat. When comparing this front with the flat reference front many non-straight fringes can be seen in the interference pattern because large phase differences can be present. In principle this fringe pattern can be analyzed using dynamic fringe analysis techniques to yield the aspherical form of the mould. In practice however, phase differences can be so large that electronic cameras as used in this technique cannot resolve all these fringes. A good measure for the necessary resolving power of the camera is the maximum fringe density.
The fringe density is determined by the slope (angle) of the aspherical surface relative to the slope of the reference sphere. Let the diameter of the asphere be D, r the radial coordinate and θ the slope. Then the local fringe density is given by:

$$LFD(r) = 2\frac{\theta(r)}{\lambda} \tag{1}$$

in units of fringes / m. For a proper sampling there should be at least 3 - 4 pixels per fringe so the required number of pixels of the camera is given by:

$$n_{pxs} \geq 3FD_{\max} = 3D \times LFD_{\max}$$

where $FD_{\max}$ is in units of fringes/aperture. The maximum slope strongly depends on the radius of the reference sphere. Let the radius of this sphere be R. The deviations of the asphere with respect to this spherical front are then given by:

$$Z = Z_{asphere} - Z_{sphere} = \sum_i a_{2i}r^{2i} - (R - \sqrt{R^2 - r^2})$$

$$= (a_2 - \frac{1}{2R})r^2 + (a_4 - \frac{1}{8R^3})r^4 + (a_6 - \frac{1}{16R^5})r^6 + (a_8 - \frac{5}{128R^7})r^8 + \ldots$$

$$= b_2r^2 + b_4r^4 + b_6r^6 + b_8r^8 + \ldots. \tag{2}$$

The slope is given by:

$$\theta(R,r) = \frac{dZ}{dr} = 2b_2r + 4b_4r^3 + 6b_6r^5 + \ldots \tag{3}$$

(See figure 2). The radius of the reference sphere can be varied by moving the asphere closer to, or away from, the objective lens. The value of the radius at which the maximum slope of the asphere is minimal is called the least slope radius. As an example see figure 3, where for a large number of aspherical designs the fringe density is given as a function of $D/2R_s$ where R_s denotes the least slope radius. All these designs need a submicron form accuracy. Partly the designs are from Philips, and partly from Cerco. As can be seen from this figure, fringe densities can be as high as 10000 fringes/ aperture. In this figure two lines have been drawn: the lower line represents the limit when using two dimensional cameras (512*512 pixels), the upper line represents the limit using one dimensional arrays (4096 pixels). Of the designs given, about 20% can be directly measured using ordinary two dimensional cameras and about 80% using the one-dimensional cameras. So when using 2-dimensional cameras special techniques have to be used to decrease the fringe density. One possible solution is to use radial shear interferometry.

3 Vignetting.

Rays traveling outside the field of view (FOV) of the used objective lens are lost. For a measurement of the complete aspherical surface we get the demand:

$$W < D_{FOV}/2$$

Where D_{FOV} is the diameter of the field of view and W the maximum deviation of the reflected beam from the optical axis (i.e. transverse radius of the waist of the caustic). This deviation can be approximated by (see figure 4):

$$W = 2R\theta(R,r) \qquad (4)$$

with θ as in the previous paragraph. The maximum deviation is minimum at least slope radius. The diameter of the field of view of microscope objective lenses can be approximated by

$$D_{FOV} = D_t/M.$$

where D_t denotes the tubus diameter of the microscope and M the magnification of the objective lens. In practice the field of view strongly depends on the numerical aperture of the objective. The numerical aperture needed to measure the asphere is given by (see figure 4):

$$NA = n \sin\alpha = n\frac{(D/2)}{R} \qquad (5)$$

where n is the refractive index of the medium between objective and asphere (normally air : n = 1) and D is the diameter of the aspheric surface to be measured. In table 1 fields of view of some objective lenses are given as a function of numerical aperture. It should be noted that these fields of view are relevant to microscopy applications based on illumination by extended white light sources. When using laser sources the situation is somewhat more complicated but the numbers give a good indication.

The fields of view needed to measure the aspheres of fig. 3 as a function of the ratio $D/2R_s$ are given in figure 5. The fields of view for the standard and super wide field series of microcope objectives lenses are indicated by the drawn curves. Aspheres beneath the two curves can be completely measured with standard or super wide field objective lenses.From the designs shown here about 50% can be measured using the standard series and about 70% using the super-wide field objective series. For the remaining 30 % of the aspherical surfaces vignetting hinders the measurement of the complete aspherical surface.

4 Aberrations of the objective used.

As the wavefront comparison is made after passage through the objective lens, aberrations of this lens are included in the measurement of the aspherical surface. The on-axis aberrations of the objective lens can be removed by subtracting the measured form of a sphere. However, when measuring aspheres, rays reflected at steep regions will travel back off-axis and will be extra disturbed by the so-called field aberrations. These aberrations limit the accuracy of the measurement. As an example, measurements of the relevant field aberrations of a 32x/0.60 objective lens are given in figure 6. Of the other primary aberrations, spherical aberration is constant in the field and is removed with the on-axis calibration, distortion is calibrated using special reticles or checking the position of a spot moving along the path of the aspherical surface. As can be seen from this figure these primary field aberrations for this lens can add up to 3 λ over a total field of $2 \times 300\mu m$. When using a HeNe-laser ($\lambda = 633nm$) this would give errors up to $2\mu m$. Requiring accuracies down to the sub-micron region thus necessitates the calibration of the field aberrations when using high numerical aperture objective lenses.

5 Radial Shear Interferometry.

Radial shear interferometry is one of several methods to solve the fringe density problem. In a shear interferometer the wavefront to be measured is not compared with a flat reference front, as in a Twyman-Green interferometer, but with a sheared version of itself. Hence the optical path difference can be much smaller depending on the shear. In the popular lateral shear interferometers the wavefront is shifted with respect to itself. The disadvantages of lateral shearing are the necessity to record two interferograms with orthogonal directions of shear and the fact that the interferogram doesn't cover the entire pupil. In a radial shear interferometer an expanded version of the wavefront acts as reference obviating the need for two separate measurements. Furthermore the complete surface is measured.

A complete overview of these radial shear interferometers is given by Malacara (1978, Ch 5). To reach the required accuracies it is necessary to perform dynamic fringe analysis. Not all the interferometers as described by Malacara can be utilized with this technique. Therefore, it must be possible to modulate one of the two beams in phase.

A very simple radial shear interferometer based on a Mach-Zehnder interferometer, using telescopic systems to produce the radial shear, has been suggested by Murty (1964) and is shown in figure 7. This interferometer can be made dynamic by using a piezo-driven mirror for one of the two mirrors.

The interpretation of a radial shear interferogram is somewhat different from that of a Twyman-Green interferogram. The wavefront to be measured can be expressed as a linear combination of circular polynomials:

$$W(\rho, \phi) = \sum_{k=0}^{n} \sum_{l=0}^{k} \rho^{k}(A_{kl} \cos l\phi + B_{nl} \sin l\phi) \tag{6}$$

with: ρ, ϕ polar coordinates over the unit circle defining the pupil. The expanded wavefront acting as reference can be written as:

$$W(S\rho, \phi) = \sum_{k=0}^{n} \sum_{l=0}^{k} (S\rho)^{k}(A_{kl} \cos l\phi + B_{kl} \sin l\phi) \tag{7}$$

where S is the shear ratio. The measured optical path difference (OPD) is given by:

$$OPD = W(\rho, \phi) - W(S\rho, \phi) = \sum_{k=0}^{n}\sum_{l=0}^{k}(1 - S^k)\rho^k(A_{kl}\cos l\phi + B_{nl}\sin l\phi) \quad (8a)$$

$$= \sum_{k=0}^{n}\sum_{l=0}^{k}\rho^k(A^\circ_{kl}\cos l\phi + B^\circ_{nl}\sin l\phi) \quad (8b)$$

with:

$$A^\circ_{kl} = A_{kl}(1 - S^k)$$

$$B^\circ_{kl} = B_{kl}(1 - S^k) \quad (9)$$

For large shear ($S \rightarrow 0$): $A^\circ_{kl} = A_{kl}$ $B^\circ_{kl} = B_{kl}$ and the interferogram resembles a Twyman-Green interferogram (exploded shear interferogram). For small shear ($S \rightarrow 1$) the wavefronts are almost the same and only a few fringes are seen. In that case the original front can be derived from the measured OPD by a fit in circular polynomials (preferably orthogonal, for example Zernike polynomials) and a determination of the factors A°_{kl} B°_{kl} according to (8b). The original front is given by eq.6 after determination of the coefficients A_{kl}, B_{kl} using (9). In employing this technique care has to be taken to centre the wavefronts carefully otherwise also a lateral shear is introduced and this algorithm will fail. An algorithm which can handle a combination of lateral and radial shear is given by Kohler (1986). We have chosen to eliminate the lateral shear optically.
The sensitivity of the radial shear interferometer as compared to a standard (for example a Twyman-Green) interferometer depends on the shear ratio and the form of the asphere. The deviation from an asphere relative to a reference sphere with least slope radius is given by (see eq. 2):

$$Z = b_2r^2 + b_4r^4 + b_6r^6 + \ldots \quad (10)$$

In a radial shear interferometer the optical path difference $\tilde{Z}$ is measured where:

$$\tilde{Z} = b_2(1 - S^2)r^2 + b_4(1 - S^4)r^4 + b_6(1 - S^6)r^6 + \ldots.. \quad (11)$$

$$= \sum_n b_{2n}(1 - S^{2n})r^{2n}$$

For small shear ($S \rightarrow 1$) the interferometer is relatively insensitive to the low coefficients. For the higher coefficients the same sensitivity as in an ordinary interferometer is reached. This insensitivity to the lower coefficients causes inaccuracies in the shear corrected front in the presence of small errors in the shear ratio. The real coefficients b_{2n} are related to the measured coefficients $\tilde{b}_{2n}$ by

$$b_{2n} = \frac{\tilde{b}_{2n}}{1 - S^{2n}} . \tag{12}$$

When $1 - S^{2n}$ is small a small error in S will introduce large errors in the corresponding coefficients; when $1 - S^{2n}$ is large only marginal errors are introduced (see figure 8). So the shear ratio is a critical parameter in the shear reduction.

6 Experimental Set-Up.

The Mach-Zehnder type radial shear interferometer as built is shown in figure 9. The expanded beam of a (stabilized) HeNe-laser is reflected at a good quality beam-splitter and illuminates the concave aspheric surface to be measured by means of an objective lens. Light reflected from the asphere is collected by the objective lens again and passes through the beamsplitter towards the interferometer. The first beamsplitter of the interferometer splits the light into a portion to be expanded and a portion to be compressed by the two telescopes. For both telescopes the same lenses are used to balance the aberrations introduced in each telescope separately. Both beams of light are brought to interfere by means of a third beamsplitter. In order to relate the interferogram spatially to the aspheric surface the asphere can be imaged onto a camera by means of a zoom imaging system (Malacara 1985).
The shear ratio of this interferometer is given by:

$$S = \frac{M_1}{M_2} = \frac{f_1/f_2}{f_2/f_1} = \left(\frac{f_1}{f_2}\right)^2 \tag{13}$$

where M_1, M_2 are the magnifications of the telescopes. Using good quality standard achromatic lenses we could reach three shear ratios in the small shear region (S=0.759, 0.875 and 0.882). In order to reach smaller shears non-standard achromats had to be used which were not available at the time of the experiments. The interferograms are modulated in phase by means of a piezo driven mirror. The phase stepping technique employed is outlined by Bruning (1974). The modified 5 step algorithm as suggested by Schwider (1983) and Hariharan (1987) is used:

$$\phi = \tan^{-1}\frac{I_2 - I_4}{I_1/2 - I_3 + I_5/2} \tag{14}$$

$I_1...I_5$ are the intensities in the interferogram at the five positions of the piezo driven mirror having a distance of $\pi/2$ in phase. In a first order approximation this algorithm is insensitive to positioning errors of the mirror.
The images can be taken in from the camera (CCD -camera 512*480 pixels) into a computer (HP9000) with help of a framegrabber (FG100, Imaging Technology) implemented in a VME-system. The computer controls also the movement of the piezo driven mirror.
In a first step the interferograms are analyzed as if they were Twyman-Green interferograms. The front is calculated according to (8b) and Zernike polynomials are fitted giving the coefficients A°_{kl}, B°_{kl}. The original front is calculated from the corrected coefficients A_{kl}, B_{kl} using eq. 9 and 6. The shear factor is determined by carefully measuring the two magnifications.
In order to get an idea of the stability and accuracy of the interferometer a large number of measurements was carried out, all of the same objects. These stability measurements showed a rather large drift with a 24-hour cycle probably due to temperature fluctuations. The fitted polynomial terms showed a fluctuation within these 24 hours of $42m\lambda$ (standard deviation). On shorter periods (less then 1 hour) the standard deviation in the coefficients was about $10m\lambda$ in the not shear-corrected measurement and didn't decrease any further when making the periods shorter.

7 Results.

Three types of aspheric moulds were selected for the experiments. The designs were moulds for CD-objective lenses. From one type several samples were available so that comparison tests could be carried out. All moulds were tested on a stylus profilometer and were found to be correct within $0.1\mu m$ of the theoretical form. For all moulds least slope radii, fringe densities and used field diameters were determined using the formulas given in the previous paragraphs. For the selected shear ratios the reduced fringe densities and fields of view were also calculated and verified experimentally. Results as function of the measured diameter are given in table 2. From the inspection of this table it becomes clear that problems arise at the edges of the asphere. Field deviation and fringe density is highest on the edge as is very often the case.

In a radial shear interferometer with a shear ratio of 0.882, using a camera which can resolve fringe densities of about 100 fringes/diameter, vignetting limits the designs A and B to be measured completely. About 90% of the surface of the moulds can be measured. At a shear ratio of 0.882 fringe density can be resolved for 60% of the inner part of design C. A higher shear ratio should be selected to increase the measured surface.

All aspheric moulds were measured to the limits given above. An example a measurement of an the asphere is shown in figure 10. In this figure the theoretical and the measured form are given. Typical values for the errors are $900 m\lambda$ ($0.57 \mu m$) peak to valley and $186 m\lambda$ ($0.12 \mu m$) RMS. These errors are mainly caused by the field aberrations of the objective. The on-axis errors were removed by a reference measurement of a sphere.

An example typical for the reproducibility is given in figure 11. This reproducibility after shearcorrection is about $150 m\lambda$ ($0.10 \mu m$) peak to valley and $25 m\lambda$ ($0.016 \mu m$) RMS.

8 Conclusions.

Important parameters in the measurement of aspherical surfaces are the maximum slope of the aspheric surface relative to a sphere and the required numerical aperture of the objective lens used for the illumination of the asphere.

The maximum slope directly determines the expected fringe densities and necessary field of view of the objective lens. When the fringe density is too high to be resolved by ordinary sensors special techniques have to be employed to decrease the fringe density. This can for example be done using a radial shear interferometer. At high numerical apertures of the objective lens the limited field of view will limit the measurement of the complete surface when slopes are large. Special objective lenses with wide fields of view then have to be used. The field aberrations of such objective lenses will limit the accuracy of the measurement. For an objective lens with NA = 0.6 we found aberrations up to 3λ ($= 2 \mu m$) over a field of view of about 600 μm.

In order to reach a sub-micron accuracy the objective lens has to be calibrated for the field aberrations. When the objective is completely known, or completely measured, this could be done by means of ray-tracing. The expectation is that with such an calibration process, accuracy can be extended to the reproducibility of the interferometer. The reproducibility of the radial shear interferometer as described in this paper was about 25 $m\lambda$ RMS at a shear ratio of 0.882 and was

limited by stability problems due to temperature drift of components in the non-common paths of the interferometer.
It is clear that in common-path interferometers reproducibility can be much better but these interferometers often don't allow dynamic fringe analysis techniques as required for these high accuracies. To improve the reproducibility of non-common path interferometers the environmental conditions have to be controlled in order to reach reproducibilities and accuracies (with calibrated objective lenses) down to 10 nm.

References

J. G. Dil et al., Appl. Optics 17, 553 (1978).

D. Malacara, Optical Shop Testing, Ch. 5. D. Malacara Ed. (Wiley, New York, 1978)

J.H. Bruning et al., Appl. Opt. 13, 2693 (1974).

J. Schwider et al., Appl. Opt. 22, 3421 (1983).

P. Hariharan et al., Appl. Opt. 26, 2504 (1978).

M.V.R.K. Murty, Appl. Opt. 3, 853 (1964).

M.V.R.K. Murty and P.P. Schukla, Appl. Opt. 12, 2765 (1973).

D. Malacara and C. Menchacha, SPIE Southwest Conference on Optics, Vol. 540, 34 (1985).

D. R. Kohler and L. Gamiz, Appl. Opt. 25, 1650 (1986).

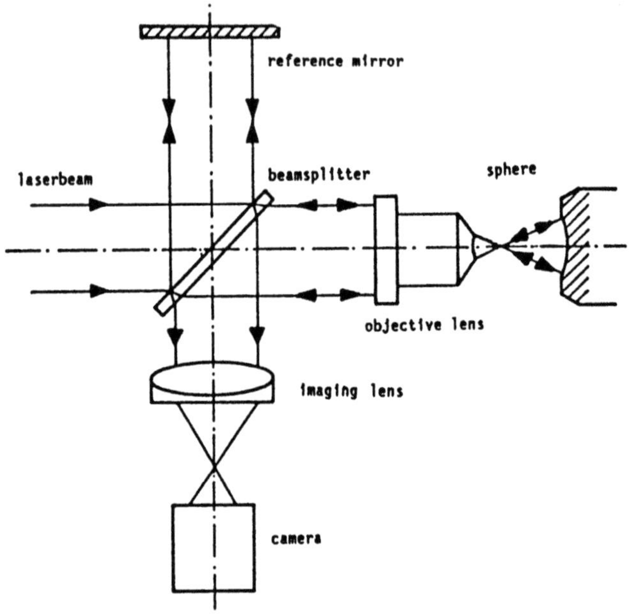

Figure 1:
Twyman-Green interferometer for measuring concave spheres. If the mould is perfectly spherical and the objective has no aberrations the wavefront emerging from the objective after reflection at the mould is perfectly flat. Then no (or only straight) fringes are seen on the camera. Deviations from the spherical form (asphericities) will give non-straight fringes.

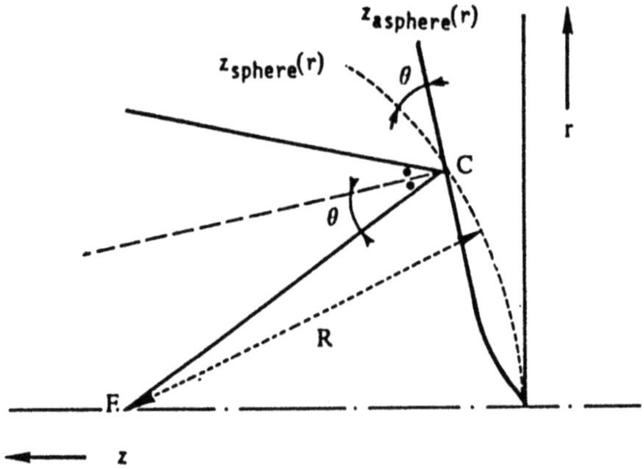

Figure 2:
Slope difference between asphere and reference sphere. $\theta \sim \frac{dZ_{asphere}}{dr} - \frac{dZ_{sphere}}{dr}$.

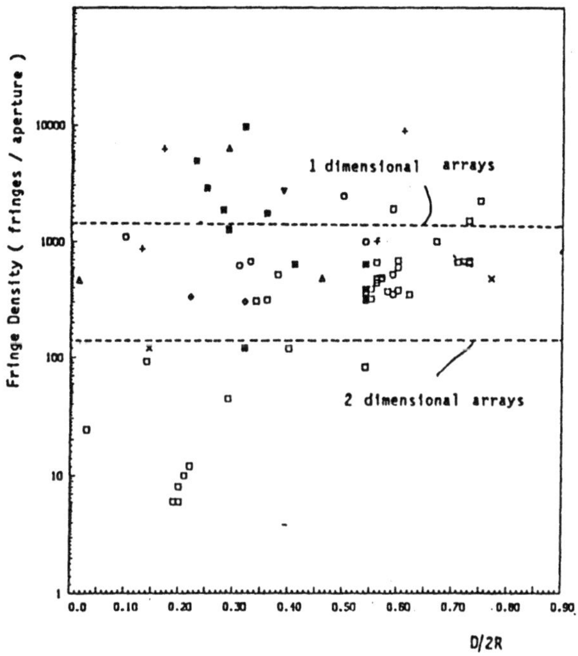

Figure 3:
Fringe densities (fringes/aperture) as function the ratio $D/2R_s$, with D = the diameter of the asphere to be measured, and R_s = the least slope radius of the reference sphere. With a standard electronic camera fringe densities can be measured up to 100-150 fringes/aperture. The different types of application are distinguished by the different markers (Designs Cerco and Philips).

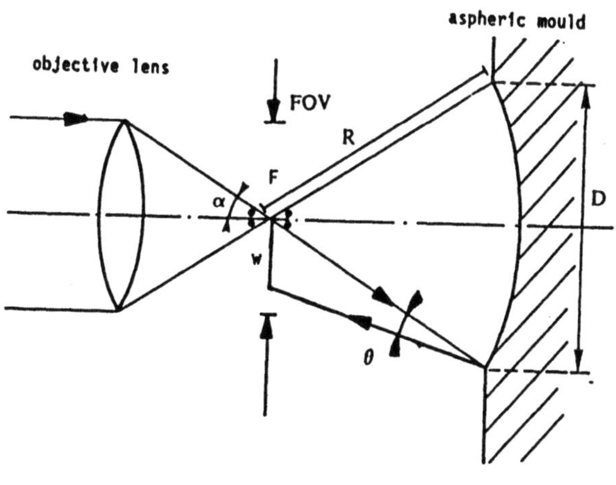

Figure 4:
Vignetting of rays traveling outside the field of view.

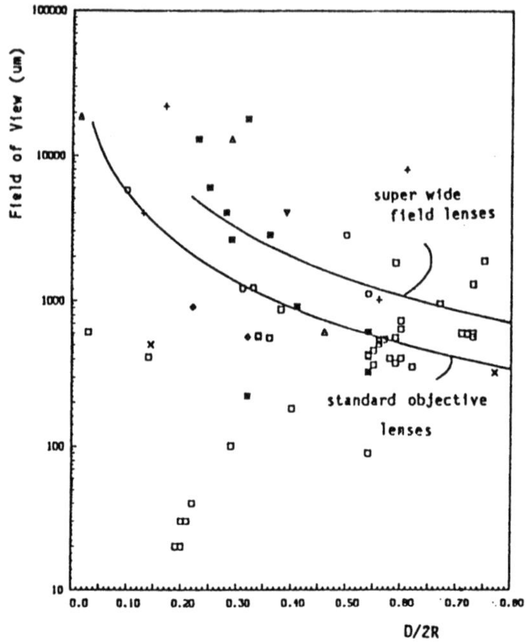

Figure 5
Necessary field diameter as function of the ratio $D/2R$ (NA) to measure the aspheres completely for the same set of designs as in figure 3. The drawn curves indicate the fields of view of standard and super wide-field series of microscope objective lenses as function of NA.

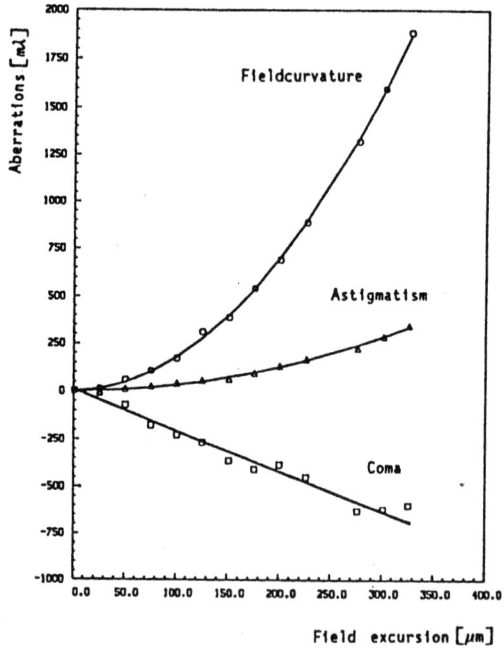

Figure 6
Example of the pure primary field aberrations of a 32x/0.60 objective lens.

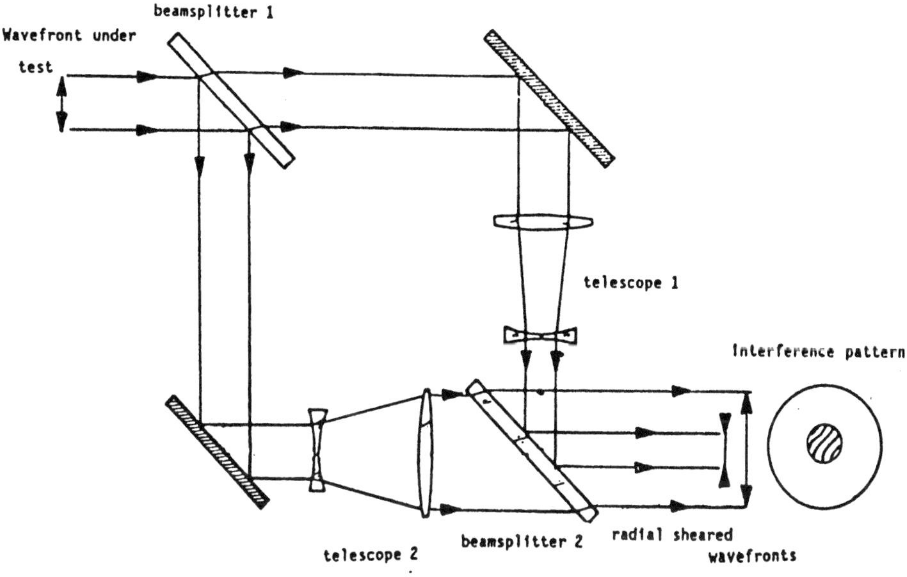

Figure 7:
The Mach-Zehnder based radial shear interferometer using telescopic systems to produce the radial shear as suggested by Murty (1964). This interferometer can be made phase-stepping by driving one of the two mirrors by a piezo moving device.

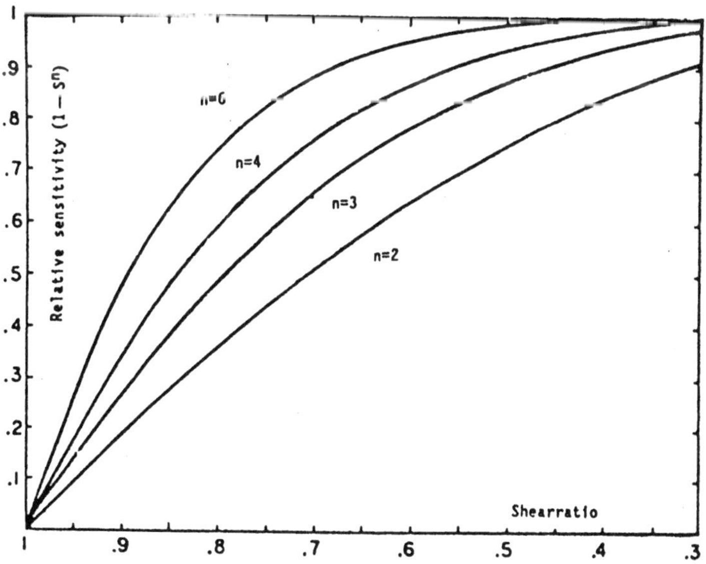

Figure 8.
Relative sensitivity ($1 - S^{2n}$) of the radial shear interferometer as function of the shear ratio S and power n. A small shear gives low sensitivity and reduced fringe densities. For small shears small errors in S give large errors in the shear correction on the vertical axis.

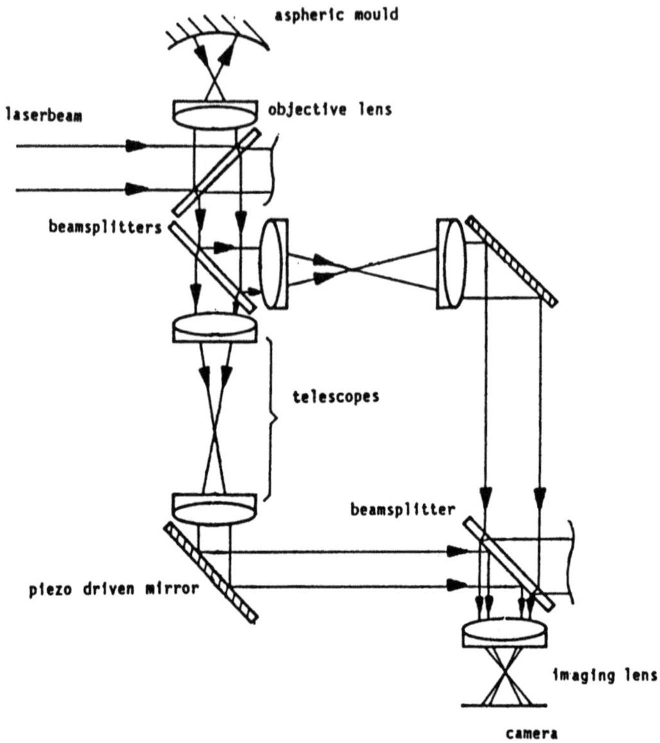

Figure 9:
The phase stepping radial shear interferometer for measuring aspheric surfaces. The shear can be controlled by changing the magnifications of the telescopes.

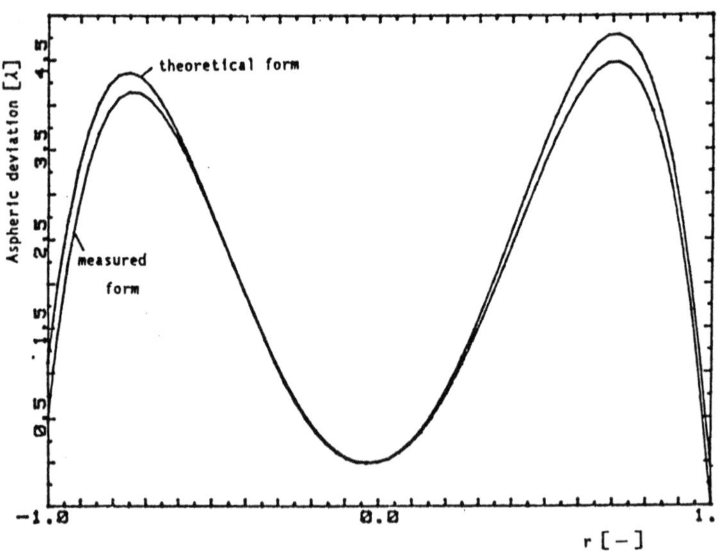

Figure 10:
Example of the difference between the theoretical form and the measured form of the asphere.

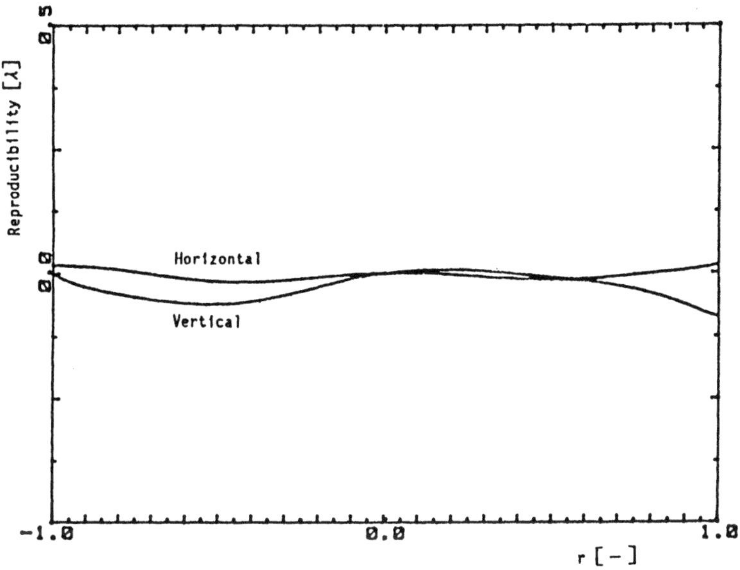

Figure 11:
Example of the reproducibility of a measurement of an asphere.

NA (--)	M (--)	FOV (mm)
	Standard obj.	
0.10	4	5
0.25	10	2
0.33	20	1
0.65	40	0.5
0.85	63	0.3
	Super wide field obj.	
0.16	4	5
0.40	10	2
0.70	20	1
0.95	40	0.5

Table 1
Fields of view of a commercially available serie of standard and super wide field objective lenses. NA: numerical aperture, M: magnification and FOV: field of view.

Design A						
	S = 0		S = 0.759		S = 0.882	
D	FOV	FD	FOV	FD	FOV	FD
[.per]	[mm]	[fr/ap]	[mm]	[fr/ap]	[mm]	[fr/ap]
10	0.00	0	0.00	0	0.00	0
20	0.00	2	0.00	1	0.00	1
30	0.01	6	0.01	4	0.02	2
40	0.02	15	0.03	10	0.04	6
50	0.03	29	0.06	19	0.08	11
60	0.06	51	0.11	33	0.14	19
70	0.10	83	0.18	53	0.22	31
80	0.15	126	0.28	80	0.34	47
90	0.22	184	0.42	115	0.50	68
100	0.32	258	0.59	160	0.72	94
Design B						
	S = 0		S = 0.759		S = 0.882	
D	FOV	FD	FOV	FD	FOV	FD
[.per]	[mm]	[fr/ap]	[mm]	[fr/ap]	[mm]	[fr/ap]
10	0.00	0	0.00	0	0.00	0
20	0.00	2	0.00	1	0.01	1
30	0.01	7	0.02	5	0.02	3
40	0.02	18	0.04	12	0.04	7
50	0.04	35	0.07	23	0.08	13
60	0.07	63	0.13	40	0.15	24
70	0.11	101	0.21	64	0.25	37
80	0.17	153	0.32	95	0.38	56
90	0.25	221	0.46	136	0.55	80
100	0.35	309	0.65	188	0.78	111
Design C						
	S = 0		S = 0.759		S = 0.882	
D	FOV	FD	FOV	FD	FOV	FD
[.per]	[mm]	[fr/ap]	[mm]	[fr/ap]	[mm]	[fr/ap]
10	0.00	1	0.00	1	0.00	0
20	0.01	8	0.01	5	0.01	3
30	0.02	27	0.04	17	0.05	10
40	0.05	64	0.09	41	0.11	24
50	0.10	125	0.18	78	0.22	45
60	0.17	214	0.32	130	0.38	75
70	0.28	335	0.54	199	0.61	115
80	0.43	497	0.77	289	0.93	167
90	0.63	699	1.13	402	1.36	233
100	0.90	955	1.62	545	1.97	318

Table 2:
Fields of view and fringe densities in a radial shear interferometer for three aspherical moulds. The values are given as a function of the percentage of the total diameter measured for three values of the shear ratio. The value S = 0 corresponds to the classical (Twyman-Green) interferogram.

Testing of Aspheric Surfaces with Computer Generated Holograms

H.Tiziani, B.Packroß, G.Schmidt

Institut für Technische Optik,
Universität Stuttgart, Pfaffenwaldring 9, 7000 Stuttgart 80, FRG

Summary
Aspherical surfaces are becoming more important and can even be mass producted. There is a need for flexible test methods of high accuracy. The paper describes the combination of a computer generated holograms with a partially compensating lens as a powerful tool. An example for testing a steep aspheric surface will be given.

1. Introduction

Interferometric techniques in combination with digital fringe analysis are frequently used for testing optical components and surfaces. Problems occur when testing aspherical surfaces with classical two–beam interferometry for the following reasons. The fringe density becomes very high and the resolution of conventional detectors e.g. CCD–arrays is too small to resolve the fringe pattern. Possible techniques are 2λ–interferometry and shearing interferometry, mechanical and optical tasters, and null–lens arrangements. Some specific disadvantages are however

- reduced sensitivity (2λ, shearing)
- damage of the surface (mechanical taster)
- no flexibility, expensive, difficult to test (null–lens).

The use of computer generated holograms (CGHs) in combination with a partially compensating lens is an alternative.

CGHs can be used for absolute testing of a wide range of aspheric surfaces. There are a number of publications concerning this subject [1–15]. For symmetrical optical systems to be tested, in–line (Gabor–type) and off–axis CGHs can be used [11,14]. Although in–line holograms have some advantages like simpler adjustment and the possibility of compensating stronger asphericities, off–axis holograms are mostly used. The spatial separation of their diffraction orders allows a simple filtering and avoids the overlap of different orders, which leads to difficulties to test the central portion of the aspherical surface when using in–line holograms.

The principle of a measurement will be described in the following. The interferometric set–up will be described as well as the calculation and manufacturing of the off–axis holograms and the digital fringe analysis for the evaluation of the interference pattern.

2.Test procedure with the Twyman–Green interferometer

For testing aspherical surfaces with computer generated holograms a two–beam interferometer in which both beams pass through the CGH is used. In this way inhomogenities of the holographic storing device are compensated. Our two–beam interferometer of the Twyman–Green type is shown in figure 1. The plane wave leaving the beam expander is separated by the beam–splitter into reference and test wave. The plane reference mirror is piezo–driven, which is necessary for the phase stepping fringe analysis to be described later.
A precision lens to be used for testing spherical surfaces is placed in the test arm of the interferometer. It was specially designed to test surfaces with short radius and high numerical aperture. An additional optical system between beam–splitter and precision lens may be required. Such a system allows to image the test surface onto the hologram to adapt the aperture and to compensate some of the asphericity of the test wavefront. The CGH is mounted in the interferometer exit. The diffraction orders of the off–axis hologram are separated spatially in order to block unwanted orders by a iris diaphragm in the focal plane of a collimating lens. Figure 2 shows a section of the Twyman–Green interferometer. For a perfect aspheric surface the hologram transforms the aspheric test wave into a plane wave in the first diffraction order (figure 3). Shape errors lead to wave deformations. The undiffracted plane reference wave remains plane when passing through the hologram in the zero order. Both waves are superimposed by tilting the reference mirror. For the fringe analysis the hologram plane is imaged onto the CCD–chip of a TV–camera connected to a frame grabber in a personal computer.

3.Calculation and manufacturing of the CGH

The calculation of the CGH starts with the data of the surface to be tested : radius, shape factor or conic constant, higher aspheric coefficients and diameter. For the ray–tracing this surface is considered as a reflecting last surface in the test arm of the Twyman–Green interferometer. The resulting wavefront in the hologram plane is obtained by ray tracing of an incident plane wave through the known collimating lens. After reflection at the surface under test the wavefront is computed back through the lens and the cubic beamsplitter to the hologram plane. For a first calculation the best fit sphere of the aspheric is evaluated. Subtracting its radius from the back focal length of the lens leads to an approximately best position of the test surface. This position is then varied by optimization routines to obtain a wavefront with minimum slope and aberration in the hologram plane. Three conditions have to be fulfilled for testing the whole diameter :

- the slope of the wavefront does not exceed a maximum given by the resolution of the photoplotter
- all rays reflected at the surface under test reach the hologram plane at a height less than the maximum plotable CGH diameter
- monotonously increasing height of rays in the hologram plane

If one of these conditions is not satisfied, either an additional compensating system has to be designed and inserted, or the testable diameter is limited up to that one, where the above conditions are satisfied.

The ray tracing delivers the optical path difference of the particular rays leading to the wavefront. For practical reasons (computation time) up to 50 rays are calculated, much less than the number of pixels to be plotted in a CGH. Thus the wavefront has to be developed into a polynomial. The number of terms depends on the slope of the wavefront. For rotational symmetric surfaces only even terms are necessary leading to even powers of the height in the hologram plane for instance. Their coefficients link the ray tracing and the plot program.

The wavefront in the hologram plane can be written as :

$$W_h(r) = A_2 r^2 + A_4 r^4 + A_6 r^6 + A_8 r^8 + A_{10} r^{10}$$

where $r^2 = x^2 + y^2$ with x,y are coordinates in the hologram plane and the A's are coefficients of the polynomial series.

For an off–axis hologram a carrier frequency is introduced by tilting the wavefront by α, which yields to a resulting transmission distribution for the CGH :

$$T(x,y) = 0.5 \left\{ 1 + \cos\left[\frac{2\pi}{\lambda} \left(W_h(x,y) + x \cdot \sin\alpha \right) \right] \right\}$$

In the real measuring procedure errors of the test apparatus need to be compensated. Manufacturing inaccuracies and misalignments of the collimating lens, the compensating lens and other optical components lead to wavefront errors $W_c(x,y)$. For this purpose the interferometer is calibrated with a very good sphere, expressing the wavefront in Zernike polynomials and adding them into the hologram calculation :

$$T(x,y) = 0.5 \left\{ 1 + \cos\left[\frac{2\pi}{\lambda} \left(W_h(x,y) + x \cdot \sin\alpha + W_c(x,y) \right) \right] \right\}$$

This function will be calculated column by column and plotted on–line.

For generating a CGH we use an Optronics P–1700 drum plotter with a maximum plot size of 250×250 mm² and a minimum spot size of 12,5 μm. This resolution is not sufficient for a carrier frequency of 50 lines/mm, which is necessary for compensating high wavefront slopes of steep aspherics. For the final use in the interferometer a photographic reduction of approximately a factor 6,4 on Agfa Millimask plates or on photoresist is done.

4.Digital fringe analysis

The fringe pattern contains information about the shape error of the surface under test. Thus an evaluation of the fringe pattern leads to a description of the surface error. Many different algorithms can be used for fringe analysis. Especially suited are phase–stepping techniques with at least three interferograms obtained by shifting the phase of the reference mirror [16–18].

We normally use two different algorithms to calculate the wavefront. The first one uses four interferograms and allows a very fast calculation of the wavefront phase. The four measured intensity distributions can be written as :

$$I_1(x.y) = I_0(x,y)\,[\,1 + m\cdot\cos(\,\varphi(x,y) - \tfrac{3}{2}\Delta\,)]$$

$$I_2(x.y) = I_0(x,y)\,[\,1 + m\cdot\cos(\,\varphi(x,y) - \tfrac{1}{2}\Delta\,)]$$

$$I_3(x.y) = I_0(x,y)\,[\,1 + m\cdot\cos(\,\varphi(x,y) + \tfrac{1}{2}\Delta\,)]$$

$$I_4(x.y) = I_0(x,y)\,[\,1 + m\cdot\cos(\,\varphi(x,y) + \tfrac{3}{2}\Delta\,)]$$

where I_0 mean intensity, m fringe contrast, φ phase of wavefront, Δ introduced phase shift. Using a phase shift of $\Delta = \frac{\pi}{2}$ we obtain the phase φ as :

$$\varphi = \arctan\left[\frac{I_4-I_2}{I_1-I_3}\right]$$

By storing the arctan values in a look–up table and using the real–time subtracting capabilities of our image processing board we obtain the wavefront in a few seconds.

If measurements with higher accuracy are required, an algorithm with five interferograms is more appropriate, because it is less sensitive to phase shift errors [19].

In this case five intensity distributions are measured. For a phase shift of $\Delta = \frac{\pi}{2}$ the phase can be calculated by :

$$\varphi = \arctan \left[\frac{2(I_2-I_4)}{2I_3-I_5-I_1} \right]$$

Figure 4 shows an example of a fringe analysis. Figure 4(A) shows the interference pattern of a steep aspheric surface measured with a CGH. The calculated phase of the wavefront is shown in a grey–scale map (B), as a pseudo 3D–plot (C) and as contour lines (D).

The deviation of the evaluated wavefront from a plane contains both surface and adjustment errors. A compensation of adjustment errors can be done by combining fringe analysis and ray tracing [15]. The measured wavefront is approximated by Zernike polynomials to have an interface to the ray tracing program. The actual adjustment state is investigated by varying the system parameters in such a way that the wavefront calculated by ray tracing has minimum deviations from the measured one.

5.Conclusions

An interferometric method for testing aspherical surfaces with the aid of computer generated holograms has been described. The combination of CGH's with partially compensating lenses is a very flexible technique, which allows the test of a wide range of asphericities. An important advantage of the synthetic holograms is the ability of absolute testing of surfaces. The digital phase measurement interferometry leads to a short measuring time and high accuracy.

References

1. Wyant, J.C.;Bennett, V.P. : Using computer generated holograms to test aspheric wavefronts. Appl. Opt. 11 (1972) 2833.

2. Fercher, A.F.; Kriese, M. : Binäre synthetische Hologramme zur Prüfung asphärischer optischer Elemente. Optik 35 (1972) 168.

3. Foulde, M.; Fercher, A.F.; Torge, R.; Wilson, R.N. : Optical testing by means of synthetic holograms and partial lens compensation. Opt. Commun. 7 (1973) 363

4. Wyant, J.C.; O'Neill, P.K. : Computer generated hologram ; Null lens test of aspheric wavefronts. Appl. Opt. 13 (1974) 2762.

5. Sirohi, R.S.; Blume, H.; Rosenbruch, K.–J. : Optical testing using synthetic holograms. Opt. Acta 23 (1976) 229.

6. Fercher, A.F. : Computer generated holograms for testing optical elements : Error analysis and error compensation. Opt. Acta 23 (1976) 347

7. Handojo, A.; de Jong, J. : Interferometer for optical testing with computer–generated holograms. Appl. Opt. 16 (1977) 546

8. Yatagai, T.; Saito, H. : Interferometric testing with computer generated holograms : Aberration balancing method and error analysis. Appl. Opt. 17 (1978) 558

9. Schwider, J.; Burow, R.; Grzanna, J. : CGH–Testing of rotational symmetric aspheric in compensated interferometers. Opt. Appl. 9 (1979) 39

10. Schwider, J.; Burow , R.: Wave aberration caused by by misalignments of aspherics and their elimination. Opt. Appl. 9 (1979) 33

11. Tiziani, H. : Prospects of testing aspheric surfaces with computer–generated holograms. Proc SPIE 235 (1980) 72

12. Schwider, J.; Grzanna, J.; Spolaczyk, R.; Burow, R. :Testing aspherics in reflected light using blazed synthetic holograms. Opt. Acta 27 (1980) 683

13. Larionov, N.P.; Lukin, A.V.; Rafikov, R.A. :Testing of aspheric surfaces by means of axial synthesized holograms. Sov. J. Opt. Technol. 47 (1980) 667

14. Mercier, R.; Lowenthal, S. : Comparison of in–line and carrier frquency holograms in aspherics testing. Opt. Commun. 33 (1980) 251

15. Dörband, B.; Tiziani, H. : Testing aspheric surfaces with computer generated holograms : analysis of adjustment and shape errors. Appl. Opt. 24 (1985) 2604

16. Koliopoulis, C.L. : Interferometric optical phase measurement techniques. Ph. D. Dissertation University of Arizona, Tucson (1981)

17. Schwider, J.; Burow, R.; Elssner, K.–E.; Grzanna, J.; Spolaczyk, R.; Merkel, K. : Digital wave–front measuring interferometry : some systematic error sources. Appl. Opt. 22 (1983) 3421

18. Creath, K. : Comparison of phase shift algorithms. Proc. SPIE 680 (1986)

19. Hariharan, P.; Oreb, B.F.; Eijn, T. : Digital phase shifting interferometry : a simple error–compensating phase calculation algorithm. Appl. Opt. 26 (1987) 2504

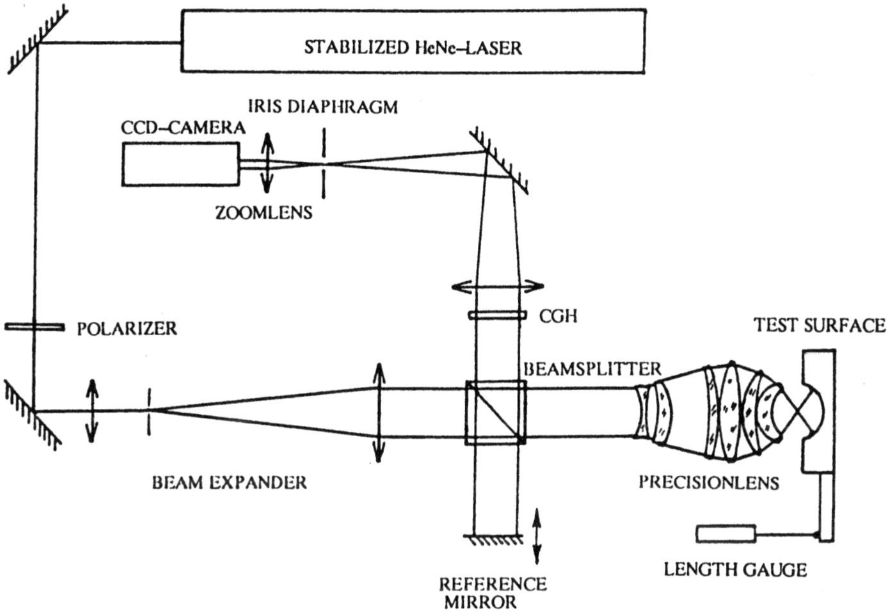

Fig.1. Twyman-Green interferometer with CGH

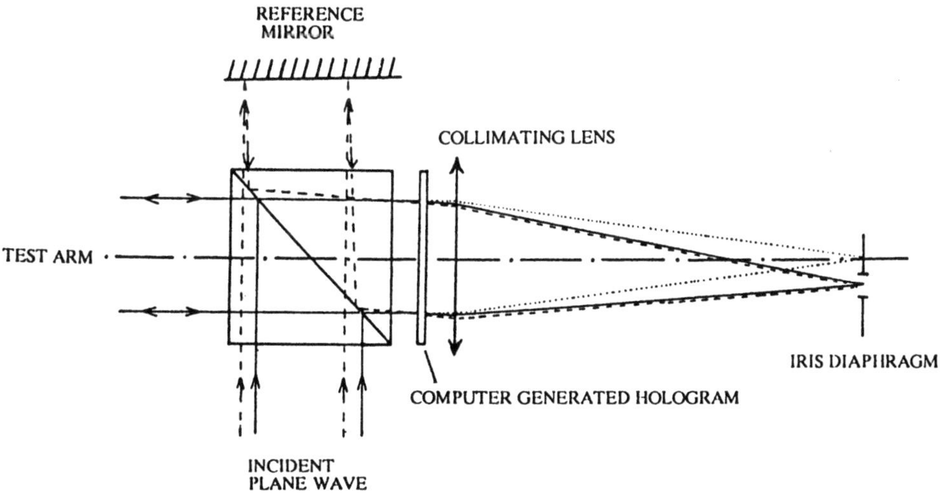

Fig.2. Superposition of diffracted waves. First diffraction order of test wave is superimposed with zero order reference wave. The other diffraction orders are blocked by an iris diaphragm.

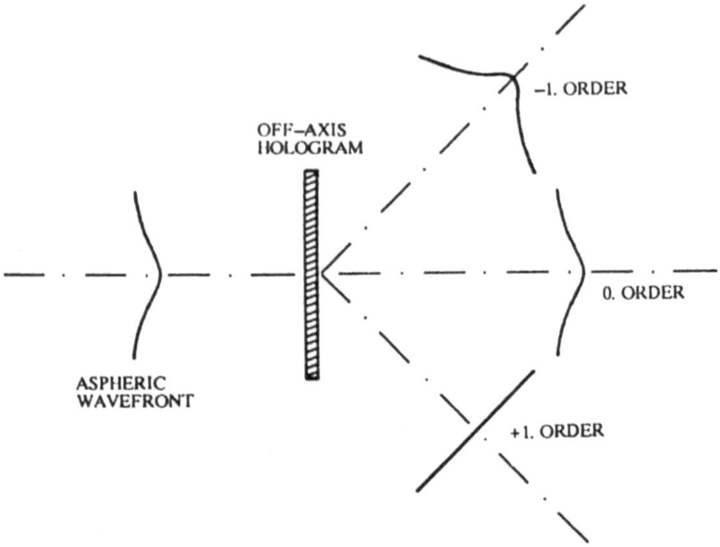

Fig.3. Diffraction of an aspheric wavefront by a CGH

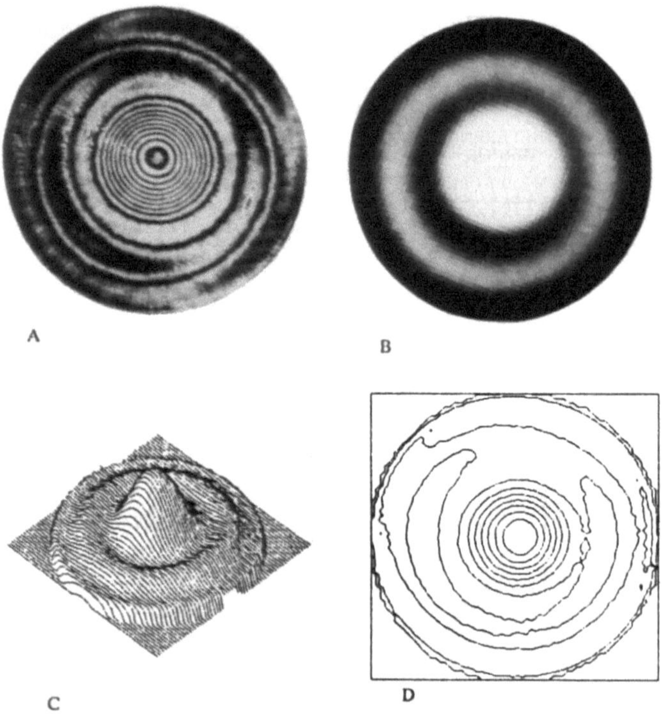

Fig.4. Fringe analysis of an asheric wavefront. Interferogram (A) and calculated wavefront coded in three different presentations:
grey level (B), pseudo 3-D (C), contour lines (D)

Synthetic Holograms for the Testing of Aspheric Surfaces

T. Pfeifer, P. Hantel, R. Tutsch

Fraunhofer-Institut für Produktionstechnologie (IPT)
Aachen, FRG

Summary

The use of computer generated holograms for interferometric testing of aspheric surfaces has become a standard measurement technique in precision engineering. In this paper the basic techniques for making computer generated holograms are pointed out and their limitations are discussed.

1. In-line and off-axis holograms

Holography can be understood as the recording of a given electromagnetic wave by coding both its amplitude- and phase-distribution onto an appropriate medium. This is achieved by superposing the object wave with a reference wave, usually a plane or a spherical wave. As shown in Fig. 1 the interference-pattern is recorded on photographic emulsion, photoresist, photothermoplastic film etc.. After processing the medium the object wave can be reconstructed by illuminating the hologram with the reference beam. Beside the undiffracted zeroth order at least two diffracted waves are formed. One of them is the reconstructed object wave, the other is the so called conjugate object wave, a wave with convergence properties inverse to those of the object wave.

If the reconstructed object wave is to be isolated from the other diffraction-orders, the direction of the reference wave has to be inclined against the optical axis. Three arrangements for recording a spherical object wave emitted from a point-source located on the optical axis 80 mm in front of the hologram are shown in Fig. 2. Plane reference waves with different angles of inclination are used. On the left the optical arrangement is plotted, on the right the shape of the interference fringes can be seen. For the chosen wavelength of 500 nm the density of the fringes is far too high to be resolved in this picture, so only every thousandth fringe is plotted.

The first example shows an in-line configuration like Dennis Gabor proposed it in 1948 [1]. Here the reconstructed object wave, the zeroth-order wave and the conjugate wave (not shown in the picture) have exactly the same propagation-direction.

In the second case the reference wave is tilted about 8 degrees against the optical axis. Still the zeroth-order beam leaves the hologram within the cone of the reconstructed wave.

The third hologram is recorded with an angle of inclination of 20 degrees. Now the separation of the different diffraction orders is easily done by setting an appropriate stop.

Three characteristic properties of holograms are illustrated by these examples:

- The separation of the different diffraction orders necessitates an increase of fringe-density in the hologram. The recording of the third hologram in Fig. 2 requires more than twice the resolution than the recording of the in-line hologram.

- The angle of inclination has to be larger than half the angle of the emission-cone of the object wave. More general it can be shown that the required angle of inclination increases with increasing maximum slope of the departure of the object wave from the reference wave [2]. Because of the limited resolution of many recording-techniques, the use of a reference wave of similar shape as the object wave is often advantageous.

- If the inclination-angle is chosen large enough to yield

separation of the different diffraction-orders, the fringes in the hologram plane will not form closed loops [3].

2. Algorithms for computing synthetic holograms

For a given reference wave every object generates a characteristic interference-pattern in the hologram plane, from which it can be reconstructed. On the other hand, every possible pattern on the hologram gives rise to a certain wavefront. For an arbitrary wave the matching pattern can be calculated and recorded on an appropriate medium. Doing so, one obtains a so called "Computer Generated Hologram" (CGH) or "synthetic hologram". Illuminating it with the reference wave the desired wavefront can be generated.

A lot of scientific research has been done in this field since the mid-60's and several techniques for making CGHs have been developed (for reviews see [3-6]). Most of the CGHs used in interferometry however are of the detour phase type or synthetic interferograms.

The algorithm of making detour phase holograms is shown in Fig. 3. The amplitude- and phase-distribution of the reference wave is computed immediately in front of the hologram plane and the amplitude- and phase-distribution of the desired object wave is computed immediately behind it. The hologram then has to be made in a way that for every point in the hologram plane the amplitude and the phase of the reference wave are altered to those of the object wave. In reality amplitude and phase can be computed only for a finite number of points in the hologram plane. Therefore the hologram plane is divided into an array of equally spaced rectangular sampling cells small enough to comply with the sampling theorem. For each sampling cell the amplitude and phase at its center is coded by drawing one or more rectangular aperture with definite size and position. Several variations of the detour phase concept have been developed to optimize the computation velocity, minimize the digitization errors and adapt the technique to special plotting devices [7-10]. Four different coding-schemes are shown in Fig. 3.

Another algorithm to compute CGHs is the "synthetic interferogram"-approach. Here the amplitude- and phase-distribution of both the reference- and the object-wave are computed immediately in front of the hologram. By superposing both waves mathematically the interference is simulated, rendering the intensity-distribution of the interference pattern. This technique is ideally suited for the computation of off-axis holograms with large angles of inclination. As mentioned before the interference fringes in this case are continuous lines without any closed loop. By use of residue arithmetic the shape of these lines can be calculated with high computation speed [11,12].

3. Techniques for making computer generated holograms

Fig. 4 shows a flow-chart for CGH making. After having computed the interference pattern the next step is to transform it into a tangible hologram. Several techniques are in use today, depending upon the required resolution. Most of them are adapted from semiconductor industry [13], where the techniques of generating microstructures are pushed forward with an immense amount of money and manpower. A rate of assessment for the different production techniques for CGHs is the resolution (that defines the maximum slope of the departure of the object wave from the reference wave) and the size of the CGH (for the maximum amount of information that can be stored in a hologram is the number of independently addressible picture elements). It should be noted that only error-free adressable elements can be taken into account. Systematic errors like non-orthogonal axis of the plotter or distortion of the reduction lens can be eliminated by the plotting-software. Thus the limits of a given plotting device are given by its construction (e.g. step width of a stepping motor) and its random errors. Typical numbers for the achievable resolution of different devices are given in Fig. 4.

Let us discuss the techniques in use today in detail. First there is a group of large-size medium-resolution devices. Computer-controlled plotters typically have plotting areas of up to DIN A1 (850x600 mm^2). Pen plotters are best suited for tracking the fringes of synthetic interferograms while photoplotters with rectangular exposition masks are to be preferred for printing the pattern of detour phase holograms.

The large-size original copy is photoreduced to the desired size. Standard photographic equipment can be used for a resolution of about 200 L/mm and a hologram size of 60x60 mm^2. If higher resolution is desired, expensive photolithographic equipment is needed. Optics with high numerical aperture specially corrected for a definite reduction ratio and a certain wavelength are used with intensive monochromatic light-sources. According to the Rayleigh-criteria

$$W = K_1\lambda/NA \qquad \Delta = k_2\lambda/(NA)^2$$

where W = the minimum feature size,
Δ = depth of focus,
λ = Wavelength
NA = numerical aperture of the imaging lens
k_1, k_1 = proportional constants often empirically determined

the depth of focus is a linear function of the second power of the minimum feature size, becoming less than +/- 2 μm for typical microlithographical applications [14]. Thus the proper adjustment of the optics, the temperature-control and the vibration-isolation are posing problems. Moreover the exposure field of high-resolution lenses usually is smaller than 20 mm in diameter. To overcome the restrictions of exposure area, so called wafer steppers are commonly used in semiconductor industry today. By moving the wafer with laser-interferometer-controlled positioning tables wafers up to 150 mm (6") in diameter are successively exposed. Lateral positioning and alignment can be done with an accuracy of better than +/- 0.25 μm, the focus repeatability is of the same order [15].

Commercial wafer steppers are designed for generating periodic structures (many equal integrated circuits per wafer), therefore only one pattern mask is needed. CGHs in contrast, are of non-periodic structure, so the mask has to be changed for each exposure.

With the wafer stepper concept even cathode-ray tubes can be used for generating the original copy [16]. The relatively low resolution of a CRT is compensated by the high speed of image-generation.

Considering environmental influences state-of-the-art wafer steppers today can generate CGHs with a resolution of about 0.7 μm [17] and a diameter of 150 mm .

Using photoreduction techniques one generally has to cope with several problems: image distortion, resolution limitations, focus-control, magnification-control and proper alignment.

The latest developments in semiconductor technology are pattern-generators writing the mask-pattern already in the right size, thus making reduction steps that can add errors unnecessary. Laser-based systems with 0.5 μm resolution are commercially available [18,19] and their suitability for production of CGHs has already been proven [20]. Not only the accuracy but also the high speed of these systems is advantegeous.

The most powerful devices for generating microscopic structures today are the e-beam pattern generators. The exposure area measures only few mm^2, therefore a positioning table is needed. The resolution can reach 5000 L/mm [13] and is limited by the positioning accuracy.

Photoreduction of grey-level patterns raises problems of non-linearity of the response of the photographic emulsion. These problems can be avoided by using binary patterns, which are formed exclusively of opaque and transparent (black and white) elements. Moreover binary holograms in principal have a higher diffraction-efficiency, resulting in a more intense recon-structured object wave. The generation of higher diffraction orders that have to be removed by a suitable stop and an increase of noise are the price to be paid for these advantages. Most CGHs in use for interferometric testing applications are binary holograms. Photoresist is preferred, when maximum contrast is desired. If photographic plates are used as recording medium, the emulsion can be bleached after processing, transforming the amplitude pattern into a phase pattern. The diffraction efficiency of binary phase holograms can reach 40 % in comparison to the limit of 10 % for binary amplitude holograms and 6 % for grey-level amplitude holograms [4].

Conventional holograms can have dimensions exceeding 100 cm^2 at a resolution of up to 7000 L/mm. By no means this can be reproduced by a CGH today. As mentioned before the required resolution can be reduced by transforming either the object wave or the reference wave with a suitable lens to get a better fit.

4. Application of computer generated holograms in interferometric testing

Fig. 5 shows a Twyman-Green phase-shifting interferomotor for the testing of concave or convex spherical surfaces. For commercially available interferometers the departure of the surface under test from the best fitting sphere is limited to of less than 10 μm [21]. For the testing of aspheric surfaces the object wavefront therefore has to be deformed in a definite way. This can either be achieved by a special master-lens, correcting nearly exactly the aberrations of the asphere or by a CGH.

An extension of the Twyman-Green interferometer for the application of a CGH is shown in Fig. 6. Typical wavefront-departures in interferometric testing of aspheric wavefronts are 100 wavelengths across the radius of the hologram. Neglecting practical sources of error a hologram with an error-free resolution of 500 L/mm and a radius of 20 mm would allow the testing of the aspheric wavefront to 1/100 wavelength. For the testing of steep aspherics the required space-bandwidth of the hologram becomes a problem. Generally a combination of a simple master lens for a coarse correction and a CGH of moderate space-bandwith will be the best choice.

References

1. D.Gabor: A new microscopic principle
   Nature 161 (1948) 777-778

2. J.C.Wyant, V.P.Bennett: Using computer generated holograms to test aspheric wavefronts
   Appl.Opt. 11 (1972) 2833-2839

3. W.-H.Lee: Computer-generated holograms: techniques and applications
   Progr. in Optics 16 (1978) 119-232

4. D.Schreier (ed.): Synthetische Holografie (German)
   Physik-Verlag, Weinheim, 1984

5. G.Tricoles: Computer generated holograms: an historical review
   Appl. Opt. 26 (1987) 4351-4360

6. W.J.Dallas: Computer-generated holograms
   in B.R.Frieden (ed.): The computer in optical research
   Springer-Verlag, Berlin, Heidelberg, New York 1980
   p. 291-366

7. B.R.Brown, A.W.Lohmann: Computer-generated binary holograms
   IBM J.RES.DEVELOP. March 1969, 160-168

8. R.A.Gabel: Reconstruction errors in computer generated binary holograms: a comparative study
   Appl.Opt. 14 (1975) 2252-2255

9. T.Takahashi, K.Konno, M.Kawai, M.Isshiki: Computer generated holograms for testing aspheric lenses
Appl. Opt. 15 (1976) 546-549

10. W.Freude, G.K.Grau, W.Liebler, B.Wüppermann: Computer-generated holograms with error compensation
Appl.Opt. 27 (1988) 138-146

11. W.-H.Lee: Binary synthetic holograms
Appl.Opt. 13 (1974) 1677-1682

12. W.-H.Lee: Binary computer-generated holograms
Appl.Opt. 18 (1979) 3661-3669

13. S.K.Ghandhi: VLSI fabrication principles. Si and GaAs
Wiley & Sons, New York 1983
Chapter 10: Lithographic processes

14. B.J.Lin: Where is the lost resolution?
Solid State Technology, January 1987, 63-68

15. J.D.Buckley: An advanced H-line stepper
Solid State Technology, January 1987, 87-93

16. H.J.Tiziani: Prospects of testing aspheric surfaces with computer-generated holograms
Proceedings of the SPIE Vol.235 (1980) 72-79

17. H.L.Stover: Frontiers of optical lithography
Solid State Technology, January 1987, 53

18. LASERSKAN II - Product information
Micronic lasersystem AB, Danderyd, Sweden

19. H.Ulrich,R.W.Wijnaendts-van-Resandt,C.Rensch, W.Ehrensperger: Direct writing laser lithography for production of microstructures
Proc. Microcircuit Eng., Paris, 22.-25.9.1987

20. K.Biedermann, O.Holmgren: Large-size distortion-free computer-generated holograms in photoresist
Appl.Opt. 16 (1977) 2014-2016

21. A.Ono: Aspherical mirror testing with an area detector array
Appl.Opt. 26 (1987) 1998-2004

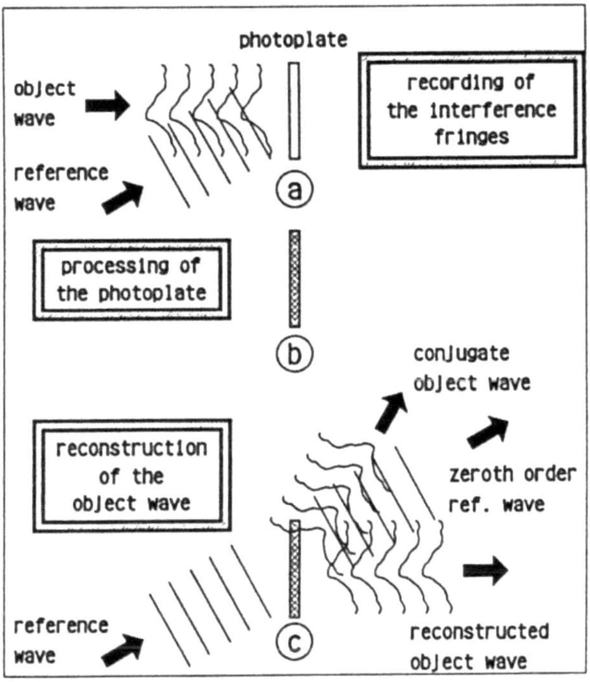

Fig. 1 Recording and reconstruction of an object wave in a hologram

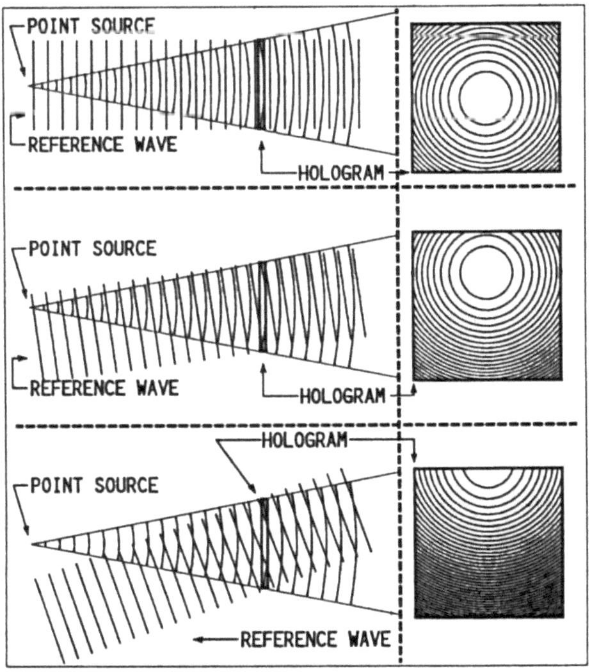

Fig. 2 Recording of a spherical wave with variable angle of inclination of the reference beam

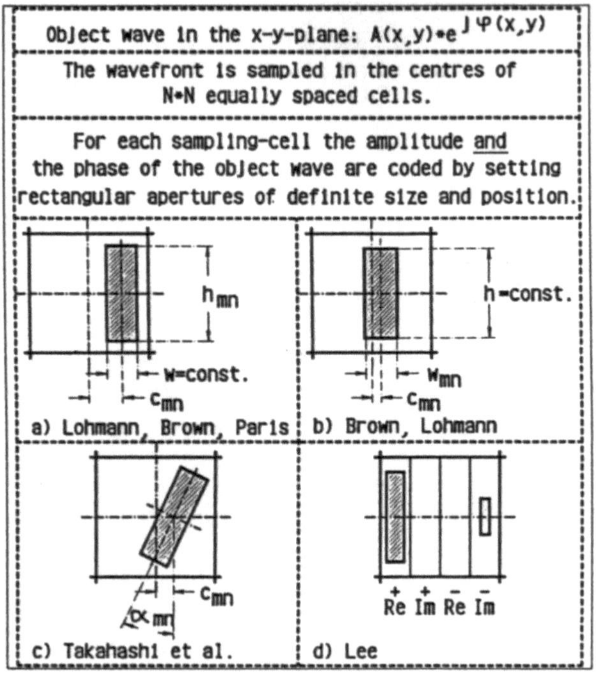

Fig. 3 Techniques for making binary detour phase holograms

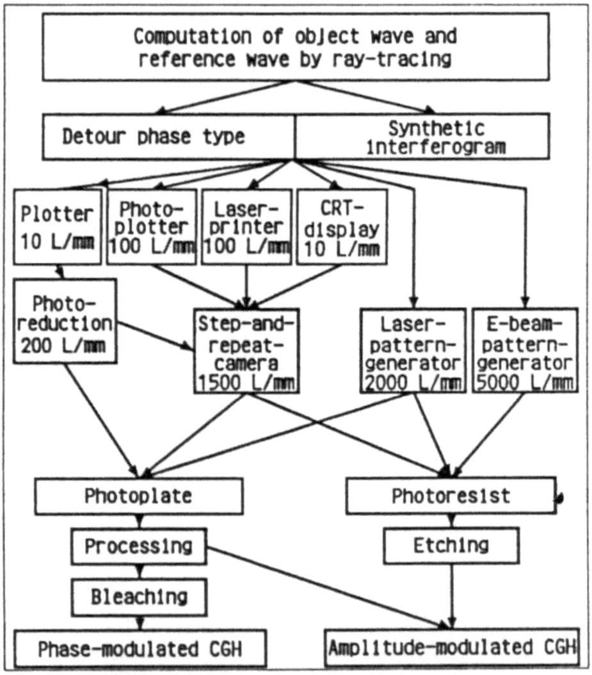

Fig. 4 Techniques for making Computer-Generated Holograms (CGHs) Resolution in Lines / mm

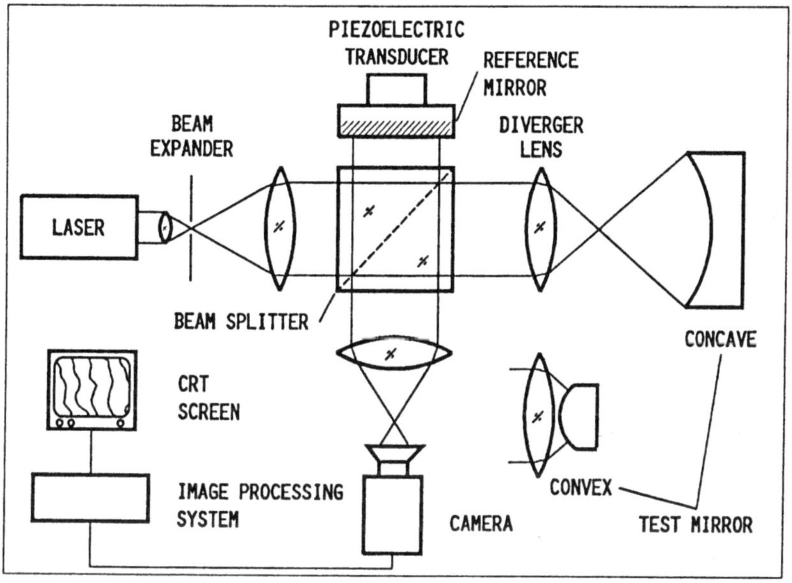

Fig. 5 Twyman-Green phase-shifting interferometer for the testing of spherical mirrors

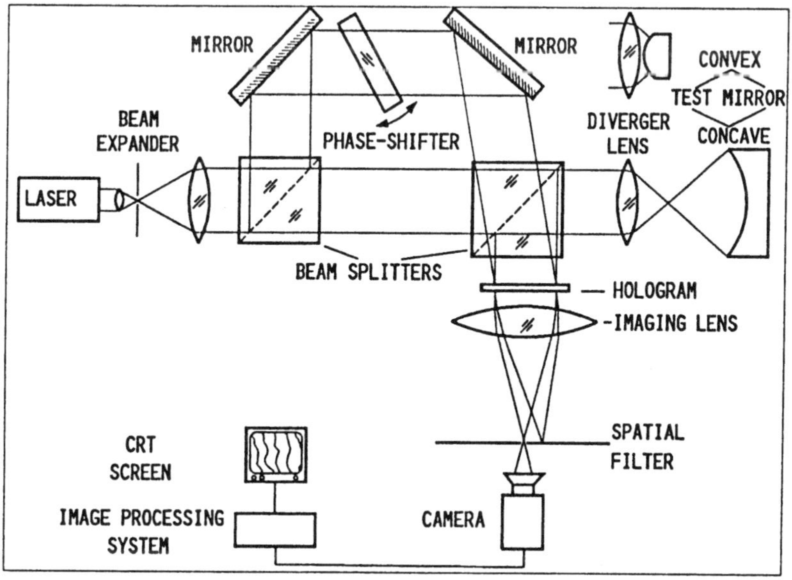

Fig. 6 Modified Twyman-Green phase-shifting interferometer for the testing of aspherical mirrors

Experimental Measurement of Optimum Depth of Cut

P. Langenbeck
D. Müller

Ingenieur-Agentur für Neue Technologie in Optik und Precision Engineering, D-7771 Frickingen 2, FRG

Syntec GmbH, D-7859 Eimeldingen, FRG

These notes relate to micromachining with diamond tools and inspection of signature of the tool and the machine as opposed to material response.
We had reported about plunge cutting and micro-interferometry of the tool path as a viable method of tool validation /1, 2/. Microscopy and micro-interferometry reveal continuous traces in the plunge cut, which may be related to damages on the edge of the tool. These traces may be measured also by sensitive stylus traversing the grooves. Machining of a homogeneous material is supposed.

The astonishing finding was that a tool initially cutting well at a few µm depth of cut (meaning a clean groove) will produce traces at increasing depth of cut which, obviously, result not from damages at the tool edge, but rather on the trailing sole behind the edge. Else, the traces should appear immediately even at the minutest depth of cut. We expect to obtain information about the tool "sole", under actual cutting condition.

The common plunge cutting with diamond tools involves the multiple passage of the tool over the same trace, as it is accomplished by raising the sample, or by sinking the fly cut spindle. This cannot be done for a sufficiently short duration, such as to observe one passage only. The oscillations in the tool path observed and reported in /2/ are, thence, difficult to be related to either tool defects, material inhomogeneity or to lack of machine stiffness.

The plunge cutting procedure is therefore modified such as to produce a multitude of plunge cuts with one passage only. Variable depth of cut is obtained additionally by a procedure explained in Fig. 1 a). The first cut surface of the sample is shown in the interferogram Fig. 1 b) (Syntec SY 01).

A sample, machined flat initially, is inclined relative to the machining plane such as to produce wedge-shaped material removal upon a second cut, starting at a minimum being indicative of "the last cut". This second machining cut, however, is produced with a feed rate high enough as to produce widely spread grooves on the surface. Flat ridges may be (but must not be) left between the grooves, these ridges being representative of the initially machined surface finish, and also serving as a reference for subsequent interferometric inspection. Different from the stationary plunge cut the tool will have different specific load on the leading and on the trailing edge, due to the fast feed. Normally, the less loaded edge is running in the free (material already moved).

The wedge plunge cut sample of AlMgSi 05 was micromachined on the ultra-precision machine described in /4/ and the sample was inspected on an interference microscope as described in /5/. Fig. 2 shows four typical locations. Fig. 2 a) shows, in white light interference, the transition from the flat initial surface to the first groove, here of 2 µm depth. Magnification is indicated. Figs. 2 b), c), d) show a groove each at successive depths of cut indicated by the number of fringes. Feed rate was about 1 mm per revolution of the fly cut spindle. Normally, as in the flat field, feed rate is about 7 µm to 15 µm. There is a 10 µm mark on each photo. One recognizes an increasing number of traces on the bottom of the groove: some continuous and indicative of the tool damage, and others interrupted short streaks, resulting from material inclusions.
The same sequence recorded in green light, Fig. 3, reveals an ondulation along the cut in the order of 0,2 µm, so obvious that not even an automatic evaluation is attempted. This ondulation is due to "material response", /3/, it is not a tool or

machine signature. It is also the residual surface microroughness after micromachining.

An increase of magnification from 20 x to 50 x reveals more details about the bottom of the groove, Fig. 4; magnification indicated, green light.
Comparison of Fig. 4 b) (2 µm) and 4 d) (10 µm) clearly indicates a useful depth of cut to exist for a given material/tool combination. This parameter can be found on the machine - and it should be part of setting-up procedure. It might also be part of machine acceptance testing.

At increasing depth of cut the attainable surface finish will be determined more by the interaction of the diamond tool sole with the material. Here, the diamond tool polishing process leaves much yet to be learned: Literature and experts communications indicate the presence of certain relief features on polished tool soles. They are not detected upon normal tool edge inspection - but they evidence their action in the variable depth of cut test.

Finally, the increase of depth of cut will lead to tool forces great enough to interfere with the stiffness of the machine assumed to be infinite for small depths. This, too, is necessary information for the machinist attempting to produce a best possible mirror surface. The wedge sample will quickly indicate the limiting depth of cut, from which on flatness cannot be obtained for lack of stiffness.

Since, obviously, observation of these effects is essential to micromachining, we have developed a special tool head fitting standard spindles. This tool head is shown in Fig. 5. It incorporates a microscopic inspection situated on the same radius as the tool. It allows in-situ inspection of the cut. As the spindle is its own stroboscopic device, it also allows in-process quick assessment of surface quality.

A sample holder was designed, Fig. 6, in order to convenience the wedge plunge cut test on ultra-precision machines. It allows comparative performance of the test both in the turning and the milling mode. The plane parallel sample is mounted on the holder which has a mounting surface inclined for 0,5 degree relative to the base. After first machining, the sample is rotated for 180 degrees and reclamped, thus providing a wedge both for the Roof Edge Test and for the wedge plunge cut test.

References

1. Langenbeck, P.; Müller, D.: Smooth low scatter metal optics by repeated machining. Prec. Engineering Oct. '85, Vol. 7, No. 4

2. Müller, D.: Testing of diamond tools. SPIE Conference April 1987, on "In Process Optical Metrology for Precision Machining", paper 802-02.

3. Stadler, H. J.; Freisleben, B.; Heubeck, C.: Response of metallic material to micromachining. SPIE Vol. 802, April 1987, Den Haag, p. 67.

4. Haberland, R.: Steife von Spindeln und Schlitten zur spanenden Herstellung optischer Flächen, Feinwerktechnik und Meßtechnik 93, 1985, p. 257.

5. Cencic, B.; Barut, M.; Cepon, F.: Converting a commercially available Linnik microinterferometer into a fringe scanning optical profiler. SPIE Vol. 802, April 1987, Den Haag, p. 182.

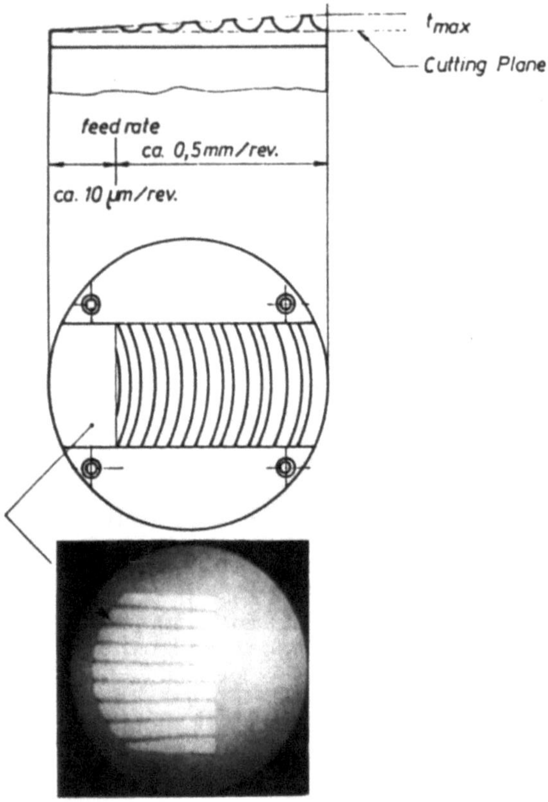

Fig.1. Wedge plunge cut test
a) standard substrate
b) interferogram of 1st cut

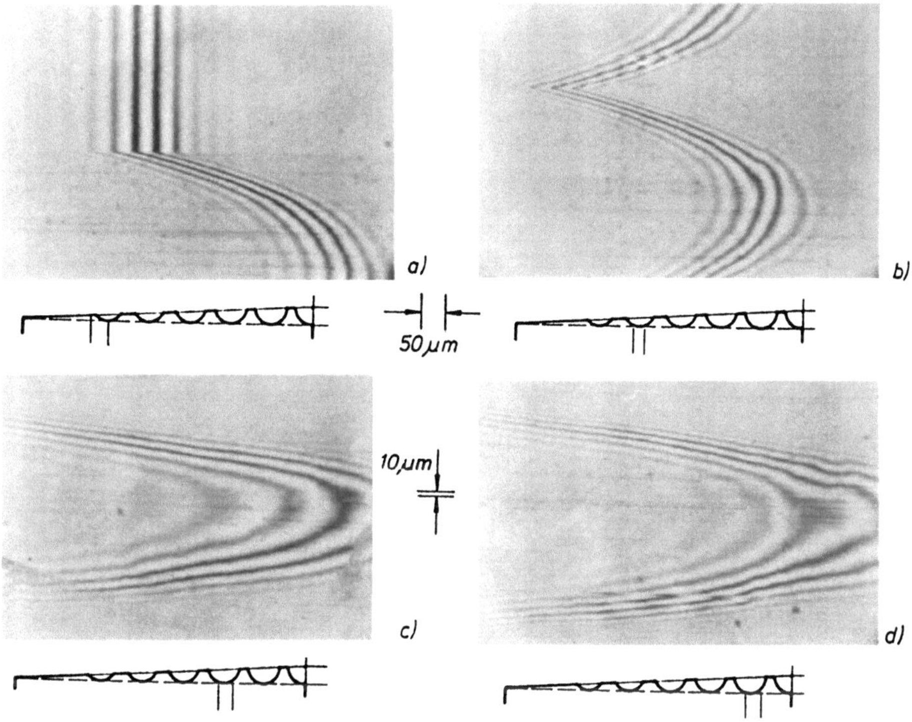

Fig.2. Wedge plunge cut - white light interference
a) 1st groove, starting wedge
...d) last groove on wedge
substrate ALMGSI 0.5, Nd-Tool 10 mm radius

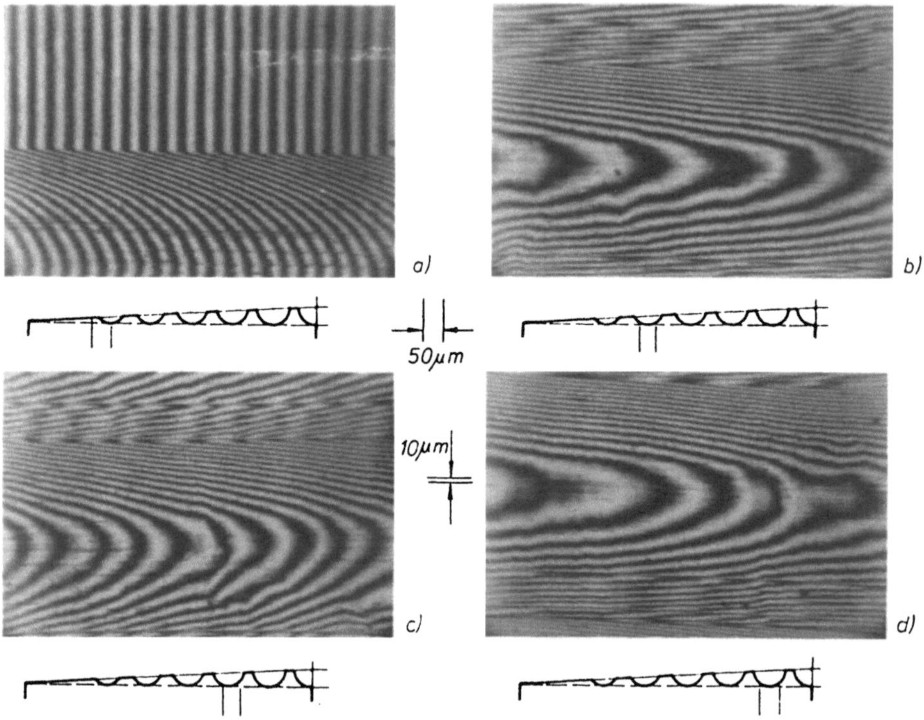

Fig.3. Wedge plunge cut - green light interference same locations as Fig.2.
Figs.2.3.4. recorded on syntec instruments

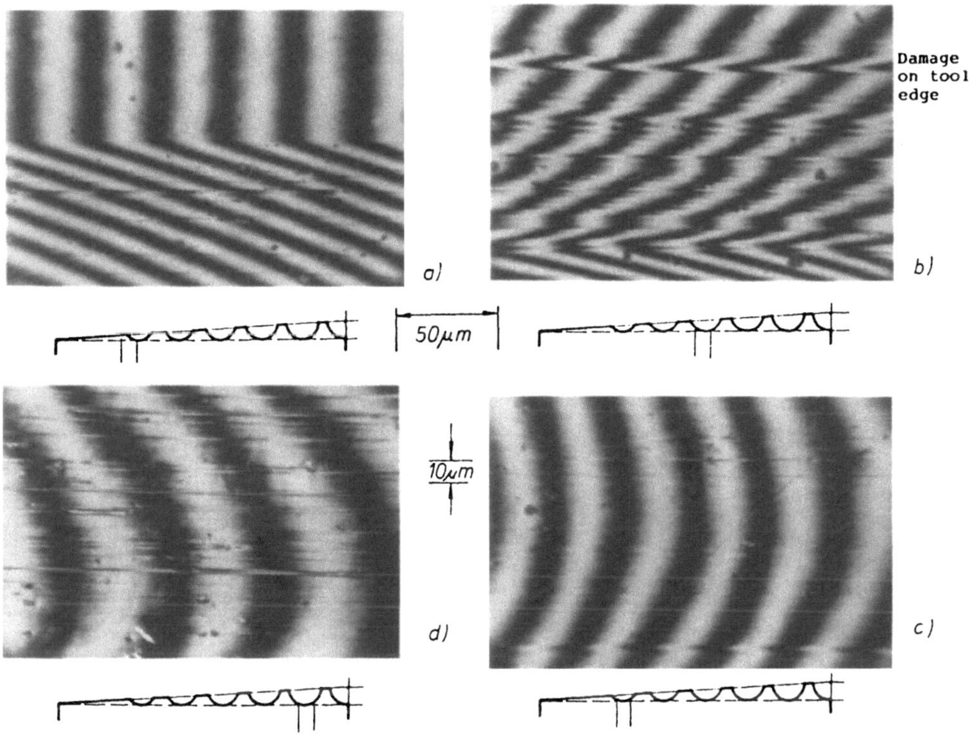

Fig.4. Wedge plunge cut - green light

a) 1st groove, starting wedge

c) 2nd groove, favourable depth of cut micromachining

d) possible but unfavourable depth of cut

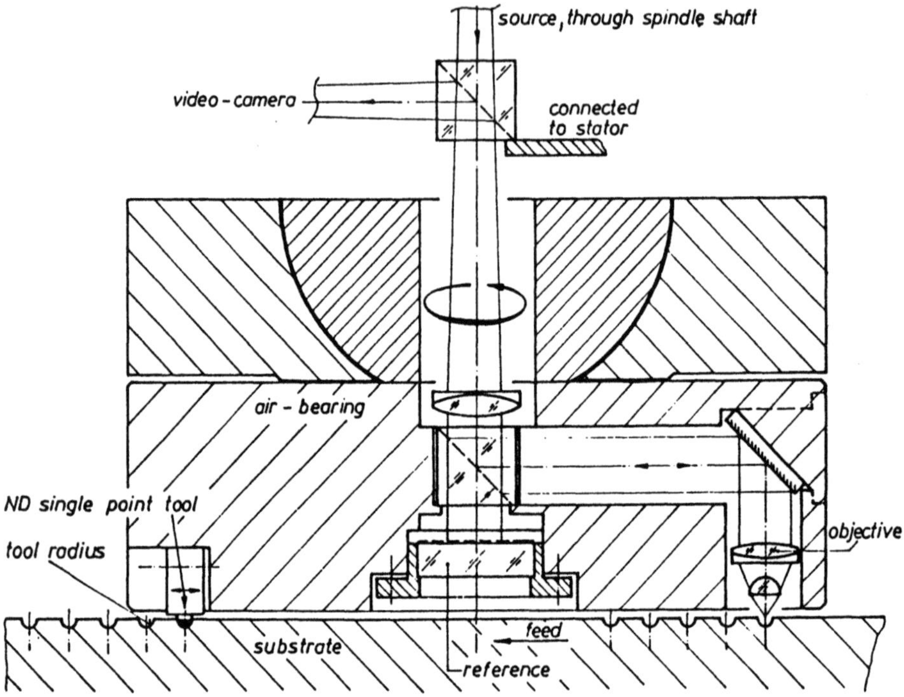

Fig.5. Fly cutting tool head with integrated microscope interferometer for wedge plunge cut inspection

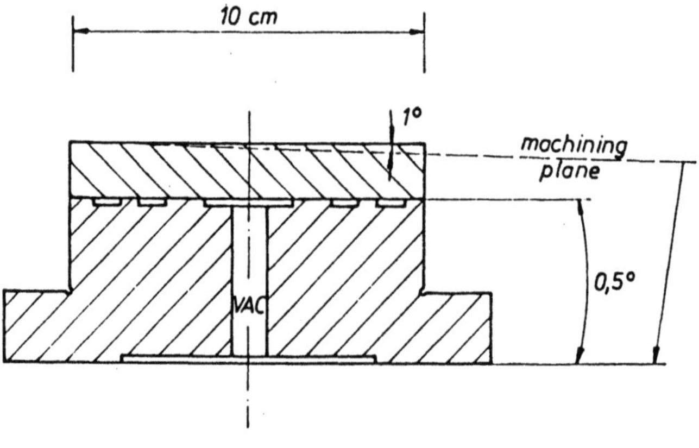

Fig.6. Sample holder for roof edge test and wedge plunge cut test

FSC
www.fsc.org
MIX
Papier aus verantwortungsvollen Quellen
Paper from responsible sources
FSC® C105338

If you have any concerns about our products,
you can contact us on
ProductSafety@springernature.com

In case Publisher is established outside the EU,
the EU authorized representative is:
Springer Nature Customer Service Center GmbH
Europaplatz 3, 69115 Heidelberg, Germany

Printed by Libri Plureos GmbH
in Hamburg, Germany